Living Culturally Responsive Mathematics Education with/in Indigenous Communities

Living Culturally Responsive Mathematics Education with/in Indigenous Communities

Edited by

Cynthia Nicol, Jo-ann Archibald Q'um Q'um Xiiem,
Florence Glanfield and A. J. (Sandy) Dawson

Foreword by Sharon Nelson-Barber

BRILL
SENSE

LEIDEN | BOSTON

Cover illustration: Golden Spruce Trail, Haida Gwaii, photograph by TJ Watt, Ancient Forest Alliance (www.ancientforestalliance.org)

All chapters in this book have undergone peer review.

The Library of Congress Cataloging-in-Publication Data is available online at http://catalog.loc.gov

ISBN 978-90-04-41574-4 (paperback)
ISBN 978-90-04-41575-1 (hardback)
ISBN 978-90-04-41576-8 (e-book)

This book is printed on acid-free paper and produced in a sustainable manner.

ADVANCE PRAISE FOR *LIVING CULTURALLY RESPONSIVE MATHEMATICS EDUCATION WITH/IN INDIGENOUS COMMUNITIES*

"Given the insufficient amount of existing research that has been conceptualized from the perspectives of Indigenous peoples, this delightful collection of papers furthers our thinking about the strengths and competencies that students develop in their own contexts and how such assets can effectively serve as bridges to learning.

This place-based perspective is particularly useful in diverse Indigenous contexts where indicators of success must extend to broader notions of self-determination and nation building. The complex elements that support learning demand innovative approaches, and the approaches presented here are fine models for others seeking transformative change in mathematics education."
– **Sharon Nelson Barber, Culture and Language in Education, WestEd. California, USA**

"This book highlights multiple ways to re-story both mathematics education (through culturally responsive pedagogies) and understandings of colonization, mathematics, and Indigenous knowledges. In doing so, the reader is introduced to the promises, possibilities and struggles in coming to critically understand how meaningful mathematics education for Indigenous students and communities must be rooted in political, social, historical, linguistic, and cultural realities. The community- and place-based research in this book not only nudges the reader out of a complacent, colonialist view that 'there is a one way to know the world mathematically,' but it draws one into a 'radical hope' for a mathematics that is respectful, responsive, sustaining and revitalizing. To me, reading this book is an act of decolonization; one that demands the reader to listen, listen well."
– **Kathleen Nolan, Professor, University of Regina, Saskatchewan, Canada**

To the late A. J. Sandy Dawson,
teacher, scholar, writer, adventurer, activist, believer, and friend

CONTENTS

SHARON NELSON-BARBER

FOREWORD

This volume centres on issues of diverse knowledge systems, use of local funds of knowledge, and the value of community understandings as foundational for creating high-quality and sustained learning and improvement in mathematics classrooms that serve Indigenous students. The rich examples brought to life in each chapter come from a range of Indigenous cultural communities in Alaska, Australia, Canada, Hawai‘i, New Zealand, and other island entities across the Pacific. Through research, development and reflection, the authors have cultivated effective ways to collaborate with Indigenous and non-Indigenous professionals—Elders, community members, evaluators, assessment developers, and educators. Collectively, they work to contextualize Indigenous students' educational experiences amid often competing conflicts between the assumptions and expectations of local communities and the expectations and assumptions of schools. Given the insufficient amount of existing research that has been conceptualized from the perspectives of Indigenous peoples, this delightful collection of papers furthers our thinking about the strengths and competencies that students develop in their own contexts and how such assets can effectively serve as bridges to learning.

The curricular and instructional approaches described in each chapter resonate with my own research as I too work alongside communities as they look inward to capitalize on local intellectual and cultural heritage, and to bring this knowledge to bear on the critical and enduring needs of diverse Indigenous students. This socio-cultural stance involves not only curriculum content, but also organizational structures, approaches to interpersonal communication, and how instructional tools reflect the world views and values of the school and the students served. The recurring themes revealed in these essays offer math educators additional access to the kinds of images, experiences, and roadmaps that can make a difference for Indigenous student success in mathematics.

Robyn Jorgensen's chapter, "Culturally Responsive Pedagogy: Addressing 'Shame' for Aboriginal Learners," uses the cultural construct, "shame," as a focal point to explore culturally responsive practices that contribute to Indigenous student math success in remote locations across Australia. Most teachers in these settings do not come from the communities of their students and do not speak the local language(s) and/or dialect(s). The author, herself non-Indigenous, first discusses the need for outside teachers to participate in ongoing learning about each community's cultural norms and values, and for these educators to come to understand how conceptions, such as "shame," can be defined very differently across ethnic groups. With a level of understanding of at least the possibilities of such distinctions, Jorgensen believes teachers are more likely to envision potential disconnections between some of the cultural practices in remote Australia and the widely understood pedagogies used in modern schooling. For example, the values and morals that underlie "shame" in Aboriginal communities and the values and morals of conventional Australian schooling can differ to such a degree that consequences arise around issues of attendance, how students show respect for authority, understanding group roles, how one monitors and manages behaviour, the role of homework, how learning is assessed, and so forth. When teachers devote substantial time to exploring contrasts of this nature, they are in a better position to work collaboratively with communites to acknowledge, show respect for, and integrate Aboriginal language and cultural practices into mathematics and the social norms of their own classrooms.

Florence Glanfield, Gladys Sterenberg, and Dwayne Donald report similar findings in their chapter, "Understanding the Landscape of Cultural Responsiveness within a Community-Driven Mathematics Education Research Project." At first, on a journey to better understand culturally responsive mathematics education, this research team ended their project having discovered an Indigenist framework guided by particular philosophies and systems of knowledge understood and lived in the Cree community. Invited to participate in community-based research, these authors were charged with finding ways to ensure student success in mathematics in the local school. Working alongside community professionals and others, Indigenous and non-Indigenous, they quickly identified inherent contradictions between the assumptions and expectations of the community and those of the school.

It was first important to acknowledge ways that the community's colonial past continues to impact community success. For example, school policies are still derived from outside perspectives. Mathematics education, in particular, adheres to mathematical standards and systemic structures that use high-stakes tests to promote accountability. Pedagogically, this implies that there is a prescribed pathway to excellence for all students. Theoretically, it suggests that mathematics is universal and culture-free. As the authors engaged with others around mathematical knowledge, they discovered the highly relational nature of social interaction and learning in this Cree community, and imagined that a flexible theory of mathematics was needed. They saw parallels between Cree ways of being and a theory of mathematics that sees effective mathematics teaching and learning not as absolutist, but instead as a relational and cultural experience. The work of this research collective underscores the foundational importance of local norms and values and how these phenomena can be used not only to build pathways to appropriate mathematics education, but also to address some of the effects of a colonial past that continue to subvert local teaching and learning approaches in favour of conventional practices.

In "Bicultural Mathematics Teacher Education and Research: Supports and Challenges," experienced New Zealand pre-service mathematics teacher educators and researchers Robin Averill, Pania Te Maro, Dayle Anderson, Herewini Easton, and Marama Taiwhati offer insights about equitable mathematics education processes and outcomes that acknowledge, reflect, and nurture Māori culture. In theory, they say, educators of Indigenous students in New Zealand should be in a good position to enact bicultural mathematics education, since New Zealand Ministry of Education policy dictates that schooling be responsive to and inclusive of Māori culture and identity. However, many teachers lack the necessary knowledge about bicultural approaches and resources to adapt or enact Māori-focused mathematics education. What then is needed to fill this bicultural mathematics teaching professional learning gap? First, the authors say to make space for bicultural consultation and partnership that places culture at the centre. Cross-cultural teams invested in developing reciprocal and respectful relationships, and who are open to others' ideas, are more willing to take risks and to invest the time it takes to create deeper understandings of Māori culture. The authors conclude that teams who share understandings about programs, pedagogy, resources, and priorities are best prepared to advocate for and deliver culturally responsive mathematics education that is truly authentic for Māori.

Roberta Hunter, Jodie Hunter, and Trevor Bills further illuminate the need for understandings of this nature in their chapter, "Enacting Culturally Responsive or Socially Response-able Mathematics Education." Many schools in Aotearoa, New Zealand serve diverse populations of Pāsifika learners—students from a range of Pacific Indigenous cultures: the Cook Islands, Fiji, Niue, Samoa, the Solomon Islands, Tokelau, Tonga, Tuvalu, and other countries. These students are often disenfranchised, and many have had negative experiences with schooling, including disconnection from mathematics education. To better meld Pāsifika learning preferences with high-quality mathematics education, the authors advocate using an assets-based strategy like that found in Developing Mathematics Inquiry Communities (DMIC). This approach encourages teachers to make use of the strengths of students' home cultures and values as bridges to the specialized participation and communication practices embodied in mathematics teaching and learning. For example, they posit the family or *fānau* to be at the core of Pasifika lifeways. If so, they suggest that beliefs and practices associated with extended family and community can be called upon to promote student motivation and achievement. DMIC, which makes use of family/community protocols around relationships, trust, respect, reciprocity, collectivism, communalism, and service, not only engages students effectively in high-order mathematics learning, but also contributes to solidifying self-efficacy and identity development processes that support success in school.

Julie Kaomea's chapter, "Hawaiian Math for a Sustainable Future: Envisioning a Conceptul Framework for Rigorous and Culturally Relevant 21st-Century Elementary Mathematics Education," reinforces these processes, introducing a framework for conceptualizing Hawaiian mathematics curricula that are culturally relevant and academically rigorous, and that also draw upon students' everyday experiences in science, technology, engineering, and mathematics (STEM). Of deep concern is the charge to develop future leaders who are prepared to address persistent environmental challenges in the islands—some related to changing climate and weather that threaten low-lying coastal regions and diminish native sea life, plants, and animals. Native Hawaiians—and, for that matter, other Pacific Islanders—are decidedly under-represented in STEM courses and careers even when the livelihoods and spiritual well-being of many depend on sustainable environmental practices. The author believes that to adequately address this pressing need mathematics education must explicitly connect school mathematics with students' lived experiences.

Historically, the subsistence activities of daily and seasonal life honed over hundreds of years enabled Hawaiian peoples to maintain a viable lifestyle. The activities they engaged in as they managed and monitored a complex system of resources from the mountains to the sea entwined processes of mathematics with aspects of spirituality. According to Kaomea, too often conventional mathematics content is decontextualized to the point that students do not see linkages to the mathematical accomplishments of their own peoples. The mathematics curricular framework introduced in this chapter is intended to guide the development of curricula that not only prepare students for the mathematical challenges of the present and future, but also integrate important elements of Hawaiian heritage beliefs and practices.

Along similar lines, "Symmetry and Body Proportional Measuring: Contributions of Yup'ik Elders to Mathematics Education," by Jerry Lipka, Dora Andrew-Ihrke, and Eva Evelyn Yanez, illustrates some of the specific contributions to mathematics education made by a cadre of Alaska's Yup'ik Elders as part of a long-term collaborative project with the Elders, teachers, and Alaska school districts. Because many Yup'ik communities continue to practice a traditional subsistence lifestyle, youth in these communities are exposed early on to deep local ecological knowledge and ways of knowing and solving problems. As they practice their lifeways they gain experience with a number of mathematical processes implicit in subsistence activities: number and operations, patterns, functions and algebra, geometry and spatial sense, measurement, data analysis, statistics, probability, reasoning and proof, representation, and so on. This is advantageous for connecting school math with their own lived experiences, and the authors show this to be optimal for bringing formal mathematics education to all students.

Unlike curricula that often ignore the socio-cultural dimensions of student knowledge, this body of empirical work centring on everyday Yup'ik knowledge related to mathematical thinking has been delivered to students of all backgrounds across the state of Alaska. Based on positive gains made by Indigenous and non-Indigenous students, the authors conclude that teaching mathematical principles gleaned from Yup'ik everyday practices offers a systematic way for educators not only to strengthen the education of Indigenous youth, but also to teach powerful, interrelated mathematical concepts, practices, and processes to all students.

Given these more fine-grained understandings about the mathematics embedded in traditional practices, one might ask, What are the particular features of mathematics curriculum development that can be codified so that developers can be sure to respond to student understandings in the most culturally appropriate ways? In addition, what components of professional learning best contribute to meaningful mathematics education in particular cultural contexts, especially for rural teachers and for remote rural schools with limited resources? The following two chapters address these prevailing questions.

In "'Remember the Time We Set up Our Tipi?' Relational Curriculum Design for Culturally Responsive Mathematics Education," Maria Jose Athie Martinez uses examples of professional learning about the geometry of the tipi as a springboard for inviting meaningful collaboration among mathematicians, Elders, and teachers as developers of appropriate and accessible school curriculum. Her approach places Indigenous knowledge at the centre, capitalizing on its suppleness not only for creating subject matter in mathematics, but also for building strong human relationships based on key cultural tenets such as respect, trust, confidence, and dialogue. Robust cultural content of this kind is advantageous well beyond mathematics teaching. It can be an important catalyst for teaching diverse themes across numerous fields, grade levels, and subjects, as it simultaneously affirms student heritage and identity.

Cynthia Nicol, Jo-ann Archibald Q'um Q'um Xiiem, and Joanne Yovanovich examine a related program of work that draws from their long-standing collaborations on Haida Gwaii, an Indigenous community in Canada's Pacific Northwest. "Being Guided by Kugann Jaad Mouse Woman for Culturally Responsive Mathematics Education," documents how a team of Indigenous and non-Indigenous educators, researchers, teachers, students, and community members engaged in participatory curriculum design, co-determining how mathematics education on Haida Gwaii might be improved to better meet the needs of students. Through a series of collective projects, this multidisciplinary group developed methods and resources for classroom and community use. Key elements central to the creation of these processes parallel those identified by other authors. Each cohort of Indigenous and non-Indigenous collaborators sought to contextualize mathematics education by first building relationships with one another as spaces of trust and dialogue. This was a rare bonus for educators of all backgrounds, who previously had few opportunities to work together. They sought to enhance their own cultural competence by collectively gaining foundational understandings about local culture, history, and place. They also explored the mathematics implicit in local cultural practices of canoe building, weaving, and carving, drawing on these contours as resources for learning. Developing deeper connections with community and bringing local knowledge together with pedagogies designed around student strengths and interests paved the way for giving back to community far beyond the classroom.

Consistent with previous chapters, in "Mathematics and Culture in Micronesia," A. J. (Sandy) Dawson, Beatriz Aguon Camacho, Tauvela Fale, Dora Borja Miura, and Amanda Fritzlan describe the philosophical underpinnings of a culture-focused, place-based program devised for Indigenous Pacific education professionals completing advanced degrees in mathematics education. The chapter recounts the early development of a program designed to be distinctly "Pacific" in nature. Program participants came from the islands of Hawai'i, American Samoa, and entities located in the northwest Pacific Ocean: Guam, the Commonweath of Northern Mariana Islands, the Republic of the Marshall Islands, the Republic of Palau, and the Federated States of Micronesia. Each of these jurisdictions experienced historical processes of education as colonization and assimilative teaching practices that resulted in language and culture loss, and community alienation from schooling. The program was intended to provide opportunities for participants to begin to recover various aspects of lost knowledges and practices and to do so on their own terms. Its goal was not to develop a one-size-fits-all Pacific approach, but instead to draw on participants' shared values of collaboration, cooperation, and collective action to identify ways to examine traditional mathematical practices in their own cultures and to translate this knowledge into elementary-level curriculum units that reflected local mathematical thought and experience. This distinctly Pacific approach has yielded effective community-based processes of mathematics education, and it has produced a cadre of master educators who are now preparing other teachers to deliver culturally responsive mathematics instruction, and influencing policy as well.

In "Show Me Your Math," Lisa Lunney Borden, David Wagner, and Newell Johnson explore the processes of delivering decolonized mathematics education to Mi'kmaw students in Canada. Here, as in other contexts described above, the cultural values of student communities do not necessarily align with those embedded in their schools' mathematics programs. The authors attribute Mi'kmaw students' lack of math engagement in formal schooling to learner perceptions that doing so essentially displaces their own ways of knowing. To combat this the program positions key individuals in students' lives—Elders, community leaders, parents, grandparents—at the core of developing math understandings, which makes use of the Mi'kmaw language and local ethnomathematics distinct. Although this strategy for decolonizing math curricula and instructional practices is designed to emphasize the power of local knowledge and culture, attention is also paid to helping students understand these processes in broad enough ways for them to acknowledge a range of mathematical views.

Finally, "Bending an Ear in the Quest of 'What is it?'" by Jennifer S. Thom, uses a series of fictional conversations about navigation practices or wayfinding as vehicles for exploring what it means to live culturally responsive mathematics education. With navigation and navigating used as metaphors for mathematics and mathematical activity respectively, living culturally responsive education as wayfinding opens the door for examination of what we know mathematically, how we know it, and a recognition of differences in our ways of knowing. It values collaborations and living relations that are constantly changing, and changing us. By examining varied perspectives on these topics, the reader can interrogate the issues raised and develop new ways of thinking about mathematics, the nature of it, and mathematical ways of knowing.

Each of these chapters examines community members' deep understandings of the social ideals, values, and behavioural standards that have shaped approaches to successful schooling for Indigenous students in their localities. They also reinforce the notion that when content areas are taught or learned in defined cultural contexts, students have increased opportunities to relate to them and find them meaningful. We would do well to extend to all educators of Indigenous youth the notion that such foundational knowledge can and should be leveraged to promote the formal learning that facilitates development to advanced levels. This place-based perspective is particularly useful in diverse Indigenous contexts where indicators of success must extend to broader notions of self-determination and nation building. The complex elements that support learning demand innovative approaches, and the approaches presented here are fine models for others seeking transformative change in mathematics education.

Sharon Nelson-Barber
Culture and Language in Education
WestEd

ACKNOWLEDGEMENTS

We acknowledge the support of many who helped to make this book possible. We begin by acknowledging the Indigenous peoples of Musqueam First Nations (Vancouver, Canada), the Indigenous peoples of Treaty 6, including the Cree, Dene, Métis, Nakota Sioux, and Salteaux, (Edmonton, Canada), and Native Hawaiians (Hawaiʻi, United States) as theirs are the lands from which we are writing and editing this book. We extend our deepest appreciation to the students, teachers, families, and communities with whom we work and to the work of the extraordinary scholars and associated communities who contributed to this volume.

This collection began many years ago with three American Educational Research Association (AERA) Symposia focused on exploring the nature of culturally responsive education with/in Indigenous communities across international contexts. At these symposia, many of the contributors to this book presented their work to large audiences with critical dialogue lead by symposia discussants including Dr. Ray Barnhardt of the University of Alaska, Dr. Liz McKinley of the University of Auckland, Dr. Marta Civil of the University of Arizona, and Dr. Christine Sleeter Emerita of California State University. These symposia brought us together as teachers, scholars, and Elders across geographical contexts to share ideas, extend our thinking, and make new connections to this growing field. As this has been an on-going multi-year project we are particularly grateful to all contributors to this volume for their patience, support, and belief in making this book a reality.

Along our journey from book conception to reality, we lost co-editor Dr. A. J. Sandy Dawson in the winter of 2015 in Hawaiʻi. We dedicate this book to Dr. Dawson who worked with Indigenous peoples throughout his academic career. Dr. Dawson's "innovative and imaginative" doctoral work in the early 1970s "looked at the implications of Lakotas' work for school mathematics" (Kieren, 2015, personal communication). Dr. Dawson was a

> selfless human being whose life work at its core was about cultural exchange, equity, and mutual respect. Mathematics education was his vehicle. He was not only intellectually curious but also profoundly interested in people and listened well to appreciate the life experiences of those with whom he worked. He always gave in his interpersonal relations and, as a consequence, received so many insightful gifts about people and their world. (Powell, 2015, personal communication)

We are truly grateful for all we learned as we travelled along side Dr. Dawson in this project and in our careers. Dr. Dawson was a leader, mentor, and friend in learning what it means to engage in culturally responsive practices.

We are indebted to the work of graduate students Amanda Fritzlan, Heather Commodore (now Dr. Heather Commodore), Jeff Baker (now Dr. Jeff Baker at the University of Saskatchewan), Maria Jose Athie Martinez, Rob McDonald, Tathali Urueta (now Dr. Tathali Urueta at the University of British Columbia, UBC), and Janice Novakowski. We valued their participation on research teams related to moving the research papers to book chapters, as well as their critical questions, feedback and innovative ideas. We are also grateful for the contributions by artist Kirsty Robbins of the UBC Department of Curriculum and Pedagogy. We also thank the reviewers, in particular Dr. Lester Rigney, who provided a number of suggestions and insights that improved individual chapters and book coherence. We are grateful to Brill Sense, and in particular to Michel Lokhorst for his continued support and for providing information in the process of preparing the book for publication. We also acknowledge Evelien van der Veer and Jolanda Karada who supported the final stages of the book to publication.

I (Cynthia) acknowledge the place of Haida Gwaii, British Columbia Canada and the many people who have mentored me and supported the making of this book. I thank my Haida family in Gaaw, Old Masset, including Aggie Davis and Helen Davis, for their ongoing teachings that continually help me to be a better listener to others, the land, my ancestors, and myself. I'm also thankful to my colleague and friend Joanne Yovanovich and her family in Skidegate, Haida Gwaii for making me a better learner and teacher by bringing me closer to Haida ways of knowing through connections to family, story and humour. I'm grateful to my UBC colleagues Dr. Ann Anderson, Dr. Susan Gerofsky, Dr. Samson Nashon, and Dr. Jo-ann Archibald for their teachings and unending confidence and encouragement. Finally thank you to my family Gord, Elyse, Chai, Monty, and Annie for their continued support and patience, and most importantly for providing me with the strength to expand my learning.

I (Jo-ann), in Coast Salish tradition, raise my hands in thanks and respect to my UBC colleagues, graduate and undergraduate students, and Indigenous Elders with whom I have learned from and with over the years, regarding Indigenous knowledge systems, mathematics education, and community-based research. Thanks to my partner, John, for engaging in countless discussions about Indigenous education and Indigenous storywork that helped me to clarify and deepen my ideas. I am grateful to my daughter, Corry, and her husband, Brian, and their two children, Sawyer and Bennett for inspiring me to share Indigenous teachings and stories about the heart, mind, body, and spirit. Thus, the circle of learning, caring, and sharing continues.

I (Florence), as a member of the Métis Nation of Alberta, acknowledge my ancestors and Elders who have walked before me. It is those ancestors and Elders who have provided the generative spaces for me to live and engage in this work in mathematics education. I am grateful for the teachings about Indigenous knowledge systems, about mathematics education, and about working with communities that I've received from Indigenous and non-Indigenous peoples who have walked alongside me. Finally, I am grateful to my husband, Andy, my father, Oliver, my sister, Brenda, and my nieces, Lara and Valisa, for their never-ending curiosity, questions, and support. The family stories we tell and live by provide me with the courage and strength to live my life-calling.

Finally we thank you, the reader, for selecting this book, and we hope that the chapters inspire engagement with these ideas and actions for living responsive, relational, and respectful mathematics education with/in your communities.

FIGURES AND TABLES

FIGURES

TABLES

NOTES ON CONTRIBUTORS

Dayle Anderson is a senior lecturer in primary mathematics and science education at Victoria University of Wellington. She is a committed pre- and in-service primary teacher educator with experience working as a scientist and in primary teaching. Her research interests in mathematics and science education have a sociocultural and culturally responsive focus. Her research includes studies of primary science education, the use of rehearsal and coaching in initial mathematics teacher education, and a number of classroom based, collaborative projects. Current projects include research into primary teachers' beliefs about science, and participatory action research into the use of drama in developing children's epistemological and substantive science understanding. As well as acting as a reviewer and consultant for the New Zealand Ministry of Education and other organisations, Dayle has written primary teaching resources and textbook chapters for tertiary teaching students in both mathematics and science.

Dora Andrew-Ihrke has been a bilingual aid, classroom teacher, and program coordinator of bilingual/bicultural education at Dillingham City Schools. Her teaching was recognized in Alaska and nationally: 1985 a recipient of Dillingham City Schools' Excellence in Teaching, 1990 a Milken Educator Award and Alaska Teacher of the Year, and a 2001 recipient of the Alaska Federation of Natives education award. She is currently a faculty member at the University of Alaska Fairbanks, School of Education in the *Math in a Cultural Context* (MCC) program. From childhood on she has been learning from and subsequently working with Yup'ik elders. During the past years through the MCC program, she has been sharing the lessons she learned from her mother and other elders and how these lessons apply to teaching school mathematics. From her mother, she learned a powerful "folding algorithm" which is foundational to MCC's new approach to teaching mathematics. She is also a co-author to journal articles and for some of the published MCC modules. She teaches *qaspeq* (dress, literally a covering) making class in Anchorage.

Jo-ann Archibald Q'um Q'um Xiiem, from the Stó:lō and St'at'imc First Nations in British Columbia, Canada, is Professor Emeritus in the Educational Studies Department at the Faculty of Education, University of British Columba (UBC). She was the former Associate Dean Indigenous Education and the Director of UBC's Indigenous Teacher Education Program (NITEP). She received a Bachelor of Education (B.Ed.) degree from the University of British Columbia, a Master of Education (M.Ed.) degree and Doctor of Philosophy (Ph.D.) degree from Simon Fraser University. Over a 45-year educational career, Jo-ann has been a school teacher, curriculum developer, researcher, university administrator and professor.

Jo-ann's scholarship relates to Indigenous knowledge systems, storywork/oral tradition, transformative education at all levels, Indigenous educational history, teacher and graduate education, and Indigenous methodologies. From 1992–2018, Jo-ann edited an annual theme issue of the Canada Journal of Native Education.

Maria Jose Athie Martinez is a current PhD student in Curriculum Studies at the University of British Columbia (UBC). She is a mixed race Mexican mestiza. She has worked alongside Indigenous people mostly in her home country, Mexico. Her working area is finding pathways to bring in dialogue, respect and dignity, Indigenous Knowledge and Western Knowledge in a curriculum. She has also worked on training teachers on continuing education courses trying to make culturally and environmental responsive curriculum mostly in Indigenous schools. She has worked in all educational levels, but mostly in higher education, in private and public Universities in Mexico. For her Master's thesis, also at UBC, Vancouver, Canada, she worked developing math lessons from K-12 to graduate studies, based on a tipi experiential story.

Robin Averill is Associate Dean (Teacher Education) at Victoria University of Wellington in New Zealand. Robin's research interests focus on enhancing equity of access to mathematics achievement. Projects include investigations into culturally responsive classroom and teacher education practices, particularly in relation to New Zealand's Indigenous learners, using coaching to promote ambitious mathematics teaching, and effectiveness of national professional development projects. Robin has published widely on the teaching and learning of elementary, secondary, and middle school mathematics and statistics in articles, edited books, and has written a range of resources for teaching. Robin is a member of MERGA (Mathematics Education Research Group of Australasia) and is the New Zealand representative on ICMI (International Commission on Mathematical Instruction).

Trevor Bills comes from a background of teaching and middle management. He is currently employed at Massey University, New Zealand as a mentor to teachers in the Developing Mathematical Inquiry Communities within a wide range of diverse schools in high poverty schools across New Zealand. The focus of Trevor's research interests includes culturally responsive mathematics teaching and learning, the development of mathematical rich and culturally appropriate tasks, mathematical inquiry classrooms, ethnomathematics, critical mathematics, and the relationship between education, social change, equity and social justice.

Beatriz Aguon Camacho is currently employed as a school guidance counselor with the Guam Department of Education. She works at Southern High School in Santa Rita. In addition to her role as a School Guidance Counselor, Ms. Camacho is actively involved in mathematic curriculum for the district as well as holding a management position with Asmuyao, a local community school. Ms. Camacho holds an M.A. in School Guidance and Counselling from the University of Guam and an M.Ed. in Educational Leadership from the University of Portland in Oregon. Her doctoral research focuses on teachers as they go through the process of developing Service Learning projects. Ms. Camacho has several accomplishments, one being she was the recipient of the Presidential Award for Excellence in Mathematics and Science Award for the U.S. Territories in 2011. Her work with Project MACIMISE (Mathematics and Culture In Micronesia: Integrating Societal Experiences) and with the Macimisers has made her appreciate and value connections with mathematics, culture, and community that allow for making sense of information and lead to a deeper understanding and knowledge base.

A. J. (Sandy) Dawson, Professor Emeritus, was a faculty member at the University of Hawaii Mānoa College of Education in the Department of Curriculum Studies and with the Institute of Teacher Education. His chapter is based on the work that he began and led through the Mathematics and Culture in Micronesia: Integrating Societal Experiences (MACIMISE) collaborative research project funded by the National Science Foundation. Through this project Dr. Dawson sought to improve elementary school mathematics education in eight Micronesian Island groups by drawing upon mathematical practices in Island communities everyday and cultural experiences. The author of many mathematics education and mathematics teacher education books and articles, Dr. Dawson's contributions to the field are felt both internationally and locally.

Dwayne Donald is a descendent of the amiskwaciwiyiniwak and the Papaschase Cree and is an Associate Professor in the Faculty of Education at the University of Alberta. His work focuses on ways in which Cree and Blackfoot wisdom and sacred ecologies can enhance understandings of curriculum and pedagogy.

Herewini Easton, previously lecturing at Victoria University of Wellington, has continued to develop his aroha for Māori education through a range of tertiary and community institutions and initiatives. Herewini was Kaiārahi (Programme Manager) and kaiako (lecturer) for Te Korowai Akonga-Bachelor of Teaching (Primary), Te Wānanga o Aotearoa and an Academic Development Lecturer for Maia Maori Development Centre, Unitec Institute of Technology, Te Whare Wānanga o Wairaka. During this time he was also involved with the National Centre of Literacy and Numeracy for Adults at Waikato University. Herewini is currently a Māori Academic Equity Lead at Auckland University of Technology, a Kaitautoko Māori at Te Pataka Korero Auckland Central Library, and an Advocate Māaori at Auckland Theatre.

Tauvela Fale occupies the position of Director of Adult Education, Literacy and Extended Learning at the American Samoa Community College. He is also adjunct faculty for the teacher education program at ASCC. He obtained a BEd in elementary education University of Hawai'i Mānoa, and an MEd with an emphasis on mathematics from University of Hawaii at Mānoa. As a member of the MACIMISE, his doctoral research focus has been the design and implementation of units for teaching mathematics at the elementary level using indigenous practice. Tauvela Fale continues to pursue his commitment to improving students' mathematical knowledge on their island, through working with the teacher education program to instil within prospective teachers the knowledge of how to teach mathematics using cultural connections especially for English language learners. He also works within the larger community through partnerships with the private sector and courses offered to adults in the work place for improvement of mathematics and language skill.

Amanda Fritzlan is a doctoral student in the Curriculum Studies program at the University of British Columbia in Vancouver, Canada. Previously, she taught for over ten years in local public schools at the middle school level. Her Master's of Education research focused on teaching Aboriginal art as a non-Aboriginal person as well as student-centred museum experiences. Her current research centres around pedagogies in culturally diverse urban classrooms. She is particularly interested in issues of reconciliation, land, and materiality in culturally responsive mathematics education.

Florence Glanfield is Vice-Provost (Indigenous Programming & Research) and Professor of Mathematics Education in the Department of Secondary Education at the University of Alberta. Florence is a member of the Métis Nation of Alberta and has had the honour of working with rural and urban Indigenous communities in Canada and internationally. Florence's research interests emerge in the interaction among mathematics teacher education, teacher education, relational inquiries, Indigenous perspectives and complexity science. Florence is a recipient of the Iris Carl Leadership and Equity Award from TODOS: Mathematics for ALL (2014) and the Kay Gilliland Equity Lecture Award from the National Council of Supervisors of Mathematics (2017).

Jodie Hunter is a mathematics education researcher of Cook Island/Pakeha descent within the Institute of Education at Massey University. She began her career as a primary teacher in New Zealand before working as a Research Fellow at the University of Plymouth in the United Kingdom. She completed her doctorate shortly after returning to New Zealand to work at Massey University. Recently Jodie has been involved in a large-scale project funded by the New Zealand Ministry of Education which focuses on developing culturally-responsive teaching to address under-achievement in mathematics for Pasifika and Māori students at low socio-economic schools.

Roberta Hunter is currently an Associate Professor of Mathematics Education in the Institute of Education at Massey University. Over the past 15 years she has actively engaged in research and professional development projects which support teachers to develop equity focused culturally responsive mathematics pedagogy particularly with Pāsifika and Māori students and other diverse students in high poverty areas in New Zealand. The focus of her research and practice is on honouring students' culture and language, so that they build positive cultural and mathematical identities. She works with teachers to draw on student values and cultural background to make mathematics real and meaningful but also to engage their participation in mathematical reasoning and communication. Other research interests include the development of mathematical practices within communities of mathematical inquiry.

Newell Johnson is from Eskasoni First Nation. Her education background is diverse: she has a Bachelor of Arts, Bachelor of Science, Bachelor of Education, Masters in Special Education, and Masters in Mathematics. Johnson began her teaching career as a high school science teacher in 1999 and gradually became a high school math teacher. She was also a vice principal for six years before becoming a full-time principal in the last four years. She is a fluent Mi'kmaq speaker (English is her second language) and enjoys learning about her Mi'kmaq culture. Johnson takes pride in her culture, which she tries to instil in her students. It is because of her wonderful, caring, compassionate and understanding staff at Chief Allison Bernard Memorial High School (formerly known as Eskasoni High School) that they are able to provide a culturally rich and academically challenging curriculum.

Robyn Jorgensen (Zevenbergen) is Professor of Education: Equity and Pedagogy at the University of Canberra. Spanning nearly three decades, her work has focused on issues of equity and access for marginalized learners. In particular, her work has focused on issues of social class, rurality and Indigenous students as these are the cohorts of students most at risk of failing school mathematics. She is the leader on a number of large research grants and is currently completing a national study on successful numeracy practices in Remote Indigenous contexts (from which the chapter herein is drawn from). She employs a socially-critical lens, often using the work of Bourdieu, in her work and seeks to explore the practices to contribute to, or address, the marginalization of particular groups of students.

Julie Kaomea, PhD, is an Associate Professor in the Department of Curriculum Studies at the University of Hawai'i–Mānoa. She has worked and researched collaboratively in a number of Hawaiian educational settings, including the initial preparation and professional development of elementary educators who are committed to teaching in predominantly Native Hawaiian school communities.

Jerry Lipka, during the past 35 years, has had the privilege and opportunity to work with Yup'ik teachers and elders in Alaska as a faculty member at the University of Alaska Fairbanks, School of Education. He has published journal articles that reflect an ethnographically-oriented approach. These publications include Yup'ik pedagogy, qualitative case studies of teachers implementing Math in a Cultural Context curriculum materials, and quantitative efficacy studies. More recently, he and his colleagues have become aware of the elegance and power of a few fundamental Yup'ik concepts, measuring as comparing and halving, as a way to build an alternative mathematical learning trajectory. This developing framework is the foundation for our new elementary school curriculum and professional development work as well as working internationally with Indigenous Knowledge holders in Greenland, in the Carolinian Islands, and Kamchatka.

Lisa Lunney Borden is an Associate Professor of mathematics education at St. Francis Xavier University in Canada and currently holds the John Jerome Paul Chair researching Equity in Mathematics. Having taught 7–12 mathematics in a Mi'kmaw community, she credits her students and the community for helping her to think differently about mathematics teaching and learning. She is committed to research that focuses on decolonizing mathematics education through culturally based practices and experiences that are rooted in Aboriginal languages and knowledge systems. Lisa is equally committed to mathematics outreach through programs such as *Show Me Your Math* that was developed with David Wagner, Newell Johnson, and a team of teachers from Mi'kmaw Kina'matnewey schools. This program invites Indigenous youth to find the mathematical reasoning inherent in their own community context. Lisa is a sought-after speaker on Indigenous mathematics education, working with mathematics educators across Canada as well as internationally.

Dora Borja Miura, as a mathematics teacher at Saipan Southern High School in the Commonwealth of the Northern Mariana Islands (CNMI), promotes building community and relationships in the classroom for ongoing student and teacher growth. In 2015, she received the US National Science Foundation Presidential Award for Excellence in Mathematics and Science Teaching. She was named teacher of the year by CNMI Public School System in the same year. Dr. Miura received a bachelor's degree from Georgetown University as well as a master's degree and a doctorate degree in education as a Macimiser from University of Hawaii at Mānoa. She has served the mathematics education community as a mentor for other public school mathematics educators in CNMI and as an administrator for CNMI Mathcourt. She has worked for the CNMI PSS as an elementary school teacher, a distant education teacher, and secondary school teacher of advanced algebra and geometry. She has also been employed as an adjunct instructor at North Marianas College. Her professional aims include having all students in the Commonwealth appreciate mathematics and sharing her research on relationship between mathematics and culture.

Sharon Nelson-Barber directs Culture and Language in Education at WestEd. Her research centers on understanding how the sociocultural contexts in which students live influence the ways in which they make sense of schooling in mathematics and science. She also focuses on understanding how aspects of cultural knowledge can become visible in assessment and evaluation to ensure that schooling is equitable for all students. She is co-founder of POLARIS (Pacific/Polar Opportunities to Learn, Advance and Research Indigenous Systems), a research and development network that fosters healthy communities in healthy environments; encourages social and educational transformation; and brings a uniquely Indigenous world view to new frontiers of knowledge. Her work spans the lower 48 states, Alaska, Micronesia, and many areas of Polynesia. She earned a doctorate in human development from Harvard University, and completed postdoctoral work at Stanford University as a Spencer Fellow.

Cynthia Nicol is Associate Professor in the Department of Curriculum and Pedagogy in the UBC Faculty of Education and the David F. Robitialle Professor in Mathematics and Science Education. She lived and taught on Haida Gwaii in Canada's Pacific North coast before moving to Vancouver to pursue her doctoral studies. With teachers and communities she is exploring new ways of making mathematics responsive to all learners by connecting math, community and culture, emphasizing place and community-based education, and exploring social justice issues through mathematics. Her current projects focus on researching ways to support teachers interested in more culturally responsive teaching practices. This includes practices to better understand Indigenous and non-Indigenous relationships. This also includes working with teachers in the Dadaab refugee camp, Northeast Kenya, to better understand what it means to live, learn and teach in Dadaab the largest protracted refugee camp in the world. As Principal Investigator of this project she is committed to working across cultures and contexts to better understand and improve opportunities for ongoing teacher learning in some of the world's most challenging conditions.

Gladys Sterenberg is a Professor in the Department of Education at Mount Royal University, Canada. Her program of research is focused on relational ethics and can be described as encompassing three interrelated research interests: mathematics education, Indigenous ways of knowing, and overlapping communities of practice within teacher education. She taught K-12 for fifteen years before joining the academy.

Marama Taiwhati is a former lecturer and advisor to kura/schools from Victoria University of Wellington. With experience teaching in Māori immersion schools, Marama contributed as advisor and investigator within a range of research projects, including evaluations of national professional development projects aimed at responsiveness to Māori learners.

Pania Te Maro, Ko Hikurangi te Maunga, Ko Waiapu te Awa, Ko Ngāti Porou te Iwi, Ko te Whānau a Pōkai te Hapū, Ko Te Kapa o Hineauta te Tūrangawaewae, Ko Pōkai te Marae, Ko Pōhatu te Wharekai. Tēnā rā tātou. A senior lecturer/

researcher in education previously at Victoria University of Wellington and now at Te Whare Wānanga o Awanuiārangi, Pania Te Maro lectures in mathematics/pāngarau, literacy and the general curriculum for Early Childhood Educators through to students who may teach in Wharekura/Māori immersion secondary schools. Pania's research focuses on the development of self-determined curriculum for Māori immersion educational settings in relation to pāngarau (mathematics). Projects include evaluations related to Māori junior and senior literacy, Te Poutama Tau curriculum implementation, and an aromatawai/assessment framework for Māori immersion schools. Retention of beginning Māori teachers in Māori schools and identifying success factors for adult students gaining high literacy achievement have been further research projects.

Jennifer S. Thom is an Associate Professor of mathematics education and curriculum studies at the University of Victoria. Her work, which is rooted in ecological thinking, seeks to understand the systemic nature of knowing as well as the co-specifying manners in which agents and their environments bring forth effects and potentials. Jennifer's research interests include embodied cognition, collective understanding, and eco-culturally responsive curricula, all of which aim to engage a deeper sense of what it means to teach and learn.

David Wagner is Associate Dean (Graduate Programs) in the Faculty of Education at the University of New Brunswick. His interest in human interaction in mathematics and mathematics learning inspires his research, including: identifying positioning structures in mathematics classrooms by analyzing language practice, ethnomathematical conversations in Aboriginal communities, and working with teachers to interrogate authority structures in their classrooms. He serves as associate editor of *Educational Studies in Mathematics*, and on the Nonkilling Science and Technology Research Committee, the International Committee of *Mathematics Education and Society,* and the editorial board of *Mathematics Education Research Journal*. His wide ranging international projects include work with UNESCO's Mahatma Gandhi Institute of Education for Peace and Sustainable Development to promote peace, sustainable development and global citizenship in school mathematics curricula. He has taught grades 7-12 mathematics in Canada and Swaziland.

Eva Evelyn Yanez is a former Yup'ik teacher, bilingual coordinator, and a State of Alaska recognized bilingual educator. She has been involved in education for over 30 years. She is a life-long learner of traditional Yup'ik stories as well as other cultural activities. She has extensive experience working with Yup'ik Elders from collecting, recording, transcribing and translating traditional Yup'ik stories to retelling stories for publication. In professional development settings, she shares the cultural practice of storyknifing and dance, and connects aspects of these cultural activities to foundational mathematical concepts. She is also a co-author for some of the modules in the *Math in a Cultural Context* series. More recently, she has worked with community members to revitalize the Yup'ik language.

Joanne Yovanovich was born and raised in the Ts'aahl Eagle Clan of Skidegate on Haida Gwaii, Her Haida name is Taanud Jaad. She has worked with School District #50 since 1995, beginning as the Full Day Kindergarten teacher and Vice Principal then becoming Principal of Sk'aadgaa Naay Elementary School. Currently she has the position of Principal of Aboriginal Education in Haida Gwaii School District. She is rooted in her community and the place of Haida Gwaii, She strives to connect the worlds of cultural and school knowledge. She is deeply committed to making a difference in student success rates on Haida Gwaii. She believes that culturally responsive education is a key to student success. Culturally responsive education can be transformative, inspiring and validating for students and staff that embrace and practice the philosophy.

CYNTHIA NICOL, JO-ANN ARCHIBALD Q'UM Q'UM XIIEM AND FLORENCE GLANFIELD

INTRODUCTION

Making a Difference with/in Indigenous Communities

INTRODUCTION

In 2016, four international organizations released a joint statement recognizing the international rights of Indigenous peoples and, in particular, the right to appropriate, sustainable, and self-determined education as outlined in Article 14 of the United Nations Declaration on the Rights of Indigenous Peoples (UNDRIP). The announcement called on the member nation states of the UN to:

> Work with indigenous peoples in a spirit of partnership to restore forms of education based on indigenous languages, beliefs, values and culture and increase efforts to address discrimination in education that has the effect of impeding indigenous peoples' rights to education. It is imperative that educational institutions are built on a human rights framework that is inclusive and respectful of indigenous peoples' cultures, worldviews and languages. (United Nations, 2016, para 7)

This statement speaks to the international call for "culturally sensitive" or culturally responsive education systems that not only recognize but also actively engage people, institutions, and systems worldwide to offer the universal and fundamental right of accessible and appropriate education. Although adopted by the UN General Assembly in 2007, not all UN countries supported the Declaration at that time. Canada, for example, only recently in 2016 announced its intention to implement the Declaration of Rights, requiring changes to its federal legislation and continued relationship-building between government and Indigenous nations. While other countries such as Australia, New Zealand, and the United States had by this time reversed their original decision and supported UNDRIP to some degree.

Internationally, countries are also working to address a legacy of human-rights abuses against Indigenous people and to reconcile and improve relationships with Indigenous communities. Canada is one of the first G7 countries to develop a formal commission and forum to redress the intergenerational trauma of residential schools and reconcile relationships between Indigenous and non-Indigenous peoples. Australia, New Zealand, and the United States are also engaged in implementing forums to review, restructure or initiate federal government policies that work

 DOI: 10.1163/9789004415768_001

toward aspects of reconciliation intentions (Geboe, 2015). In 2015, the Truth and Reconciliation Commission of Canada (TRC) released its recommendations in the form of 94 *Calls to Action* that include calls for the Canadian government to increase funding for education of Indigenous students for "improving education attainment levels and success rates" (TRC, Calls to Action 10.ii, p. 2) and for "developing culturally appropriate curricula" (Calls to Action 10.iii, p. 2).

These international conventions, declarations, and commissions for recognition and action seek to address, challenge and change educational systems that systemically discriminate against Indigenous peoples, assimilate Indigenous people into the dominant society, and as a result eradicate culture, language, literacies and identity as well as displace Indigenous people from their lands for the purposes of natural resource extractions. Most often these educational systems, policies and curriculum are developed without the participation, input or consent of Indigenous people. Indigenous scholar Battiste (2013) reports that about "600 million Indigenous people in the world exist in various landscapes and ecologies" (p. 160). It is through the land and ecologies that generations, Battiste continues, "passed this knowledge and experience to succeeding generations through its language," creating "the foundation for their knowledge, culture, and heritage" (2013, p. 160). This means the right to education is inextricably connected to all human rights, including the rights to lands. However, calls for access to a high-quality education that is culturally respectful, responsive, sustaining and revitalizing, and that is relational involving communities and youth in decisions of curriculum and pedagogies, are a challenge to actualize. Indigenous peoples are still challenging new forms of colonization and neoliberal actions through decolonizing efforts (Battiste, 2013).

A decade ago mathematics education researchers Brian Greer, Swapna Mukhopadhyay, Arthur B. Powell, and Sharon Nelson-Barber published their edited book *Culturally Responsive Mathematics Education* (2009), which stems from a 2004 conference exploring culturally responsive mathematics education in the United States. Following others before them (e.g. Bishop, 1991), these authors challenged the idea of mathematics being culture-free and argued for mathematics as a human activity "grounded in human interactions with the environment and with one another, in the solving of practical problems and the human desire to transcend matters of simple survival, in universal human traits such as the use of physical and mental tools, the decorative impulse, and pleasure in intellectual play" (p. 1). Their book was ground-breaking in its discussion of culturally responsive mathematics from political, historical, cultural and social perspectives for Native American, African American and Latino/a students in the United States marginalized by systemic inequalities such as ethnicity or language.

This book, *Living Culturally Responsive Mathematics Curriculum and Pedagogy: Making a Difference with/in Indigenous Communities* began many years ago with three American Educational Research Association (AERA) Symposia focused on similar issues as those discussed by Greer and colleagues (2009). Scholars also explored the nature of culturally responsive education with/in Indigenous communities, across international contexts. At these symposia many of the contributors to this book presented their work to large audiences with critical dialogue lead by symposia discussants such as Dr. Ray Barnhardt of the University of Alaska, Dr. Elizabeth McKinley of the University of Auckland, New Zealand, Dr. Marta Civil of the University of Arizona, and Dr. Christine Sleeter Emerita of California State University. These symposia brought us together as teachers, scholars, community members, and Elders across geographical contexts to share ideas, extend our thinking, and make new connections to this growing mathematics education field.

Following these Indigenous education mathematics gatherings, the four original editors who had either worked cooperatively or who had known each other's scholarship began collaboration on a new book to extend the work of Greer et al. (2009). The result is this book, a collective involving 30 authors, that focuses specifically on mathematics education with/in Indigenous communities, across international contexts, involving Indigenous and non-Indigenous authors, human and more-than-human, natural and supernatural, responding to calls for improved education for all Indigenous students.

To begin, we follow a common Indigenous protocol that involves introducing ourselves in relation to place, community, and experience. As mentioned in the Acknowledgements, one member of our editing team, Dr. Sandy Dawson passed away during the book's publication process.

Cynthia Nicol is a seventh generation Canadian of European ancestry rasied on the Kootenay territory of the southern interior of British Columbia (BC), and learned to teach mathematics on Haida Gwaii in BC's Pacific northwest coast. The land, community and her students were, and continue to be, her teachers. Cynthia teaches mathematics education with connections to community, culture, and place at the University of British Columbia. Her experiences learning and teaching mathematics in rural and Indigenous communities ground her commitment to expanding the circle of what it means to do mathematics, who can do it, and what counts as success for students, teachers and teacher educators.

Jo-ann Archibald Q'um Q'um Xiiem is a member of the Stó:lō Nation located in southwestern British Columbia (BC), Canada. She also has ancestry from the St'át'imc Nation in the interior of BC. Her Indigenous name, Q'um Q'um Xiiem, means strong clear water. Q'um Q'um Xiiem, recently retired from the University of British Columbia where she taught Indigenous education to undergraduate and graduate students, engaged in community-based research, and developed Indigenous programs and student services. Her work has placed Indigenous knowledge systems at the core of learning. Regarding mathematics, she welcomed the opportunity to work with colleagues who challenged conventional mathematics approaches and who sought to find mathematics in Indigenous worldviews and knowledge systems.

Florence Glanfield is a member of the Métis Nation of Alberta, was born and raised in Northeastern Alberta, and is a Professor of Mathematics Education and Chair of the Department of Secondary Education at the University of Alberta. Florence teaches courses in mathematics education, Indigenous perspectives, and curriculum studies to undergraduate and graduate students. Her work has lead her to working with teachers in all regions and territories of Canada and internationally. Florence's work was situated in her early university experiences when she met people who were surprised that she, a 'native' person, was studying mathematics. Throughout her professional life, Florence has searched for coherence and relationships among her enjoyment of teaching and learning mathematics and the teachings of her family.

The concept of culturally responsive pedagogies in relation to Indigenous students' cultures and their communities is introduced next to set a context for culturally responsive pedagogies in relation to mathematics education. This introductory chapter concludes with a discussion about how the book chapters exemplify the theoretical principles of Indigenous storywork (Archibald, 2008).

CULTURALLY RESPONSIVE PEDAGOGIES

> Meaning comes from the context rather than the content. Knowledge holders spend a great deal of time setting up the conditions for engagement rather than trying to predict or control the outcome. Ultimately we access knowledge through the quality of our relationships and the personalized context we collectively create—the meaning comes from the context and the process, not the content. (Simpson, 2014, p. 112)

Indigenous writer and scholar Leanne Simpson writes of the power of place or context for meaning making focused on relationships. Similarly, more than 20 years ago Indigenous American scholars Oscar Kawagley (2006) and Gregory Cajete (1994, 1999) argued for the need for Indigenous students to experience a "culturally based education" that "honors holism and integrated learning"; however, they also considered how content is placed in relation to context for learning (Cajete, 1999, p. 49). Both scholars challenged earlier assumptions, prominent in the 1960s and 1970s, that Indigenous students' achievements in mathematics and science were significantly lower than their White American peers because of cultural and cognitive deficits. Theories of teaching and learning that focused on deficit approaches posited that marginalized students' languages and cultural practices were deficits or obstacles that needed to be overcome by education. In fact, for many educators working from a cultural-cognitive deficit perspective, minoritized students' home environments were considered as lacking and in need of altering in order for students to be successful in school (Cajete, 1999; McCarty & Lee, 2014). Thus, the goal of deficit approaches to education was to ignore, and even eradicate, the cultural practices that students brought to school, with the intent of assimilating students into the dominant culture. Cajete (1999) challenged such approaches to education and articulated strategies and models of curriculum development that made learning more interdisciplinary, culturally interactive, and contextulized so that the "lesson and the learning of the lesson [were] intimately interwoven within the situation and the environment of the learner" (p. 53).

Cajete (1994, 1999) and Kawagley (2006) also challenged cultural difference theories prominent in the 1970s and 1980s. Cultural difference theories recognized the differences between school cultural practices and home cultural practices not as deficits but rather as just being different. However, educators working from cultural difference approaches sought to develop bridges between home and school cultures under the intent of replacing students' home, cultural, and linguistic practices with the dominant practices of mainstream schools. The implication of deficit and cultural difference approaches was that students could succeed in school if they had increased opportunities for access to the dominant academic practices that required students to lose their heritage as well as their community, cultural, and linguistic practices.

Although many educators have taken up some form of culturally responsive pedagogy, Ladson-Billings (2014) claims that much of what she has witnessed in classrooms over the past 20 years is a "distortion and corruption of the central ideas" (p. 82) she developed some years ago. Some practices, she notes, claim to be culturally relevant yet are only enacted—for example, when celebrating certain holidays or cultural customs. And for those attempting to enact culturally responsive pedagogy, few, according to Ladson-Billings (2014), engage students in the critical socio-political consciousness aspect, such as critiquing dominant power structures. Well-meaning educators, writes Darder (2015), therefore miss the "revolutionary potential" of culturally responsive approaches focusing instead on "a fractured curriculum of cultural songs, stories, holidays, and heroes … entrapped in a deficit paradigm of difference" (p. 43).

Consciousness-raising is an important aspect of culturally responsive pedagogies. Without some form of criticality, pedagogy grounded in cultural knowledge risks reproducing a static view of culture, limited to what is already known and reproduced through a system of cultural apprenticeship. Culturally responsive pedagogies use students' cultures as resources or assets (Paris, 2012) rather than as aspects that need to be erased or replaced. Furthermore, such pedagogies recognize the fluidity of culture while acknowledging that classrooms likely privilege certain voices over others. Thus, culturally responsive pedagogies are not only responsive to students' heritage, language, and cultural practices but also to their contemporary cultural practices as well as students' imagined futures, or what Skovsmose (2011) calls their "foregrounds."

As a result, there are many different approaches to culturally relevant or culturally responsive pedagogy (Gay, 2010). What these approaches share is a focus on the resources available to various groups of students in subject areas such as mathematics, "so that curricula can be designed in a way that empowers all students to be successful in learning and using mathematics" (Enyedy, Danish, & Fields, 2011, p. 275). Culturally responsive pedagogies challenge the uniformity and neutrality of knowledge creation, recognize the central role of language, ethnicity, home, and community culture on students' learning, and advocate pedagogies of multiple discourses (Gutiérrez & Rogoff, 2003; Ladson-Billings, 1995; Phunstog, 2001).

CULTURALLY RESPONSIVE PEDAGOGIES AND MATHEMATICS EDUCATION

Culturally responsive pedagogies in mathematics education consist of complex sets of concepts that can be interpreted from multiple perspectives. For some, culturally responsive pedagogies in mathematics education are considered as ways of enacting critical mathematics education (Skovsmose, 2014). Scholars in critical mathematics education connect purposes, intentions and approaches in critical education to the field of mathematics education (Frankenstein, 1987, 2012; Gutstein, 2006; Skovsmose, 2012, 2016). Here, liberatory problem posing and problem solving education involve both teachers and learners in addressing issues of equity, inclusion and exclusion, teaching and learning for justice, and mathematical agency. For others, culturally responsive pedagogies in mathematics education are connected to the field of ethnomathematics education (D'Ambrosio, 2006). For Powell and Frankenstein (1997), a critical view of ethnomathematics includes not only the mathematics of various cultures and cultural groups such as "the mathematics of Angolan sand drawings" but also "the politics of imperialism that arrested the development of this cultural tradition" and "the politics of cultural imperialism that discounts [this kind of] mathematical activity" (p. 2).

Considered as critical mathematics education, culturally responsive mathematics pedagogies, as articulated in this volume, extend previous visions of pedagogies that attempted to support students in moving from outside to inside dominant cultural practices thereby erasing students' culture and language. Furthermore, many of the conversations on culturally responsive pedagogies have heard less from Indigenous communities and more from other groups also marginalized from institutional and systemic inequalities. Thus, this book calls for educational change and offers culturally responsive mathematics education that is grounded in place and community and speaks to communities and landscapes.

Each chapter is also a story of Indigenous and non-Indigenous authors' experience with culturally responsive education. There are many ways to work with and understand Indigenous stories, as this book demonstrates (Archibald, 2008). In the Foreword, Dr. Sharon Nelson-Barber presents her understandings of these 11 chapters and introduces each chapter in detail using a thematic structure that was important to her. Complementing this structure and honouring the flexibility of meaning-making through stories, the structure of this book emphasizes place by situating chapters geographically (see Figure 1.1). An Elder reminded Florence that the earth rotates eastward, and so we introduce the chapters following that direction. Imagine embarking on a story journey with the editors starting in the Pacific westcoast of British Columbia (Chapter 2: Nicol, Archibald, & Yovanovich and Chapter 3: Athie Martinez), moving eastwards to Alberta (Chapter 4: Glanfield, Sterenberg, & Donald)), then to Atlantic Canada (Chapter 5: Lunney Borden, Wagner, & Johnson). Next we travel further east to New Zealand (Chapter 6: Averill, Te Maro, Anderson, Easton, & Taiwhati and Chapter 7: Hunter, Hunter & Bills), Australia (Chapter 8: Jorgensen), Micronesia (Chapter 9: Dawson, Camacho, Fale, Miura, & Fritzlan), before returning to the United States (Hawaii, Chapter 10: Kaomea and Alaska Chapter 11: Lipka, Andrew-Ihrke, & Yanez) and finally back to Western Canada (Chapter 12: Thom).

The visions articulated throughout these chapters do not represent quick solutions or 'best practices' for culturally responsive pedagogies with and within Indigenous communities. Instead, this book contains chapters from Indigenous and non-Indigenous authors working together to explore the nature of culturally responsive pedagogies. Many chapters are co-written with community members and academics highlighting the importance of relationships with communities and the land. Each chapter engages critically with ideas of culturally responsive curriculum and pedagogy, exploring what this stance might mean and how it is lived in local contexts within global conversations. The authors explore projects that are making a difference with and within Indigenous communities. However, this book is not a set of tested and contained strategies. This collection offers a living pathway in which researchers, educators, youth and community members critically take up culturally responsive teachings and the possibilities and challenges that arise along the journey. Each chapter includes principles that guided the research and community-based work, which is relevant to a particular place/context. Yet, a commonality is found in how to present and work with the lived stories of experiencing culturally responsive mathematics education: Indigenous storywork (Archibald, 2008; Archibald, Nicol, & Yovanovich, 2019).

LIVING CULTURALLY RESPONSIVE PEDAGOGIES WITH INDIGENOUS STORYWORK

Embracing the chapters in this volume, reading across the contributions, reading and rereading the chapter contributions alongside different authors, and sometimes reading across great lengths of time we find many of Archibald's (2008) principles of Indigenous storywork that include respect, reciprocity, reverence, responsibility, interrelatedness, synergy and holism. For Archibald (2008), these principles are both separate and interconnected. The description of weaving a cedar basket brings these to life:

> Each principle has a separateness that is like a long flat piece of cedar bark used for basket weaving. As each piece is woven together, it may lose its separateness and become the in-between space that creates the background for a beautiful design. As the basket maker continues, she interweaves the pieces, transforming them into distinctive designs. (Archibald, 2008, p. 153)

Respect

Searching for "a culturally appropriate Indigenous research methodology … about Indigenous storytelling," Archibald (2008) writes that she "started with the principles

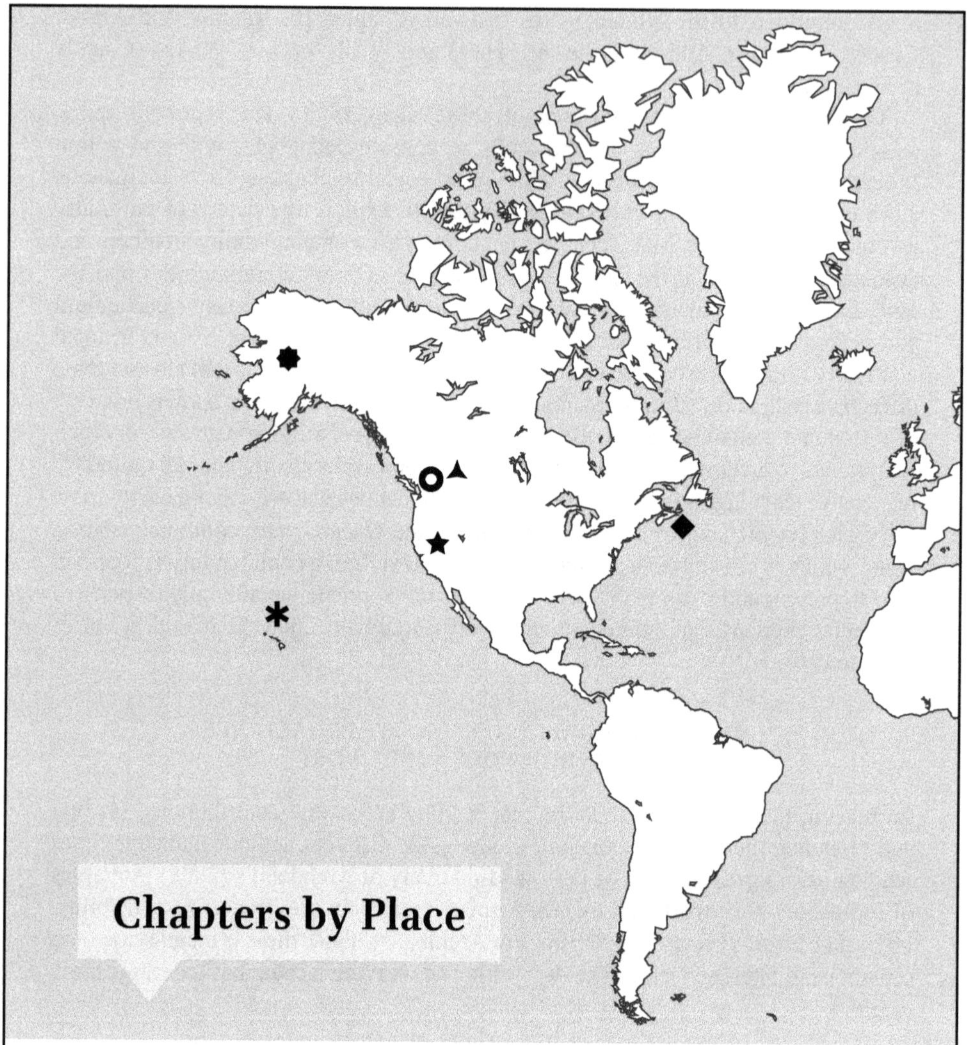

- ★ **Foreword** | Living Culturally Responsive Pedagogy
- **O** **Chapter 1** | Introduction: Living Culturally Responsive Mathematics Curriculum
- **O** **Chapter 2** | Being Guided by Kugaan Jaad Mouse Woman
- **O** **Chapter 3** | Relational Curriculum Design
- ▲ **Chapter 4** | Understanding the Landscape of Culturally Responsive Education
- ◆ **Chapter 5** | Show Me Your Math
- ◈ **Chapter 6** | Bicultural Mathematics Teacher Education and Research

Figure 1.1. Chapters by place map

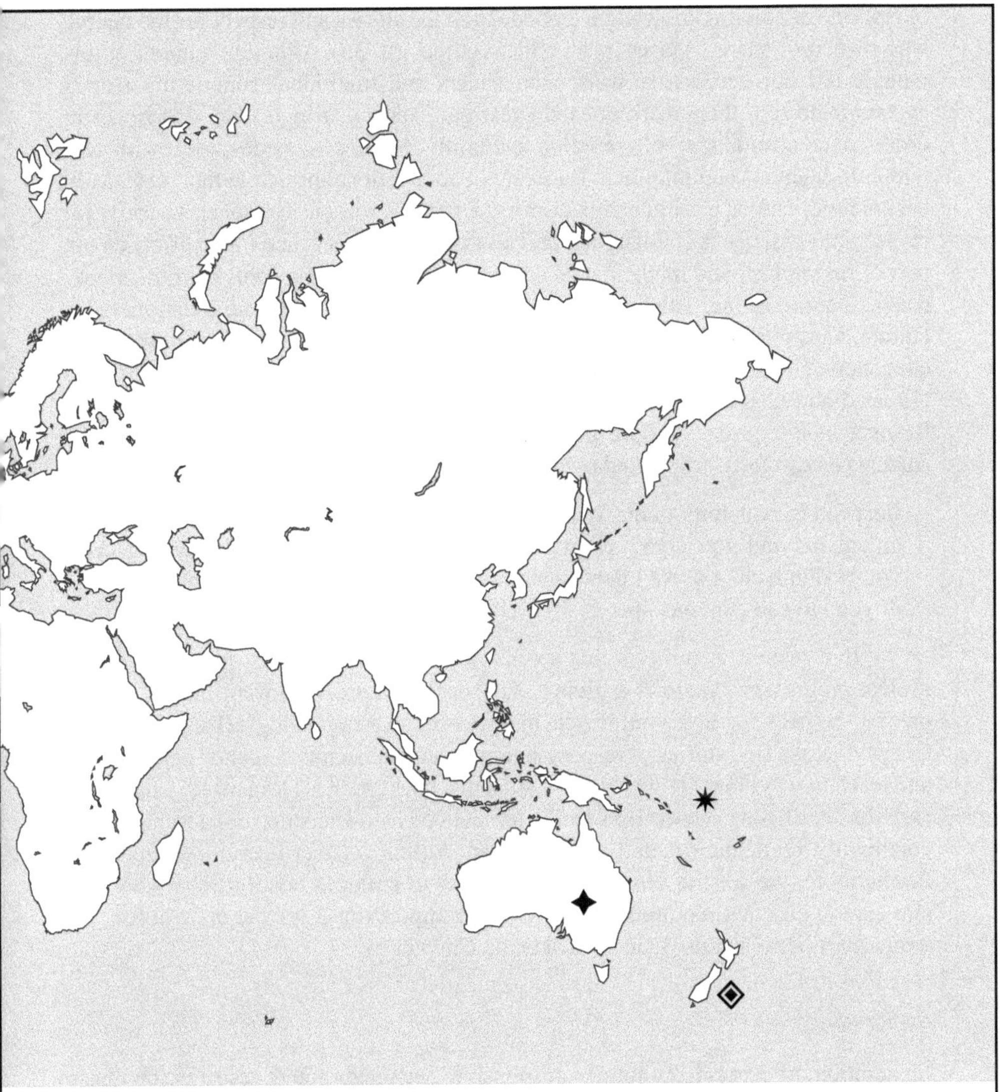

◈ **Chapter 7** | Enacting Culturally Responsive Socially Response-Able Mathematics
✦ **Chapter 8** | Culturally Responsive Pedagogy
✳ **Chapter 9** | Mathematics and Culture in Micronesia
✱ **Chapter 10** | Hawaiian Math for a Sustainable Future
✹ **Chapter 11** | Symmetry and Body Proportional Measuring
○ **Chapter 12** | Bending an Ear in the Quest of "What is it?"
○ **Chapter 13** | Conclusion: Re-Storying Mathematics Education

of respect for cultural knowledge embedded in the stories and respect for the people who owned or shared stories as an ethical guide" (p. 36). Although authors in this volume did not necessarily work with Elders and traditional community stories as Archibald did, their work does suggest the need for a high level of respect in order to conceptualize and research culturally responsive pedagogies with and within Indigenous communities. Respect is a foundational principle that may easily become a cliché or platitude and dismissed for that reason. However, to truly put respect into practice is challenging, and takes time, patience, and much effort on the part of researchers and mathematics educators. Respect in the context of this book, refers to appreciating, valuing, and honouring Indigenous knowledge that includes culture, nature/land, relationships, mathematics, communities/families/students and education. The ideas, visions, and hopes of those who fill these chapters speak to the need and challenge for respectful relationships and actions at multiple levels. Respect in this sense is more than good intentions. It includes respect between students as carefully articulated by one student, Sione (Hunter et al., Chapter 7):

> Respect is real important. When you have respect you can have friendly arguments and you argue about the maths so it's not nah you're wrong or you're dumb eh, it's like I don't agree with your maths and this is what I think or you have to convince me.

We find respect in projects that work with community Elders, in which, as the project progresses, "there is a giving of respect to those who went before … and gaining respect for and confidence in the work we are doing" (Dawson et al., Chapter 9). We find stories of respect in how teachers listen to students' perspectives and teach in ways that are inclusive of students' backgrounds (Jorgensen, Chapter 8) and stories of how respect is crucial for collective ownership, collaborative and community participation, and research team building (Averill et al., Chapter 6). Furthermore, we see the importance of respect in building relationships with land and community that require listening without imposition (Lipka et al., Chapter 11; Athie Martinez, Chapter 3; and Nicol et al., Chapter 2).

Reciprocity

In addition to respect, culturally responsive pedagogies and research on these pedagogies involve reciprocal relationships. Reciprocity, another theme found across chapters, builds on the principle of respect and requires mutual responsiveness to the needs of students, teachers, researchers, and community members, as well as individuals and families. Reciprocity, as Archibald (2008) notes, is a giving-back, a sharing of knowing with others. It is an act that challenges unethical research interactions that historically can be one-sided and exploitive, in which the intentions and practices of the researcher are privileged over the needs, interests and involvement of Indigenous participants. Building reciprocal relationships with and within Indigenous communities requires respect for another's perspective, an openness to engage in critical questions, and a confidence to share uncertainties that arise along the way. Questions stemming from the principle of reciprocity include: Who benefits from this work? Whose questions are privileged? How are relationships between those involved fluid and changing depending on the context? Whose research agenda is followed? What should we do next, when what was tried didn't work?

This principle of reciprocity was explicitly stated in some chapters such as Athie Martinez (Chapter 3), Kaomea (Chapter 10) and Nicol et al. (Chapter 2), and in others revealed in the way that authors came together to share their research projects. Almost all chapters are collaborations written with academics, community members, or teachers. Across the projects we see how contributors and researchers are considered as participants, co-researchers, teachers, and community members—where researchers are no longer authoritative outsiders. For some teacher participants, sharing their learning with others "motivated teachers to continue their work, to recognize the importance of mobilizing their learning, and the importance of giving back to the community" while at the same time "validat[ing] the importance of their work for others" (Nicol et al., Chapter 2). The research and educational projects described are important to all those involved. Reciprocity in culturally responsive pedagogies and research in this sense addresses not only how academic researchers benefit from such projects, but also how communities benefit from creation of innovative curricula, events, or pedagogies for Lunney Borden et al. (Chapter 5), Lipka et al. (Chapter 11), and Nicol et al. (Chapter 2).

Reverence

For Archibald (2008), reverence, as an Indigenous storywork principle, is difficult to describe because it is one that should be deeply felt and at the same time is both personal and subjective. We link ethical ways of working with Indigenous knowledge-stories-people-land to reverence in order to counter the negative legacy caused by irreverent laws, policies, practices, education, and research toward Indigenous peoples. Indigenous Elders often say that we need to do our work with a good heart and mind, and this requires ethics and reverence. Thus, reverence as a principle guides ethical ways of approaching the work. Although we do not see this principle directly stated in the chapters of this book, it is a significant principle for culturally responsive mathematics pedagogies with/in Indigenous communities. Reverence can be enacted through ethical ways of relating: how we listen, how we use what participants share in the research process, how we regard those involved, and how we attend to our hearts and minds in order to create an open caring relational environment for engaging in work related to mathematics. In this reverence, space may become open to challenge the revered position of mathematics in school curriculum and the revered position of those who are perceived to know mathematics (see Glanfield et al., Chapter 4 and Thom, Chapter 12).

Responsibility

Culturally responsive pedagogies involve the Indigenous storywork principle of responsibility. Attention to responsibility can be found throughout the chapters of this book and focuses on responsibilities accepted by all involved, as well as on the shifting emerging responsibilities that appeared as the projects unfolded. The determining of research or project responsibilities with students, teachers, researchers, and community members can be challenging, as these responsibilities aren't necessarily pre-determined but are co-constructed throughout the projects. This is evident in Chapter 5, in which Lunney Borden et al., "initiated a new series of conversations" with community members in order to disrupt an interpersonal positioning in which the researchers felt participants had defined them (the researchers) "as the ultimate audience." Awareness of and attention to how researchers and participants were positioned thereby "opened up new opportunities" as students and community members participated in the community's Show Me Your Math events. Responsibility is therefore performed in the ethical encounter and interaction between those involved. Hunter et al. (Chapter 7) refer to this as ethically or socially-response-able mathematics education that support Pāsifika students in "seeing mathematics as part of the world they inhabit and something they need within that world."

Living culturally responsive mathematics education as an ethical responsibility changes the kinds of interactions between students and teachers from those focused more on teacher interests to those for which pedagogical decision-making embraces student interests and needs. Jorgenson (Chapter 8) addresses such a focus in her research that involves teachers in "celebration of the language/s students bring to the mathematics classrooms." Here, teachers are also learning to be in relation with students creating environments that highlight "synergies" between home and mathematical languages and "ensuring the students could be proud."

Other chapters remind us of our responsibility as educators to consider varied epistemologies for culturally responsive mathematics pedagogies. Hawai'i's future leaders, Kaomea (Chapter 10) argues, need to have strong foundations in "accumulated wisdom of our Hawaiian ancestors" along with "exceptional skill in complex, mathematical problem solving" in order to address "future dilemmas." Offering three lesson examples as contexts for Hawaiian students to engage with pressing local and global issues, Kaomea argues for a transformative education that supports student responsibility as future leaders. As with other chapters in this volume such pedagogies connect mathematical ways of being with students' historical backgrounds, contemporary cultural practices, and students' future possibilities or foregrounds.

Interrelatedness

Culturally responsive pedagogies, as the chapters in this volume portray, involve a dialogical interrelatedness between students, teachers, researchers, community members, mathematics and the local place, which requires a different kind of participation, a different kind of listening. Archibald (2008) writes that "an interrelationship between the story, storytelling, and listener is another critical principle of storywork" (p. 32). This interrelatedness for Archibald emphasized the potential power of stories to be used "in a cultural way that enabled people to interact with a story and each other through a storywork process" (p. 99). Similarly culturally responsive pedagogies provided opportunities for interrelationality that were sometimes surprising, challenging, and powerful.

Living culturally responsive pedagogies requires listening to students' stories and cultural histories along with what interests and motivates them. It involves listening to Elders and community members for the desired cultural, language and heritage practices that are important to the community. It involves listening to openings for multiple ways of being mathematical. Listening in these ways emphasizes the interrelatedness, the connections among and between all involved. Averill and colleagues (Chapter 6) provide examples of interconnectedness with communities; Athie Martinez (Chapter 3) highlights the possibilities of bringing Indigenous mathematicians, educators, Elders and researchers together for curriculum design; and Nicol, Archibald, and Yovanovich (Chapter 2) consider interrelatedness of human and nonhuman, animate and inanimate to conceptualize culturally responsive pedagogies.

Synergy

Interrelatedness can create "a synergistic story power" writes Archibald (2008, p. 100) with healing aspects and ability to bring a story to life. For culturally responsive mathematics education as articulated, explored and experienced in this volume, synergies were created that extended projects beyond initial intentions or expectations. Dawson and colleagues (Chapter 9), for example, created the project Mathematics and Culture in Micronesia: Integrating Society Experiences (MACIMISE) that resulted in many team members completing masters and doctoral degrees. For Nicol, Archibald and Yovanovich (Chapter 2), what began as a three-year project continues today, 10 years later. In this project, ideas for enacting culturally responsive education extend beyond mathematics to now include other subject areas, administrative practices, and staff hiring protocols within the community. These ideas are examples of synergistic connections in living culturally responsive pedagogies with and within Indigenous communities.

Holism

More than a collection of parts, holism, as conceptualized from an Indigenous philosophy "refers to the interrelatedness between the intellectual, spiritual (metaphysical …), emotional, and physical … realms to form a whole healthy person (Archibald, 2008, p. 11). Holism includes interrelations between self, family community and nation, with emphasis on a mutual balance between connections to land, place, animate, inanimate, human and nonhuman. Holism becomes a context for Indigenous story. Similarly, holism is a context for culturally responsive mathematics education and involves understanding the important contributions of students, teachers, Elders, researchers, community members, culture and land. This means that researching culturally responsive pedagogies requires understanding the local context, involving multiple perspectives and points of view, and resisting attempts to reduce such relational practices to a series of fundamental processes. Hunter and colleagues' research (Chapter 7) considers holism with examples of values, cultural/community practices, language and body language.

The principles of Indigenous storywork, which include respect, reverence, reciprocity, responsibility, interrelatedness, synergy and holism characterize the various ways in which culturally responsive pedagogies are lived with and within Indigenous communities across the world. The principles highlight the ethical and relational nature of culturally responsive mathematics pedagogies and the associated practices required for respectful research. Although we use these principles retrospectively, as a strategy to discuss research reported in this volume, we suggest that the principles could also be used to guide the initiation, creation, and development of culturally responsive mathematics curriculum, pedagogy and research.

Cynthia, Jo-ann Q'um Q'um Xiiem, and Florence invite you, the reader, to continue on your own storied journey with these chapters. We look forward to continuing conversations that result in new experiences and new stories about ways to bring place, community, Indigenous knowledge systems, and mathematics education together for good quality learning.

REFERENCES

Archibald, J. Q'um Q'um Xiiem. (2008). *Indigenous storywork: Educating the heart, mind, body, and spirit*. Vancouver: UBC Press.

Archibald, J. Q'um Q'um Xiiem, Nicol, C., & Yovanovich, J. (2019). Transformative education for Aboriginal mathematics learning: Indigenous storywork as methodology. In J. Archibald Q'um Q'um Xiiem, J. Lee-Morgan, & J. de Santolo (Eds.), *Decolonizing research: Indigenous storywork as methodology* (pp. 72–88). London: Zed Books.

Battiste, M. (2013). *Decolonizing education: Nourishing the learning spirit*. Saskatoon, SK: Purich Publishing.

Bishop, A. J. (1991). *Mathematical enculturation A cultural perspective on mathematics education*. Dordrecht: Springer.

Cajete, G. (1994). *Look to the mountain: An ecology of Indigenous education*. Durango, CO: Kivaki Press.

Cajete, G. (1999). *Igniting the sparkle: An Indigenous science education curriculum model*. Durango, CO: Kivaki Press.

D'Ambrosio, U. (2006). *Ethnomathematics: Link between traditions and modernity*. Rotterdam, The Netherlands: Sense Publishers.

Darder, A. (2015). Paulo Freire and the continuing struggle to decolonize education. In M. Peters & T. Besley (Eds.), *Paulo Freire: The global legacy* (pp. 39–54). New York, NY: Peter Lang.

Enyedy, N., Danish, J., & Fields, D. (2011). Negotiating the 'relevant' in culturally relevant mathematics. *Canadian Journal of Science, Mathematics and Technology Education, 11*(3), 273–291.

Frankenstein. M. (1987). Critical mathematics education: An application of Paulo Freire's epistemology. In I. Shor (Ed.), *Freire for the classroom: A sourcebook for liberatory teaching* (pp. 180–210). Portsmouth, NH: Heinemann.

Frankenstein, M. (2012). Beyond math content and process: Proposals for underlying aspects of social justice education. In A. Wager & D. Stinson (Eds.), *Teaching mathematics for social justice: Conversations with educators* (pp. 49–62). Reston, VA: National Council of Teachers of Mathematics.

Gay, G. (2010). *Culturally responsive teaching. Theory, research and practice*. New York, NY: Teachers College Press.

Geboe, G. (2015). *Comparative approaches to reconciliation: Canada, United States of America, Australia and New Zealand.* Policy Paper for the Reconciliation Working Group. Institute for the Study of International Development. Retrieved from https://mcgill.ca/isid/files/isid/pb-2015-03_0.pdf

Greer, B., Mukhopadhyay, S., Powell, A., & Nelson-Barber, S. (2009). *Culturally responsive mathematics education*. New York, NY: Routledge.

Gutiérrez, K., & Rogoff, B. (2003). Cultural ways of learning: Individual traits or repertoires of practice. *Educational Researcher, 32*(5), 19–25.

Gutstein, E. (2006). *Reading and writing the world with mathematics: Toward a pedagogy for social justice*. New York, NY: Routledge.

Kawagley, A. O. (2006). *A Yupiaq worldview: A pathway to ecology and spirit* (2nd ed.). Long Grove, IL: Waveland Press.

Ladson-Billings, G. (1995). But that's just good teaching: The case for culturally relevant pedagogy. *Theory into Practice, 34*(3), 159–165.

Ladson-Billings, G. (2014). Culturally relevant pedagogy 2.0: A.k.a. the remix. *Harvard Educational Review, 84*(1), 74–84.

McCarty, T. L., & Lee, T. S. (2014). Critical culturally sustaining/revitalizing pedagogy and Indigenous education sovereignty. *Harvard Educational Review, 84*(1), 101–124.

Paris, D. (2012). Culturally sustaining pedagogy: A needed change in stance, terminology, and practice. *Educational Researcher, 41*(3), 93–97.

Powell, A. B., & Frankenstein, M. (1997). *Ethnomathematics: Challenging eurocentrism in mathematics education*. Albany, NY: State University of New York Press.

Simpson, L. (2014). "Bubbling like a beating heart": Reflections on Nishnaabeg poetic and narrative consciousness. In N. McLeod (Ed.), *Indigenous poetics in Canada* (pp. 107–119). Waterloo: Wilfred Laurier University Press.

Skovsmose, O. (2011). *An invitation to critical mathematics education*. Rotterdam, The Netherlands: Sense Publishers.

Skovsmose, O. (2012). Critical mathematics education: A dialogical journey. In A. Wager & D. Stinson (Eds.), *Teaching mathematics for social justice: Conversations with educators* (pp. 35–47). Reston, VA: National Council of Teachers of Mathematics.

Skovsmose, O. (2014). Critical mathematics education. In S. Lerman (Ed.), *Encyclopedia of mathematics education* (pp. 116–120). Dordrecht, The Netherlands: Springer.

Skovsmose, O. (2016). What could critical mathematics education mean for different groups of students? *For the Learning of Mathematics, 36*(1), 2–7.

Truth and Reconciliation Commission of Canada (TRC). (2015). *Truth and reconciliation commission of Canada: Calls to action*. Retrieved from http://www.trc.ca/

United Nations. (2016). *Indigenous peoples' right to education – A transformative force for empowerment*. United Nations Human Rights Office of the High Commisioner. Retrieved from https://www.ohchr.org/EN/NewsEvents/Pages/DisplayNews.aspx?NewsID=20342&LangID=E

CYNTHIA NICOL, JO-ANN ARCHIBALD Q'UM Q'UM XIIEM
AND JOANNE YOVANOVICH

2. BEING GUIDED BY KUGANN JAAD MOUSE WOMAN FOR CULTURALLY RESPONSIVE MATHEMATICS EDUCATION

INTRODUCTION

Place and our relationship to the land defines who we are, what we notice, how we live in the world, and how we live with each other. Haida Gwaii, situated in Canada's Pacific Northwest as a collection of 150 islands, comprises bogs and lowlands, snow-capped mountains and cliffs extending to the ocean, long stretches of white sandy beach, and areas that commonly receive 10-metre-high waves. It is a land between sea and sky where the worldview that "everything is connected to everything else" means there is a symbiotic relationship between humans and non-humans, natural and Supernatural Beings. The place of Haida Gwaii reminds us that under our skins, both animate and inanimate, we are all the same—we are all connected.

> Our world began thousands and thousands of years ago. They say it was a time when it was both light and dark and a great flood covered the land. The supernatural beings had no place to rest until Xaagyah came to the surface. Haida oral histories tell of these beginnings and the events to follow, including the affairs of the notorious Raven. In 1900 and 1901, linguist John R. Swanton recorded many of these oral histories, as told to him by Haida informants. (Collison, 2006, p. 7)

Haida Elder Gwaaganad, Diane Brown, tells us that there are many oral stories of interactions between Haida and Supernatural Beings. These Beings of interconnectedness, or shape-shifters, respectfully appear in Haida designs carved into masks and painted onto bentwood boxes. Many, like Supernatural Being Kugann Jaad Mouse Woman, can appear both consciously and unconsciously, be both positive and negative space, and provide guidance or redirection to those who are about to "travel into unknown terrain" (Davidson, 2016, para. 3). It was Kugann Jaad who appeared in Haida artist WNCY Billy Yovanovich's representation of mathematics education connected to the place of Haida Gwaii (see Figure 2.1).

For Billy, Kugann Jaad is known for her call to restore balance and equity, and with her strength, wisdom, and vision she guides us, speaking with both her hands and her eyes with knowledge of what is to happen, how to prepare for what

 | DOI: 10.1163/9789004415768_002

Figure 2.1. Mouse Woman, WNCY Billy Yovanovich

will happen, or even provide resources to meet challenging situations (Nicol & Yovanovich, 2011). Billy's vision of Kugann Jaad captures our efforts in working with teachers, students, and community members of Haida Gwaii to explore how culturally responsive education might be lived in such a place—a place where Haida artist and activist Guujaaw states, "Culture actually is our relationship to the land" (Jones, 2006, p. 29).

This chapter offers a critical examination of culturally responsive education enacted with educators in contexts of professional learning. We examine the nature of culturally responsive mathematics education, drawing upon our long-standing collaborative work on Haida Gwaii, an Indigenous[1] community located in Canada's Pacific Northwest. Our guiding questions include, What are culturally responsive teaching, professional learning, and leadership being "responsive" to? We first examine conceptualizations of culturally responsive education in general, and more specifically within mathematics education for Indigenous students, and then, drawing upon our own research, provide examples of possibilities and challenges of culturally responsive and sustaining mathematics education.

CONCEPTIONS OF CULTURALLY RESPONSIVE EDUCATION

Counter to deficit and cultural difference approaches to education, Indigenous American scholars Gregory Cajete (1994, 1999) and Oscar Kawagley (2006) proposed, alongside Indigenous scholar Carol Pewewardy (1998), that education for Indigenous students needed to be culturally based, whereby culture became the context for education rather than a topic inserted into school curriculum. Barnhardt and Kawagley (2005) offer examples of culturally based curricula. Working with Yup'ik students in Alaska, they describe school curriculum that weaves Yup'ik language and linguist practices together with local community support. Yet, as Brayboy and Castango (2009) note, advocating for curriculum that is responsive to local language, and linguistic and cultural practices, does not mean that parents and communities of Indigenous students are not also interested in their children being successful in mainstream education. Instead, parents and families, "insist … that children's learning to 'do' school should not be an assimilative process; rather, it should happen by engaging culture" (Brayboy & Castango, 2009, p. 31).

Around the same time as Cajete (1994, 1999) and Kawagley (2006) African-American scholar Gloria Ladson-Billings (1995a) introduced the idea of culturally relevant pedagogy. Ladson-Billings was one of the first to turn the research gaze from focusing on what African-American learners were not able to do to focusing instead on what teachers experiencing success with these learners did. In her study of eight teachers, Ladson-Billings (1995a, 1995b) conceptualized culturally relevant teaching to describe teachers who were caring, demanding, critical, and connected to students and their communities. Based on her study with these teachers, and with a dynamic and fluid view of culture, Ladson-Billings (1995a) theorized culturally relevant pedagogy as consisting of three characteristics: (1) academic success—striving to support students' academic needs and intellectual growth; (2) cultural competence—celebrating students' cultures and community knowledge as valuable and capable resources; and (3) socio-political consciousness—extending learning beyond the classroom to solve real-world problems. Reflecting on her theory almost 20 years later, Ladson-Billings (2014), writes how culturally relevant pedagogy helped teachers to learn from their students by "focusing on student learning and academic achievement versus classroom and behavior management, cultural competence versus cultural assimilation or eradication, and sociopolitical consciousness rather than school-based tasks that have no beyond-school application" (p. 76).

Building on Ladson-Billings's conceptions of culturally relevant pedagogies, Paris (2012) and Paris and colleagues (Paris & Alim, 2014, 2017) offer culturally sustaining pedagogy as an approach that not only considers education in terms of academic success, cultural competence, and sociopolitical consciousness, but also extends this to include educational practices that sustain and support students' heritage, language, and linguistic practices of communities for the present and future. "Culturally sustaining pedagogy seeks to perpetuate and foster—to sustain—linguistic, literate, and cultural pluralism as part of the democratic project of schooling" (Paris, 2012, p. 95). Therefore, in contrast to cultural difference pedagogies that attempt to move students marginalized by systemic inequalities to become more like students of the dominant culture, Paris and Alim (2014, 2017) argue for pedagogies that foreground students' heritage and contemporary practices. Culturally sustaining pedagogies use students' culture not only as a resource for access to educational opportunities, but also, together with the evolving practices of students' communities, as a focus for more pluralist educational outcomes. As an example, Paris and Alim (2017) examine how pedagogies emerging from African-American youth's hip-hop culture can be culturally sustaining by recognizing present cultural practices as critical educational spaces for the future.

CULTURALLY RESPONSIVE MATHEMATICS EDUCATION

How might we enact a culturally responsive/sustaining pedagogy of mathematics education? Gutstein (2006, 2012) writes of his efforts to work with American Latino/a students and their communities to support connecting mathematics to students' cultural and community histories. Gutstein, like Civil (2002, 2016) drawing upon students' and community funds of knowledge, develops curriculum and pedagogical approaches that view students' home cultures and languages as "strengths upon which to build, rather than as deficits for which to compensate" (Gutstein & Peterson, 2013, p. 3). In considering connections between students' cultural and community experiences, Gutstein, inspired by Freire (1970/2000), draws upon three kinds of knowledge domains. Classical knowledge is the mathematical knowledge students need "to pursue advanced mathematics and mathematically related careers" (Gutstein, 2006, p. 203). Community knowledge involves knowledge of "community life in all its complexity, and of perspectives and interpretations of the world" (Gutstein, 2006, p. 201). Critical knowledge, the third form of knowledge, involves critical awareness of both how to read and transform the world with mathematics. Bringing these forms of knowledge together provides opportunities to consider alternative ways of envisioning school mathematics by considering home communities as sources of viewing the world mathematically and pedagogically.

A critical approach to mathematics education involves students using their mathematical backgrounds to participate in community and cultural projects and decision-making that could involve mathematics. Critical approaches also, argues Skovsmose (2011), need to consider not only what is familiar to students (i.e., their backgrounds), but also what might be possible for students (i.e., their foregrounds). A critical approach therefore draws upon students' lived experiences so that students can be involved in understanding (reading) and intervening (writing) the world. For Gutstein, Skovsmose, Freire, and Cajete, responsive teaching practices bring student and teacher in reciprocal relationships of co-learning. Cajete (1994, 2012) considers this form of co-learning as an ecology of relationships connecting humans and place, in which humans and place shape each other. In Cajete's words, "people make place, as much as place makes them" (1994, p. 84).

Cajete (1994) further suggests that with place we have a dialogical relationship in which we learn more about ourselves, our relationships with each other, and our relationships with the more-than-human world. If we consider culture as relationship to the land and sea (Guujaaw in Jones, 2006), what kinds of pedagogies might be responsive to land, sea, ourselves, and communities? To explore this question, we move now to our collaborative work on Haida Gwaii, a continuing and long-term project with university researchers and Island teachers and community members exploring what it means to live culturally responsive mathematics education (Nicol & Yovanovich, 2011, 2017; Nicol, Archibald, & Baker, 2013; Nicol, Yovanovich, & Gear, 2016).

THE CONTEXT

Haida Gwaii, which means Islands of the People, is located in Canada's Pacific Northwest. It consists of 150 islands with two main islands acting as home to those currently living on Haida Gwaii. According to Elders and oral traditions, the Haida people originated from the ocean. Before European settlers made contact in the late 1700s and the British claimed sovereignty in the mid-1880s, tens of thousands lived on Haida Gwaii. With European contact came the spread of disease, decimating the Haida population to less than 1,000 and requiring, for survival, that people congregate in two main communities: G̲aaw Old Masset in the north and HlG̲aagilda Skidegate in the south. In 1853 the British claimed Haida Gwaii as British lands and, in 1876, the Canadian government's Indian Act declared all "Indians" or First Peoples to be the responsibility of the government and made it illegal for all First Peoples to engage in cultural activities such as potlatches used by the Haida for cultural and economic governance practices. With ancestral communities concentrated in two main locations on the Islands, the settlers who arrived interpreted the vast open lands as unsettled and began claiming it for themselves, using the land and ocean for resource extraction (e.g., clam canneries, logging and sawmill camps, open-pit gold and copper mining, and whale stations). Much of this resource extraction occurred without the approval or input of the Haida people. However, beginning in 1985, the Haida protested this kind of resource removal, leading to a historical agreement for the lands to be co-managed by the Government of Canada and the Haida Nation.

Today a little more than 4,000 people, both Indigenous and non-Indigenous, live on Haida Gwaii, with less than half the population identifying as having Indigenous ancestry (Statistics Canada, 2016). The Haida Gwaii School District has six public schools, with one school located on HlG̲aagilda Skidegate reserve land. The school district serves about 450 students, of whom approximately 70% identify as being of Haida or other First Nations ancestry. Since 2005, we have worked together to create, explore, and research the development of learning environments responsive to the place and communities of Haida Gwaii. When we began, the Haida Gwaii School District was interested in improving school graduation rates for all its students but especially for its Indigenous students. For example, between 2002 and 2007, the district secondary graduation rates were between 46% and 28% for Aboriginal students, while rates for non-Aboriginal students ranged between 97% and 39%. This occurred during a time when provincial secondary school graduation rates were close to 80% for all students.

This chapter is part of a larger project involving multiple people, including Indigenous and non-Indigenous educators, researchers, teachers, students, and community members. Teaching mathematics in the Haida Gwaii School District

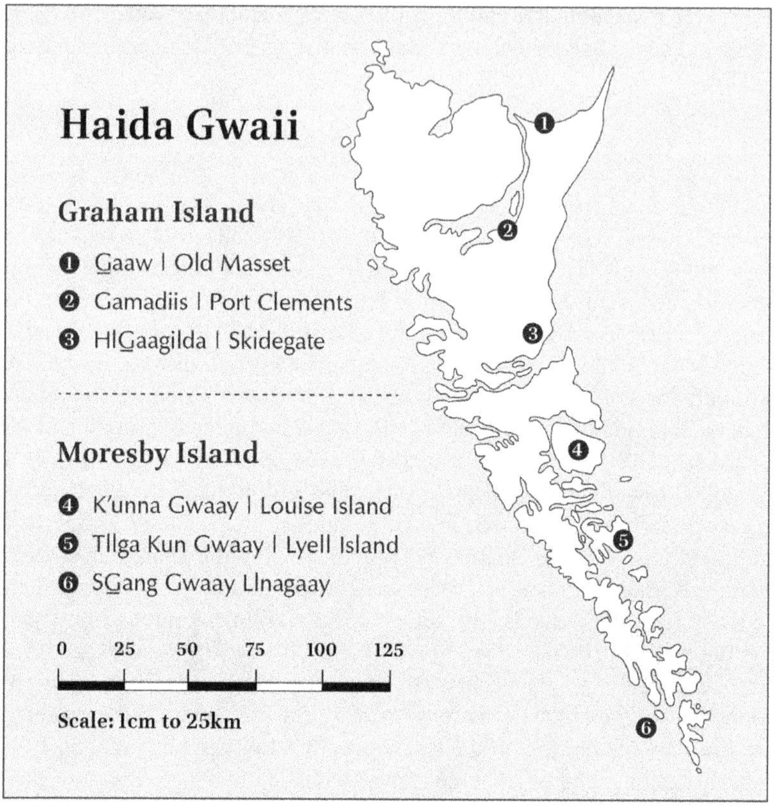

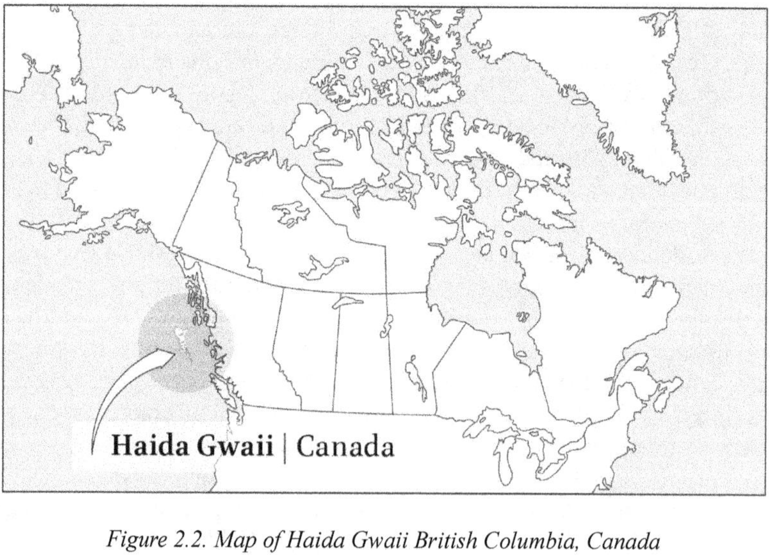

Figure 2.2. Map of Haida Gwaii British Columbia, Canada

was for me (co-author Cynthia) my first teaching position at Gudangaay Tlaats'gaa Naay Secondary School in Masset, located four kilometres from Gaaw Old Masset. At the time I taught, and until just recently, the school carried its colonial name, George M. Dawson Secondary School, after a Canadian surveyor and geologist. I taught here for eight years before moving to Vancouver to pursue graduate studies. Now at the University of British Columbia (UBC) as an associate professor, my relationship with Haida Gwaii as a non-Aboriginal teacher and researcher continues. Various collaborative projects with the Haida Gwaii School District keep me connected to the land, people, and communities. My past students, now with families of their own and children attending the school where I once taught, tell me of their desire for education that builds on their children's strength so that youth are successful in school mathematics, and have confidence in their abilities to make a difference both within and outside the community.

I, co-author Jo-ann—Q'um Q'um Xiiem, of Sto:lo and St'át'imc ancestry—recently retired from a 45-year educational career as a school teacher, curriculum developer, and university professor. I continue my affiliation with UBC's Faculty of Education as professor emeritus, Educational Studies. For 36 years, I had the privilege of teaching and working with undergraduate and graduate students, mainly Indigenous, faculty members, and many Indigenous communities throughout British Columbia, Canada, and elsewhere. My involvement with Haida Gwaii began in 2005 when Cynthia invited me to join her in a culturally responsive math research project. In this project, I helped to develop an Indigenous storywork approach (2008) that addressed making meaning of and working with Indigenous traditional and life experience stories, using the heart, mind, body, and spirit.

For me, co-author Joanne—Taanud Jaad—Haida Gwaii is my ancestral home. I was born into the Haida Ts'aahl Eagle Clan[2] of Skidegate and have worked as both a teacher and principal in one of the district's elementary schools since 1995. Currently I am the district's principal of Aboriginal education, a position I have held for the past 10 years. As a district we are working to improve our graduation, participation, and success rates for students. We are also very committed to incorporating Haida culture and language in all schools throughout the district. It is my hope that this work will inspire others to connect with Haida culture and language and consider what is mathematical in our everyday activities.

As a team of teachers and researchers, we began meeting approximately every six to eight weeks during the school year, inviting various community members to join when they were available. In order to better experience the place of Haida Gwaii, we often met in out-of-school settings such as the Haida Gwaii Museum at Kay Llnagaay, the Tluu Xaada Naay Longhouse at Gaaw, local cafés, and the District Board Office. Our group most often consisted of six to eight people, sometimes more, and we engaged in mathematical activities often focused on pattern, transformations, and relationships. From community members and Elders, we learned more about cultural practices and protocols such as canoe building, cedar weaving, wool weaving, mask carving, design drawing, storytelling, harvesting, and celebrating.

The role of culture, social practices, and mathematics, particularly questions around school mathematics and out-of-school mathematics, were also an integral part of our discussions. We sometimes struggled to bring perceived competing goals of improving what Gutstein (2006) calls classical mathematics achievement while at the same time creating experiences for students to deepen their connection to the place of Haida Gwaii. These conversations challenged our own assumptions, biases, differences, privileges, and power related to mathematics, community, and culture. Developing mathematics problems together and trying these problems in classrooms with students gave us shared experiences of what might be possible and the challenges that accompanied those possibilities.

To examine our learning, we audio-recorded and sometimes video-recorded all team meetings, resulting in over 300 hours of recordings collected to date. In addition, researcher field notes helped us locate critical discussion points for transcribing and sharing. Photos of classroom activities, videos of students working on developed mathematical activities, audio-recorded conversations with community members and Elders, interviews with teachers, along with some samples of students' work formed the data for our project. Following each project meeting, those who could listened to the audio or video recordings of the meeting and took notes of salient events and conversations. These notes and our researcher field notes guided decisions around which audio excerpts to transcribe.

The composition of our team, over the years, reflects the interests and focus of those involved. In total at least 35 teachers and administrators together with 15 community members such as artists, museum curators, Haida Elders and community knowledge holders have been involved in the project from 2005 to 2017. We have engaged in participatory curriculum design, co-creating how mathematics education on Haida Gwaii might be improved to better meet the needs of students, and how teaching could use students' cultures as resources (Paris, 2012) rather than as aspects to be replaced or rewritten. In the following sections, we draw upon examples related to place to discuss the nature of culturally responsive mathematics education in the context of professional learning.

EXPLORING CULTURALLY RESPONSIVE MATHEMATICS EDUCATION ON HAIDA GWAII

As our project began with teachers, we begin this discussion of culturally responsive mathematics with a focus on professional learning. Learning to incorporate students' cultural knowledge into mathematics curriculum or to see such cultural knowledge pedagogically continues to be a complex process. Teaching in a rural school district is demanding. There are few opportunities for professional development, mainly due to the expense of transportation costs on or off Haida Gwaii. Teachers have access to limited professional development funding to attend professional conferences. Furthermore, the geographic distance between schools in the north of Haida Gwaii and those in the south make it challenging to bring teachers from different schools together even for short meetings.

Yet teachers were keen to participate. In an initial survey given to all participants involved at the project's beginning, teachers rated their enthusiasm for the project as high and hoped the project would provide them with "resources that are applicable, easy to use, accessible, relevant and connected to the curriculum" (project elementary teacher) or "creative methods for passing on skills to varied levels of learners" (secondary teacher). Like many teachers in small rural schools, participants were heavily involved in school activities demanding much of their time. Therefore, ready-made materials they could easily try in their classrooms were perceived to ease their work commitments. However, as the project progressed, the call for ready-made materials lessened for some, particularly teachers participating over multiple years. As noted by one teacher two years into the project, "The opportunity to meet with colleagues was a highlight." She continues:

> I feel so privileged because meeting with colleagues just doesn't happen for the majority of teachers here. To work with a group of individuals over the last two years and meet regularly and discuss and talk and go back and try things [in our classrooms] and come back and discuss how different people approached it. Just meeting to discuss the lessons we created, kids and math. For example, we were just measuring with shells. Simple enough. Just different shells found on the beach so we can explore non-standard units of measurement, and the stories that came out from the kids!
>
> - "Oh, that is a razor clam, and I go digging for razor clams" or
> - "I've seen that shell before—there's lots of those," or
> - "The scallop shell. Oh, we eat scallops."
>
> It's one of the most beneficial things that has stood out for me. (Miki)

The interests and questions of those attending and the interests and needs of the community directed the focus and content of our meetings. Meeting topics included:

1. exploring students' mathematical thinking to learn more about students' strengths, interests, and cultural histories;
2. understanding student emotions, identity, and agency to consider how we might teach mathematics lessons that are emotionally engaging;
3. investigating mapping and oral stories to examine the interrelatedness of mathematics education and cultural, social, political, and historical issues;
4. taking and studying digital images of Haida Gwaii to inspire mathematical problem solving in the culture and community; and
5. learning from community members by exploring cultural practices of canoe building, weaving, and carving and the mathematics involved in these through collaboration with museum curators, Elders, and community artists.

As a group we collaborated on what we call "collective projects," bringing us together to develop resources for classroom and community use. These collective projects began small, were not predetermined, and were shaped by those participating in the group at the time. One collective project involved creating photo mathbooks with images, taken by participating teachers of Haida Gwaii, linked to cultural history, current community practices, and mathematics. To date, the group has created three photobooks along with numerous lessons involving local food harvesting, cultural and contemporary practices of canoe building, bentwood box design, mask and pole carving, as well as Haida legends.

A second collective project involved bringing the stories of Haida Elders together with mathematical adventures in a book, *Tluuwaay 'Waadluxan Mathematical Adventures* (Nicol & Yovanovich, 2011). In this project, participating teachers each developed mathematical problems inspired by Elders' stories of the place and importance of the Haida canoe as interconnected to Haida ways of being and living on Haida Gwaii.

Creating a living model of culturally responsive mathematics education became a third project for teachers. They called it PAIRS, an acronym for a series of questions teachers ask themselves when designing culturally responsive mathematics (Nicol, Yovanovich, & Gear, 2016; Nicol, Archibald, & Baker, 2013). The model puts emphasis on Place (How is my lesson responsive to this place of Haida Gwaii?); Action (How does this lesson engage students and myself in critical social consciousness, for example, making changes to improve mathematics teaching and learning, or engaging students in using mathematics to, as Gutstein [2006] says, read and write the world with mathematics?); Inquiry (How does my lesson provide opportunities for both social-cultural and mathematical inquiry?); Relationships (How does my lesson support building relationships with students, communities, teachers, families, mathematics, and Haida Gwaii?); and Storywork (How does my lesson consider Archibald's (2008) Indigenous storywork principles of respect, reciprocity, responsibility, relevance, synergy, interrelatedness, and holism?). We are currently working on a fourth project guided by Haida Traditional Ecological Knowledge and the Haida Marine Harvesting calendar. This project involves working with community members with extensive marine knowledge to share harvesting experiences, protocols, and connections to mathematics.

Participating teachers also worked on individual projects within their own classrooms. During our group meetings, we often developed a mathematical problem together and agreed to try the problem in respective classrooms. Teachers collected copies of student work, took photos, and sometimes video-recorded their classrooms in order to share examples with group members of how their students responded to the tasks. They also "field-tested" many of the problems and lessons included in our lessons, photobooks, and the Tluuwaay 'Waadluxan Mathematical Adventures book. When possible we also visited each other's classrooms, co-teaching, observing, and interacting with students doing mathematics. In each of these activities, the goal was to learn more about culture, community, mathematics, ourselves, and the interrelatedness of these in order to improve opportunities for teaching and learning mathematics.

Teachers began with few visions of teaching culturally responsive mathematics. They mentioned neither culture nor community as resources for teaching or learning mathematics. However, with time and involvement in the project, teachers referred to their participation as "life-changing" and "transformative." They spoke differently about their teaching while articulating their own views of what culturally responsive mathematics education could be. For example, three years into the project, Kit, a non-Aboriginal teacher, stated:

> I used to think Haida culture was scary and something I couldn't touch and wasn't allowed to talk about … now I think … I can talk about [it] as long as I'm respectful. [Before the project] I was afraid of offending somebody … now I realize that it is something that you can talk about and you can learn from the students because they want to tell you about it. They want to tell you what they know, and they want to share it with you. And it's OK if you make a mistake, they will correct you and laugh at you [in a good way]. That's OK. So, I'm not afraid of talking about [Haida culture], of bringing it into the classroom anymore. (Kit)

Similarly, for this teacher of Haida ancestry, participation was also significant. Here he speaks of the importance of teaching his own relatives, his family, and of the personal dedication to learning Haida culture through his own cultural engagements:

> [Participating] is making me a better teacher, a more responsible teacher in my approach because I am teaching my own family. I want the kids I teach, especially kids in my family, to have the best and the most success that I could possibly give them. I am spending a lot time doing my own artwork and practising my own. The whole culturally responsive aspect is key—I'm living it. (Darren)

Generally, teachers reported learning more about Haida culture. They spoke of gaining strategies for developing deeper connections with the community, designing pedagogies around their students' strengths and interests, and questioning their own assumptions of what is possible in teaching and learning mathematics.

LEARNING CULTURALLY RESPONSIVE AND SUSTAINING MATHEMATICS EDUCATION

What was it about these meetings that may have supported such professional learning? We discuss professional learning for culturally responsive mathematics education as opportunities to enhance cultural competence, build relationships as spaces of trust and dialogue, and participate in social/cultural action.

Opportunities to Enhance Cultural Relations

Keen to learn and enhance their own understanding of local culture, teachers recognized the possibilities that local culture offered for deepening their own connection with students and mathematics teaching. But, as Kit notes, they were also somewhat hesitant:

> I look at Haida culture as a belief system—like a religion—so you just don't touch it. It is not surprising then that most teachers reported satisfactory to poor knowledge of local culture for teaching as they entered the project. (Kit)

All teachers self-reported building community connections through the project. The project gave a "purpose," as one teacher stated, for intervening in or making their way into community events and activities. This was the case both for teachers who were relatively new to the school district (had arrived within three years) and those who were long-time residents (had lived over 12 years in the community). For example, various well-known Haida artists and Elders attending project meetings shared personal histories, cultural stories, and community protocols as well as their personal experiences with mathematics through carving, weaving, and food harvesting. Although these were well-known community members, this was the first time teachers had personally met these artists and Elders. This was also the case for community members. One well-known community artist remarked how meeting with project teachers was the first time the school district had invited him to share his knowledge in this way. He was honoured and pleased to make such a contribution.

If we consider culture as connections to the land as Guujaaw (in Jones, 2006), past president of the Council of the Haida Nation stated, then perhaps teachers' interactions with the land, living in the rural communities on Haida Gwaii, could also provide opportunities to deepen cultural competence. Some teachers interacted with the land for recreational purposes while others did so for food-gathering purposes. Surf kayaking the waves provided intimate connections with the ocean above and below, while food gathering, such as fishing, provided quite a different connection. One teacher spoke of this as the difference between cultural and community responsiveness, stating, "I'm not Haida myself, I didn't grow up here but I live across the street from the ocean and I cannot help but be impacted by the ocean every day of my life … it just becomes part of you" (Carol). This teacher referred to such connections as community responsive, while defining cultural responsiveness as more generational, consisting of Haida art, language, story, literacies, and traditions. Not all teachers agreed, but it was a way for some to draw upon their own experiences with the land as resources for considering mathematical contexts and noticing historical differences in interpreting connections to the land.

More recently, project teachers extended their thinking about connections to the land through an openness to learn more about Haida Supernatural Beings such as Wasco (Nicol & Yovanovich, 2017) or Kugann Jaad Mouse Woman. Meeting with community Elders and artists provides opportunities to learn more about cultural practices, Haida worldviews, Haida language, cultural protocols for learning and sharing local knowledge, and how such protocols are often changing rather than fixed and immutable. For example, at a meeting in May, 2015, Haida Elder Gwaaganad, Diane Brown, shared a story of Haida creation with participating teachers. She spoke of three different waves of Haida people who came to Haida Gwaii. The first wave was through air, the second through clay, and the third through the ocean. In her recounting of the story, she emphasizes this connection to the ocean during a previous presentation at Gaaysiigang: An Ocean Forum for Haida Gwaii:

> There were many, many spots all over Haida Gwaii … where our ancestors came out of the ocean. I and many of us have been brought up to respect the ocean … we're all brought up that these guidelines of living with the ocean are a matter of life and death or a matter of us losing a certain seafood for disrespecting it. In the beginning of time, there were supernatural beings that lived in the ocean and could come out to be with us and go back to the ocean at will. That is how close we were to the supernatural beings, to all the beings that live in the ocean. So in a sense, we're all related. And that's how you should treat the ocean, as your relative. It's what gave you life … so we should all revere the Tang.Gwanaay—the ocean. (Brown, 2009, cited in Haida Marine Traditional Knowledge Study, 2011, Vol. 1, p. 23)

Teachers at the meeting, most of non-Aboriginal ancestry, discussed with Elder Gwaaganad, Diane Brown, the interconnectedness of Haida life and wondered further about the nature of the relationships between natural and Supernatural Beings. At previous meetings, when Supernatural Beings were introduced or mentioned, teachers considered the stories more similar to European-style fairy tales for children, or requested scientific proof of supernatural-human connections rather than as a way of living in relation to the ocean and land. Nonetheless, engagement with these ideas provided opportunities for teachers to challenge their own assumptions and better understand the community in which they worked and the histories and backgrounds of students they taught.

Developing this kind of cultural awareness, understanding, and respect in the context of teaching mathematics proved challenging for teachers and community members. We often debated the role and presence of culture in learning and teaching school mathematics. Community members and Elders shared their experiences of using mathematics in cultural and economic practices such as canoe building, cedar weaving, wool weaving, mask carving, Haida design drawing, dancing, and storytelling, along with marine harvesting practices such as salmon and halibut fishing and shellfish and seaweed gathering. Mathematical processes of patterns, measuring, cycles, symmetry, transformations, orientations, and relationships arose in these discussions (Nicol & Yovanovich, 2011). However, a goal of teaching school mathematics often made it difficult to reconceptualize what mathematics is and what it means to do mathematics. It also made it challenging to be open to examining ways mathematics can be a context for learning more about Haida culture while at the same time how culture can be a context for learning mathematics. As one teacher stated during an interview after two years participation in the project:

> I used to think that math was something you did from a textbook, that it was something you did, that it had a right way to do it and a wrong way to do it. And now I think that that is not the case, that math is everywhere, that mathematical opportunities are everywhere. (Miki)

This teacher also questioned the inclusion and exclusion of particular mathematical shapes in the provincial elementary mathematics curriculum. Missing, she suggested, were those shapes found in the natural environment on Haida Gwaii and represented in Haida art, such as a razor clam shell or the eye of Kugann Jaad Mouse Woman (ovoid shapes) or the tail of an orca or the wings of an eagle (U shapes). These elements, the ovoid shape and the U-shape, are foundational aspects of Haida art and very familiar to most students on the Islands, both Haida and non-Haida (see Figure 2.3).
Miki states:

> The two-dimensional, three-dimensional shapes in a mathematics IRP [Integrated Resource Package for BC curriculum] doesn't include an ovoid and a U shape [shapes used as elements in Haida art forms]. So I will be spending time with these shapes the same way I would with other shapes. Finding the shapes [ovoid and U] in our natural environment. So doing a shape walk around that .... And then maybe looking through Haida art and finding those shapes and comparing those shapes to see that nothing has right angles. So sort of do a comparison of those shapes to a rectangle and a square, and how do they differ and drawing them. (Miki)

Similarly, during a meeting hosted at the local Kay Llnagaay Museum, one community member and artist spoke of the opportunities the project offered for exploring mathematical activities of Haida cultural practices:

> I'm an artist and talking with all of the artists and the carpenters and people that create canoes—the intuitive math that it takes to create a canoe is phenomenal. In one of our interviews we talked with an Elder, and he talked about how the

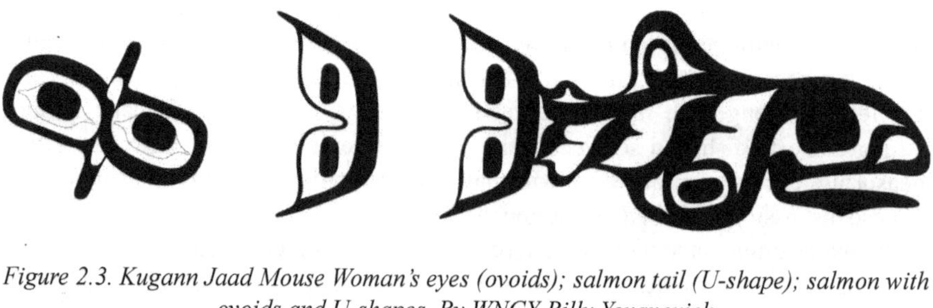

Figure 2.3. Kugann Jaad Mouse Woman's eyes (ovoids); salmon tail (U-shape); salmon with ovoids and U-shapes. By WNCY Billy Yovanovich

> people of Haida Gwaii made these canoes that were so perfect that they were only 1/16th of an inch off on either side … they never knew how to read or write … so how did they create these markings to make these perfect vessels? … I think it accesses a whole new kind of math that you don't actually get in the formal education system. (Kit)

This community member was fascinated with what she refers to as a different kind of mathematics, a mathematics different from school mathematics, yet powerful and precise enough to create powerful seaworthy vessels. This form of mathematics wasn't a lesser mathematics, a subordinate version of what is now taught in schools. Instead, for this community member, the mathematics needed for canoe building was "phenomenal," a wisdom intimately connected to culture that

> teaches you not only about creating a beautiful object, but it teaches you about the math, it teaches you how to be environmentally responsible. It teaches you about integrity. It teaches you about a kind of holistic kind of feel to it … It's a holistic approach to what you are looking at. (Kit)

Discussions of culture and connections to the place of Haida Gwaii also supported teachers decolonizing their own views of themselves as teachers. This teacher's comments speak to her awareness and the challenge of teaching students who are different from herself:

> These meetings have made me think about relevance and connection. I didn't grow up in this beautiful area. I grew up in the suburbs as a non-Aboriginal person. I think that I constantly need reminding of this. This place offers connections to the land and nature that is so part of Haida culture …. I need continued reminding that my background and learning may be different from my students. I have to remember that my students, my children, are coming from a different perspective. (Carol)

Dion (2009) refers to the "perfect stranger" stance as a coping mechanism of most non-Aboriginal Canadian teachers: they mean to do well, but they endlessly claim no knowledge, no relation or connection to Aboriginal peoples (Dion, 2009, pp. 179–181). In this way, non-Aboriginal teachers often abdicate any responsibility in actively knowing or discriminating against Aboriginal peoples and students. However, as can be seen in these comments, the perfect-stranger stance had shifted somewhat with teachers' involvement in the project and opportunities to deepen cultural awareness. This teacher's words highlight the shift further:

> I do think there is always more to learn about the culture. I could learn more about the language, and that shows respect … but I have to be careful …. Sometimes I stereotype to income and lifestyle correlation … I put on those stereotypical hats … So I need to catch myself when I make those judgments because I think I do [make judgments], even though I say I don't. (Carol)

In summary, participation in project meetings and activities provided teachers with opportunities to enhance their cultural awareness. This awareness was evident in teachers' interest in learning more about students' heritage, interest in deepening their own understanding of Haida culture, and their attempts to move beyond Dion's (2009) perfect-stranger stance. Deepening cultural awareness is also seen in teachers challenging what mathematics could be or should be taught in schools, suggesting that mathematics practised and familiar to those on Haida Gwaii are worthy of study. Finally, through connections made with community artists and Elders, some teachers through their own initiative reconnected with community members to plan further activities for their students.

Opportunities to Build Relationships

Intertwined with deepening cultural competence are opportunities to build relationships among teacher participants, between teachers and community members, and with us as facilitators. Meetings drew teachers across schools and grade levels from kindergarten to Grade 11, and across years of professional work from two to 14 years of teaching experience. Teachers valued this diverse composition, stating that it was "unique" and "fascinating" to be part of conversations that involved both elementary school and high school perspectives. This is evident in one task that involved learning more about the community through creating visual community maps. Teachers were inspired to create this task after listening to a story as told by community Elders and knowledge holders (Worthy, 2007). In this story, Raven is tasked with taking two stones while flying over the ocean. For the first stone, the black stone, Raven was asked to bite on it and spit it down so that it would become the seaward land, North America. For the second stone, the speckled one, Raven was told to bite it and spit it down so that it would become Haida Gwaii. However, Raven mixed up the stones, dropping the speckled stone first and then the black stone, making Haida Gwaii much smaller than North America. "That's the kind of thing Raven would do" (Worthy, 2007).

We explored both the content of the story and the characters for their potential to inspire mathematical inquiry (Nicol, Archibald, & Baker, 2013). With guidance from a Haida language educator, teachers developed a task that involved seeing the world from different perspectives: from a bird's-eye perspective (as Raven had), a ground perspective, or a human perspective. The task involved students in creating maps of community places important to them from more than one perspective (e.g., eagle's perspective and frog's perspective). With permission to share the story in their classrooms, teachers engaged students from kindergarten to high school in lessons around the mathematics of mapping, perspective taking, decisions around whose view is valued, and around the trickiness and mischievousness of Raven as a problem maker and problem solver.

Tasks such as these provided opportunities to build relationships with community members, learn more about Haida culture through stories, and strengthen professional trust and care among teachers themselves. Teachers developed and explored these tasks collaboratively before trying them in their classrooms. The meetings were, therefore, crucial in providing a space for teachers to share their concerns, develop ideas about mathematics and culture, and ask critical questions about how the task could be structured in the classroom. They also contemplated how it might be received by students, and later, reflected on how students actually did respond. Some teachers voiced concern that taking students outside for community mapping would not be possible due to student management issues. One teacher voiced her concern and hesitancy to move outside the classroom "because there would be lots of behaviour problems. [My students] aren't very independent. They're good at teacher-guided stuff … and very reluctant to do anything challenging …. In math, they like their student workbooks" (Maria). This point recurred during multiple project meetings:

> I can't take them out of my room or try something different, because they would be all over the place. They'll see it as a chance to goof off or goof around …. What I'm trying to do is to establish a routine. So that they know what is going to happen when and what is expected of them. Every day there are detentions. (Maria)

Setting foot outdoors risked losing the management this teacher felt she was establishing with her class. In response, another offered ideas for what she planned for her students outside:

> I have an idea of creating a book with my class of math on the land. I want to take students outside, to the beach, to the forest, and ask students to find pattern materials. We'll take photos of what they've found and together create a book that we can share with families …. But to get outside there is an element of, maybe initially risk taking, but also trust building. Or maybe initially it seems too scary because they [students] are wild in the class and we think, "Oh, they're not going to listen to be outside," and so we don't even try and then we don't even know. (Miki)

Later in the discussion, another teacher revealed her challenges of developing trust with her students:

> What I am finding is that there are some students that I really want to ensure they have some positive successes and positive experiences in their math class … and I am having a hard time getting to that place …. I am realizing too that it probably has to do with trust issues. They don't trust me yet. Because they don't really know me yet. I am only really new to the school though I am not new to Haida Gwaii … so … is there any brainstorming we can do about achieving some trust and some positive math experiences that incorporate all of these ideas? (Stacy)

The fact that such issues were raised in this meeting speaks to the trust developed among participants, where teachers felt comfortable bringing such issues forward for collective dialogue. Our dialogues included working together to move beyond simplistic ways of thinking about culturally responsive education and challenging each other to reconsider what Sleeter (2012) refers to as "cultural celebration, trivialization, essentializing culture, and substituting culture for political analysis and inequalities" (p. 568). For example, in Haida culture ceremonial button robes depicting family crests are worn as regalia on special occasions for, as Haida artist Dorothy Grant states, "part of our way of recording history" (1986, p. 60). This visual record of cultural and generational heritage is still celebrated at feasts and special events such as high school graduation. Grade 12 Haida graduates are presented a button blanket to mark their accomplishment, acknowledge the transition to adulthood, and celebrate their Haida ancestry. Given the robe's prestige, students' pride in receiving one, and the artistic work involved in designing and making it, the button blanket could be a cultural context through which to explore mathematical ideas. One teacher participant in preparation for her son's high school graduation ceremony used the opportunity of designing and making a button blanket to pay attention to her mathematics in designing, making, fitting, and sewing a blanket (Yovanovich, 2009). Like Sleeter (2013), this teacher was conscious to not simplify culture by trivializing it as a context for school mathematics. Instead, she recorded all of the mathematical questions she asked herself through the months it took to make a finished blanket and created a photobook documenting the process. Her questions included: How will the design be enlarged/decreased? The original design is 27.5 cm x 17.5 cm; if enlarged, what scale should it be? What type of sewing stitch will be used for the design and the border? What tension gives the best seam, and why is tension important?

Creating a photobook of the process provided this teacher and the group with a more critical and respectful view of both mathematics and culture. When sharing her book with community members, and discussing her mathematics with others, this participant reiterated that many had not considered their practices as mathematical. In retelling this story she said, "I used to think that we didn't use a whole lot of math, we as being Haida people, now I think we use lots of math, but we just never called it that." (Jess) In fact, there is no Haida word for the discipline of mathematics, yet Haida language is filled with words that describe measuring, relationships, comparing, time, and movement. In terms of culture, the photobook and other activities transformed other participants' understanding of culture:

> My view of culture was more of a token thing. You could get a token image here and there of Haida whatever. And now it is more rich and strong in how I have used it [culture] and how I have integrated it .... We live on these beautiful islands, the land, that is so tied to Haida culture. It is almost impossible to discern when it does not become Haida and it becomes the land. So that is part of the culturally responsiveness here on these particular islands. (Carol)

Teachers also spoke about the challenges of bringing culture into mathematics in meaningful ways that do not essentialize culture. For example, this teacher asks:

> How do we bring [culture] into the classroom in a way that the children can identify with, but also work with? I think that is the responsive part, how do you or I bring in the culture—because it's living, right? It's today. It's something that is alive. (Miki)

And in response, another teacher suggested the need for teachers to be learners alongside their students:

> It means we should mainly be willing to jump right in there and do it. Drawing on your own personal interest in the area and culture. Through this approach you end up becoming a learner just as much as you do a teacher. (Darren)

These teachers consider culture not as fixed and immutable but instead as living. They nudge each other in moving from simplistic views of culture and responsive pedagogies to more interactive, living approaches that preserve language and art but also reclaim cultural identity.

At project meetings, teachers shared their experiences through telling their stories, showing samples of student work, photographs, and videos of teaching episodes. Teachers discussed their understandings of culturally responsive pedagogy for this context and with these students on Haida Gwaii. As facilitators, we recognized the need to create open spaces of dialogue in which participants felt comfortable sharing questions, critically challenging ideas, and discussing possible courses of action. Meeting together within the community over several years, including at the museum and in longhouses and carving houses, with field trips to beaches and forests provided shared experiences over time for building trust and caring relationships among participants.

Opportunities for Social Action

In true spirit of reciprocity and respect, project teachers shared their work on the project with their colleagues during school staff meetings, district professional development days, and community meetings. Over the years, teachers also presented their understanding of culturally responsive mathematics education at local, provincial, and national conferences. Although teachers understood the value of their work on the project for themselves and their community, they did not expect that others outside Haida Gwaii would also see such value. The group's discussion following a provincial presentation highlights the teachers' shift in understanding the impact of their work:

> Miki: I had no idea people would respond in that way. They actually called our work revolutionary. Can you believe it?
>
> Darren: I'm amazed .... It was really interesting that they could see how they could take these ideas and apply it to their own culture and community. I hadn't thought about it like that before.

Jess: It's so powerful.

Darren: I still can't believe the reaction .... That people are really interested in what we're doing.

Maria: We even had people ask to buy our book [math photobook].

Sharing their understanding of culturally responsive education in the context of mathematics education motivated teachers to continue their work, and to recognize the importance of mobilizing their learning and giving back to the community. Living culturally responsive education included sharing their understanding with others, and that, in turn, seemed to validate the importance of their work for others. Their keen interest led the University of British Columbia to offer a university-level course on Haida Gwaii for interested teachers, providing further opportunities to explore the nature of culturally responsive mathematics education. Almost all project teachers were able to enrol, along with a number of community members. One assignment in the course involved working with interviews of Haida Elders and using Archibald's (2008) Indigenous storywork principles. This assignment evolved into a book, *Tluuwaay 'Waadluxan Mathematical Adventures* (Nicol & Yovanovich, 2011), with sales across the province, Canada, and internationally. Community-based mathematics education led the community and the group to greater trust, rapport, confidence, and credibility of the importance of this work.

As a team we have presented our developing ideas and work at local gatherings and School District conferences, at provincial professional learning conferences, and at national academic conferences. Aspects of our work are included in a MOOC (Massive Online Open source Courseware) for teachers titled *Reconciliation through Indigenous Education* with thousands of students each term. As discussed elsewhere, our living model of culturally responsive education, PAIRS, is used not only to guide further curriculum development with participating teachers but is also introduced to teacher candidates in university teacher education programs (Nicol, Yovanovich, & Gear, 2016; Nicol, Archibald, & Baker, 2013; Nicol & Yovanvoich, 2017).

In summary, participation in the project provided opportunities for teachers to engage in social action through giving back to the community and their colleagues what they were learning in the project. Although all teachers were excited to participate in changing the culture of school mathematics on Haida Gwaii, teachers still considered their work to be ongoing. Some remained hesitant to embrace the land as teacher, to attend community events, to reach out to community members, and to move further from Dion's (2009) perfect-stranger stance. Others, on the other hand, were actively connecting school classes with community members, moving classrooms to the outdoors, and recognizing their own position as Haida or non-Haida in teaching culturally responsive mathematics. A fourth opportunity for professional learning of culturally responsive mathematics education came through our opportunities for mathematical engagement.

DISCUSSION

Contrary to deficit approaches to education that tend to ignore or replace the cultural practices that students bring to school, culturally responsive, sustaining, and relational practices focus on developing a pedagogical stance that is interdisciplinary, culturally interactive, and contextualized so that, as Cajete (1999) states, the "lesson and the learning of the lesson [are] intimately interwoven within the situation and the environment of the learner" (p. 53). Ladson-Billings (1995a, 1995b, 2014) argues that culturally relevant teachers expect academic excellence of their students, draw upon students' cultural histories and current experiences as resources for learning, and prepare students to participate and act for social change. Ladson-Billings (2014) also writes of the challenges and complexities of teachers to enact culturally responsive, sustaining educational practices.

Our research provides insight into the professional learning of teachers interested in developing more culturally responsive practices in their mathematics classrooms. In this chapter, we highlighted four aspects of professional learning that supported teachers in developing awareness and actions as culturally responsive mathematics teachers. First, if teachers, both Aboriginal and non-Aboriginal, are to draw upon students' histories and current experiences as resources for learning, then teachers need opportunities to become culturally competent themselves. Developing cultural competence requires both awareness and engagement in deepening understanding of the place (in our case, Haida Gwaii), its people (both human and supernatural), and the land (animate and inanimate), as well as critical analysis of personal assumptions about mathematics, culture, and what is involved in teaching, as Delpit (1995/2006) states, "other people's children" to challenge Dion's (2009) perfect-stranger stance.

Second, we found that professional learning for culturally responsive mathematics education requires opportunities for teachers to build relationships with each other, with community members, and with facilitators. Creating environments sustained over many years can support teachers in asking difficult questions and sharing personal assumptions and biases. These were key to supporting teachers in challenging simplistic interpretations of culturally responsive mathematics education. A third aspect considers opportunities for teachers to engage in social action, such as sharing their developing understanding of culturally responsive mathematics education at local, provincial, and national levels through presentations, classroom lessons, and professional resource books. This sharing was not only a way of giving back to the community to rethink how mathematics might be taught on Haida Gwaii, but also provided feedback to teachers on the importance of this work, developing their confidence and credibility.

Finally, a fourth aspect for learning culturally responsive mathematics education is seen in opportunities for teachers to work on mathematics together. This aspect was not viewed as a separate feature but instead as something that permeated all meetings. Mathematics and culture were considered together, each providing an opening to learn about the other. Visiting artists carving monumental poles or masks, for example, provided teachers with opportunities to learn about others' mathematics and to challenge what counts as mathematics as well as the mathematics selected for study in the provincial school curriculum.

Learning culturally responsive mathematics education and connections between community, culture, and place, required that teachers develop understandings of students' cultural histories, backgrounds, and funds of knowledge that included connections to the land (Skovesmose, 2011; Gutstein, 2006; Civil, 2016), as well as rethink what it means to be doing mathematics across contexts. However, supporting participating teachers in teaching with a stance and pedagogies that are culturally responsive, sustaining, and relational is both challenging and complex. Our research focused on documenting the dialogue, questions, and shared work during our meetings. Our research is limited, therefore, to what teachers say they do in their classrooms and supported, to some extent, by the student work and classroom videos teachers shared during our meetings and our field notes during team-taught lessons. Nonetheless, our work deepens understanding among teachers and community members of values, cultures, and intercultural dialogue of culturally responsive mathematics education. We discuss our findings in light of three questions focused on the challenges and complexities of this work: What leadership is needed for learning culturally responsive mathematics education? How might critical consciousness be supported? What is needed to move from awareness to engaging students' histories, languages, and literacies in mathematics education?

Culturally Responsive Professional Leadership in Mathematics Education

Opportunities for teachers to develop cultural competence for culturally responsive mathematics education requires, we suggest, supportive leadership at the school and district levels. Our university–school district collaboration began with a shared interest in exploring culturally responsive education as a way of supporting teachers to address the diversity in their classrooms; build upon student strengths and improve performance; and connect school, community, and culture. We did not start with a preconceived or defined view of culturally responsive mathematics education. Instead, we met as a group—teachers, administrators, community members and Elders, and university researchers/facilitators—with a collective goal of improving education for students on Haida Gwaii by exploring the nature of culturally responsive education and how it might be lived (Nicol, 2010; Nicol, Archibald, & Baker, 2013; Nicol & Yovanovich, 2011; Nicol, Yovanovich, & Gear, 2016; Nicol & Yovanovich, 2017). The project was and continues to be supported by school principals encouraging at least one or two teachers per school to attend ongoing project meetings. The school district continues its support by providing teachers with release time to attend meetings.

However, with the success of efforts to engage with culturally responsive mathematics education, School District 50 Haida Gwaii is supporting teachers, schools, and communities through extended visions of culturally responsive education. At the district level, there are changes to its hiring practices that now consider culturally responsive pedagogies, implementation of school projects that value and honour Haida language and culture, and continued development of school-community partnerships and relationships (Nicol & Yovanovich, 2017). For example, current candidates for teaching positions on Haida Gwaii are now asked to be prepared to discuss their visions for culturally responsive education, not only specific to mathematics but to other subject areas as well.

Culturally responsive education is also found at the district level through professional learning sessions and staff meetings. For the past 10 years, all district professional days have included culturally responsive education sessions, many but not all focused on mathematics education. The sessions included cedar weaving, food gathering and preparation, bentwood box building, drum making, and Haida singing, and support teachers in including local culture and history, Indigenous knowledge, and pedagogies into school curriculum. Other sessions such as after-school gatherings during the year have focused on bringing community members and teachers together in discussions of place-based learning, Haida stories, and Haida principles of learning. These sessions are not specific to mathematics education, but as professional learning activities they provide opportunities for distributed expertise, enabling teachers to learn with and from community members about Haida knowledge, culture, and values. Culturally responsive education is also an agenda item for all school staff meetings throughout the district and the district's school principal meetings. At staff meetings, for instance, teachers report and share their culturally responsive ongoing projects for feedback and discussion.

Culturally responsive professional leadership is key for supporting ongoing efforts to move culturally responsive mathematics education from a one-off event to considering it as an integral aspect of teaching. This leadership approach is responsive in its focus on collaborating and co-constructing ideas and strategies for building relationships, in its focus on recognizing and building upon individuals' strengths not deficits, and in its focus on reclaiming Haida language and cultural practices across the curriculum, including mathematics education. Yet, as argued by Khalifa, Gooden, and Davis (2016), this field requires further theorizing and documenting, so that we can better understand the kinds of culturally responsive leadership needed to improve education for school students.

Culturally Responsive Mathematics Education and Critical Consciousness

A second question arising from our research is how to support teachers, administrators, and facilitators in deepening critical consciousness. Gutstein (2006), building upon Freire (1970/2000), writes about his efforts with Latino/a students in social justice mathematics projects, where students engage in reading (understanding) and writing (transforming) the world with mathematics. Gutstein reported the challenges of designing and implementing projects to the level that students can participate in action-oriented change. Freire calls this kind of engagement critical consciousness.

Our project did engage teachers in sharing their knowledge and understanding of culturally responsive mathematics education with others, what we call a form of social action. Yet, as non-Indigenous teachers in the project stated, they needed constant reminders that their ways of being in and seeing the world were likely different from their Haida students. Although many teachers would agree with this statement, we noticed in our meetings that some discussions revealed deficient beliefs and assumptions about students, leading us to wonder how teachers' critical consciousness can be an ongoing project.

One promising suggestion is to engage in a critical investigation of one's own family history, what Sleeter (2013) calls a "critical family history" and Haig-Brown (2012) calls a "decolonizing autobiography." Each offers strategies to help members of a dominant group analyze their own social power and positioning so that they critically reflect on how their own lives have benefited from various relationships and historical political decisions. In this way, teachers could make explicit their own stories of coming to be on Indigenous lands, as settler, immigrant, or refugee and to acknowledge and see themselves in relation to the land, each other, and the human and non-human worlds. We see these as promising strategies for teachers to continue critical self-consciousness for creating mathematical pedagogies that are responsive and sustaining to students' interests and experiences.

Our work with teachers reveals that learning culturally responsive pedagogies for teaching mathematics involves considerations of mathematics and more than mathematics. It involves socio-cultural and socio-political views that, as Gutiérrez (2009) argues, will "help us see the potential of teachers rewriting the narrative in mathematics education to more adequately prepare students for a participatory democracy" (p. 15). Teachers, including ourselves, need opportunities to critically engage with their own land-related histories so that they might be in stronger positions to decolonize or Indigenize their teaching practices. Opportunities to investigate and make more explicit how societal decisions are culturally stratified could encourage teachers to deconstruct and reconstruct (Greenwood, 2013) their relationships with land, community, students, and families, as well as opportunities to reject the resistance to act, for as Dion (2016) states, "The desire to maintain the perfect stranger position is immense" (p. 472).

Drawing upon Bonilla-Silva's (2013) work, Nasir (2016) provides reasons why mathematics educators should care about culture. One reason offered is that "high-quality mathematics instruction (potentially) disrupts unequal access to mathematics learning" (Nasir, 2016, p. 13). High-quality mathematics learning (Boaler, 2006, 2008) includes approaches for students to learn in detracked classrooms, using project-based mathematics curriculum and group work guided by Complex Instruction (Cohen & Lotan, 1997).

Teachers in our project sought to offer high-quality mathematics instruction for students. The small numbers of students per grade required that teachers use multi-ability instruction. Participation in the project provided opportunities for designing culturally responsive practices that included project-based learning when possible.

Our visits with teachers in classrooms, and our efforts to co-teach, provide evidence of teachers' willingness to work toward such high-quality mathematics instruction through a more relational curriculum. There are some positive results. Since 2007, graduation rates have improved for Aboriginal students in the district. Provincial exams in reading, writing, and numeracy have remained fairly constant but have seen modest improvements in some years. Such success, we hope, can disrupt the negative stereotypes of who is and can be successful in school mathematics. It requires unpacking what counts as success and whose norms are used. Nonetheless, to disrupt such deficit narratives requires, we suggest, a culturally response stance that includes experiences for teachers to engage in critical consciousness, making their own beliefs and assumptions a focus of inquiry—what Paris and Alim (2014) term an "inward gaze."

Placed Engagement: Culturally Responsive and Revitalizing Mathematics Education

Paris and Alim (2014) respectfully argue for a re-vision of Ladson-Billings' (1995a, 1995b) idea of culturally relevant education and offer culturally sustaining pedagogy as a way of honouring students' current cultural practices as educational opportunities. In this way, the goal of democratizing schooling is found in "supporting both traditional and evolving ways of cultural connectedness for contemporary youth" (Paris, 2012, p. 95). Paris and Alim (2014) further ask:

> What if, indeed, the goal of teaching and learning with youth of color was not ultimately to see how closely students could perform White middle-class norms but to explore, honor, extend, and, at times, problematize their heritage and community practices? (p. 86)

Recently we were asked by an administrator in a large urban school district how culturally responsive mathematics education can move beyond traditions of the past, such as those located in and connected to weaving, stories, and canoe building. The question speaks to challenging views of Indigenous cultures as trapped and fixed in traditions and the past. This administrator was perhaps speaking to a desired view of culture as dynamic, changing, and living and wondered how mathematics education might respond to such active visions of culture.

Yet, from our experience, few community members on Haida Gwaii would refer to Haida culture as static. Community members, Elders, and artists on Haida Gwaii are actively engaged in reclaiming and revitalizing Haida language and cultural practices previously forced underground by colonizer laws and ways of life. Carving and raising monumental poles, carving and dancing cedar masks, weaving cedar-bark and spruce-root hats, designing and making ceremonial robes, and building seaworthy canoes are all about participating in revitalizing cultural knowledges: a culture that looks to the past for reclamation but to the future for revitalizing.

A culture that is defined, lived, and connected to the land and ocean. A culture disrupted and displaced by colonization.

The importance placed on language revitalization is one example that speaks to possibilities and complexities. Haida Gwaii School District has recently received provincial approval for its Haida-language curriculum for Grades 5 to 12. This comes with the challenge that Haida language is recognized by the United Nations Educational, Scientific, and Cultural Organization (UNESCO) as critically endangered (Moseley, 2010). Following European contact on Haida Gwaii, two main dialects of the language remain: X̱aad Kil in the north G̱aaw Old Masset area and X̱aayda Kil in the south HlG̱aagilda Skidegate area. In efforts to assimilate Haida children and families, government-run residential day schools prohibited students to speak their language. As a result less than 40 fluent speakers now remain, most over the age of 80. With the new X̱aad Kil/X̱aayda Kil Haida language curriculum, community members hope that learning Haida language will provide opportunities for language and cultural revitalization, emphasizing connections of people to land and ocean, community, and human and non-human supernatural worlds. Perhaps engaging with place through language could offer alternatives to Western (mainly European) epistemologies that tend to separate knowledge and place, that tend to consider mathematics as placeless. We can hope that through place, language, and stories we might be in a better position to engage with, as Johnson (2012) describes, "alternative economies, politics and environmental management methods" (p. 834) and thus, we add, alternative ways of being mathematical.

Perhaps learning about place, language, and cultural histories will offer further opportunities to problematize current cultural practices, to question current school mathematics curriculum as did at least one project teacher, or to consider alternative mathematical perspectives alongside what Barton (2008) calls nearly universal and conventional mathematics. We have hope in culturally responsive sustaining and revitalizing practices for mathematics education to move forward "against an education system that contains us toward one that sustains us" (Alim & Paris, 2017, p. 15).

KUGANN JAAD MOUSE WOMAN

We end now, as we began, with Kugann Jaad Mouse Woman, a Supernatural Being who loves a good puzzle, and is willing to support and help out where she can. Haida artist Robert Davidson (November 2018, personal communication) refers to Kugann Jaad as appearing in both the conscious and unconscious, the negative and the positive space. As a shape-shifter, Kugann Jaad can slip into the skins of other beings and seek to restore balance and equity. Kugann Jaad guides our work to gain a deeper understanding of this place, Haida Gwaii. With her curiosity, she has connected us to place and people and supported efforts to shift understandings of mathematics, culture, and what it means to do mathematics—all toward creating a culturally responsive, sustaining, and revitalizing mathematics education. Alongside

Kugann Jaad, we are hopeful for continued shape-shifting, for opportunities to dwell and think more deeply, to look beyond schools and classrooms as places of learning, for as Davidson says "we are all Supernatural Beings having a human experience" (Davidson, 1974).

ACKNOWLEDGEMENTS

We thank the students, teachers, community members and Elders of Haida Gwaii for their support and contributions to this work. We also acknowledge the place of Haida Gwaii as playing a valuable part in our education, how we view and understand culturally responsive mathematics education, and how we see and interpret professional learning.

This research was supported in part by funding from the Social Sciences and Humanities Research Council of Canada, the Vancouver Foundation, and the Canadian Council on Learning.

NOTES

1 We use the terms *Indigenous* and *Aboriginal* interchangeably and in the Canadian context intend them to be inclusive of First Nation, Métis, and Inuit peoples.

2 In the Haida Nation, all Haida are members of either the Raven Clan or the Eagle Clan, whereby a clan is a social governance group.

REFERENCES

Alim, S., & Paris, D. (2017). What is culturally sustaining pedagogy and why does it matter? In D. Paris & S. Alim (Eds.), *Culturally sustaining pedagogies: Teaching and learning for justice in a changing world* (pp. 1–21). New York, NY: Teachers College Press.

Archibald, J. Q'um Q'um Xiiem. (2008). *Indigenous storywork: Educating the heart, mind, body, and spirit*. Vancouver: UBC Press.

Barnhardt, R., & Kawagley, A. O. (2005). Indigenous knowledge systems and Alaska native ways of knowing. *Anthropology & Education Quarterly, 36*(1), 8–23. doi:10.1525/aeq.2005.36.1.008

Barton, B. (2008). *The language of mathematics: Telling mathematical tales*. New York, NY: Springer.

Boaler, J. (2006). How a detracked mathematics approach promoted respect, responsibility, and high achievement. *Theory into Practice, 45*(1), 40–46.

Boaler, J. (2008). Promoting "relational equity" and high mathematics achievement through an innovative mixed-ability approach. *British Educational Research Journal, 34*(2), 167–194.

Bonilla-Silva, E. (2013). *Racism without racists: Color-blind racism and the persistence of racial inequality in America* (4th ed.). Lanham, MD: Rowman & Littlefield.

Brayboy, B. M. J., & Castagno, A. E. (2009). Self-determination through self-education: Culturally responsive schooling for indigenous students in the USA. *Teaching Education, 20*(1), 31–53.

Cajete, G. (1994). *Look to the mountain: An ecology of Indigenous education*. Durango, CO: Kivaki Press.

Cajete, G. (1999). *Ignite the sparkle: An Indigenous science education curriculum Model*. Durango, CO: Kivaki Press.

Cajete, G. (2012). Contemporary Indigenous education: Thoughts for American Indian education in a 21st-century world. In S. Mukhopadhyay & W. M. Roth (Eds.), *Alternative forms of knowing (in) mathematics: Celebrations of diversity of mathematical practices* (pp. 33–51). Rotterdam, The Netherlands: Sense Publishers.

Civil, M. (2002). Culture and mathematics: A community approach. *Journal of Intercultural Studies, 23*(2), 133–148.

Civil, M. (2016). STEM learning research through a funds of knowledge lens. *Cultural Studies of Science Education, 11*(1), 41–59.

Cohen, E. G., & Lotan, R. A. (1997). *Working for equity in heterogeneous classrooms: Sociological theory in practice*. New York, NY: Teachers College Press.

Collison, N. (2006). In the beginning Haida stories. In D. Augaitis, M. Jones, & P. Macnair (Eds.), *Raven travelling: Two centuries of Haida art* (pp. 7–22). Vancouver/Toronto. Vancouver Art Gallery/ Douglas & McIntyre.

Davidson, R. (1974). *Sea ghost*. Vancouver: Lattimer Gallery.

Davidson, R. (2016). *"Kugann jaad giidii": Child of Mouse Woman*. Douglas Reynolds Gallery. Retrieved from https://www.douglasreynoldsgallery.com/artwork/print-original/prints/kugann-jaad-giidii-child-of-mouse-woman/

Delpit, L. (2006). *Other people's children: Cultural conflict in the classroom* (2nd ed.). New York, NY: New Press.

Dion, S. (2009). *Braiding histories: Learning from Aboriginal peoples' experiences and perspectives*. Vancouver: UBC Press.

Dion, S. (2016). Mediating the space between: Voices of Indigenous youth and voices of educators in service of reconciliation. *Canadian Review of Sociology/Revue Canadienne De Sociologie, 53*(4), 468–473.

Freire, P. (2000). *Pedagogy of the oppressed*. New York, NY: Bloomsbury. (Original work published in 1970)

Gay, G. (2010). *Culturally responsive teaching. Theory, research, and practice*. New York, NY: Teachers College Press.

Grant, D. (1986). Sculpting on cloth. In D. Jensen & P. Sargent (Authors), *Robes of power: Totem poles on cloth* (pp. 60–62). Vancouver: UBC Press.

Greenwood, D. (2013). A critical theory of place-conscious education. In R. Stevenson, M. Brody, J. Dillon, & A. Wals (Eds.), *International handbook of research on environmental education* (pp. 93–100). New York, NY: Routledge. doi:10.4324/9780203813331

Gutiérrez, R. (2009). Embracing the inherent tensions in teaching mathematics from an equity stance. *Democracy & Education, 18*(3), 9–16.

Gutstein, E. (2006). *Reading and writing the world with mathematics: Toward a pedagogy for social justice*. New York, NY: Routledge.

Gutstein, E. (2012). Connecting community, critical, and classical knowledge in teaching mathematics for social justice. In S. Mukhopadhyay & W. M. Roth (Eds.), *Alternative forms of knowing (in) mathematics: Celebrations of diversity of mathematical practices* (pp. 299–312). Rotterdam, The Netherlands: Sense Publishers.

Gutstein, E., & Peterson, B. (2013). *Rethinking mathematics: Teaching social justice by the numbers*. Milwaukee, WI: Rethinking Schools.

Haida Marine Traditional Knowledge Study Participants, Winbourne, J., Haida Oceans Technical Team, & Haida Fishers Program. (2011). *Haida marine traditional knowledge study: Methods and results summary* (Vol. 1). Gaaw Old Masset: Council of the Haida Nation.

Haig-Brown, C. (2012). Decolonizing diaspora: Whose traditional land are we on? In A. Abdi (Ed.), *Decolonizing philosophies of education* (pp. 73–90). Rotterdam, The Netherlands: Sense Publishers.

Johnson, J. (2012). Place-based learning and knowing: Critical pedagogies grounded in indigeneity. *GeoJournal, 77*, 829–836.

Jones, M. (2006). Haida art and Haida Gwaii. In D. Augaitis, M. Jones, & P. Macnair (Eds.), *Raven travelling: Two centuries of Haida art* (pp. 28–39). Vancouver/Toronto. Vancouver Art Gallery/ Douglas & McIntyre.

Kawagley, A. O. (2006). *A yupiaq worldview: A pathway to ecology and spirit* (2nd ed.). Long Grove, IL: Waveland Press.

Khalifa, M. A., Gooden, M. A., & Davis, J. E. (2016). Culturally responsive school leadership: A synthesis of the literature. *Review of Educational Research, 86*(4), 1272–1311.

Ladson-Billings, G. (1995a). But that's just good teaching: The case for culturally relevant pedagogy. *Theory into Practice, 34*(3), 159–165.

Ladson-Billings, G. (1995b). Toward a theory of culturally relevant pedagogy. *American Educational Research Journal, 32*(3), 465–491.

Ladson-Billings, G. (2014). Culturally relevant pedagogy 2.0: A.k.a. the remix. *Harvard Educational Review, 84*(1), 74–84.

Moseley, C. (Ed.). (2010). *Atlas of the World's languages in danger* (3rd ed.). Paris: UNESCO Publishing. (Online version) Retrieved from http://www.unesco.org/culture/en/endangeredlanguages/atlas

Nasir, N. (2016). Why should mathematics educators care about race and culture? *Journal of Urban Mathematics Education, 9*(1), 7–18.

Nicol, C. (2010). Learning about culturally responsive mathematics education. In M. F. Pinto & T. F. Kawasaki (Eds.), *Proceedings of the 34th Conference of the International Group for the Psychology of Mathematics Education (PME)* (Vol. 3, pp. 329–336). Belo Horizonte: PME.

Nicol, C., & Yovanovich, J. (2011). *Tluuwaay 'Waadluxan: Mathematical adventures*. Skidegate, BC: School District 50 Haida Gwaii.

Nicol, C., & Yovanovich, J. (2017). Sustaining living and learning culturally responsive pedagogy. In J. Archibald & J. Hare (Eds.), *Learning, knowing, sharing: Celebrating successes in K-12 Aboriginal education in British Columbia* (pp. 83–100). Vancouver: British Columbia Principals' & Vice-Principals' Association.

Nicol, C., Archibald, J., & Baker, J. (2013). Designing a model of culturally responsive mathematics education: Place, relationships and storywork. *Mathematics Education Research Journal, 25*(1), 73–89.

Nicol, C., Yovanovich, J., & Gear, A. (2016). Considering place for connecting mathematics, community and culture. In A. Anderson, J. Anderson, J. Hare, & M. McTavish (Eds.), *Language, learning and culture in early childhood: Home, school and community contexts* (pp. 123–141). New York, NY: Routledge.

Paris, D. (2012). Culturally sustaining pedagogy: A needed change in stance, terminology, and practice. *Educational Researcher, 41*(3), 93–97.

Paris, D., & Alim, H. S. (2014). What are we seeking to sustain through culturally sustaining pedagogy? A loving critique forward. *Harvard Educational Review, 84*(1), 85–100.

Paris, D., & Alim, H. S. (2017). *Culturally sustaining pedagogies: Teaching and learning for justice in a changing world*. New York, NY: Teachers College Press.

Pewewardy, C. D. (1998, November). *Culturally responsive teaching for American Indian learners*. Paper presented at the Kansas Institute on Effective Teaching Practices for Indian Education, Lawrence, KS.

Skovsmose, O. (2011). *An invitation to critical mathematics education*. Rotterdam, The Netherlands: Sense Publishers.

Sleeter, C. (2012). Confronting the marginalization of culturally responsive pedagogy. *Urban Education, 47*(3), 562–584.

Sleeter, C. (2013). *Critical family history*. Retrieved from http://christinesleeter.org/historical-context-questions/

Statistics Canada. (2016). *Census profile, 2016 census*. Retrieved from http://www12.statcan.ca/census-recensement/2016/dp-pd/prof/index.cfm?Lang=E

Worthy, B. (2007). *Legends of the Old Massett Haida: Gaaw Xaadee Gyaahlaangaay* [Audio CD]. Canadian Broadcasting Corporation.

Yovanovich, J. (2009). *The mathematics of button blankets*. Haida Gwaii: Author.

MARIA JOSE ATHIE MARTINEZ

3. "REMEMBER THE TIME WE SET UP OUR TIPI?"

Relational Curriculum Design for Culturally Responsive Mathematics Education

THE LITTLE TIPI STORY

As Told by Blackfoot Elder Mary Ruth MacDougall

> One year in the late 60s there was an Indian Day Encampment. The people invited our family to camp there. We accepted the invitation; however, we had no tipi, so we borrowed our parents' tipi which was a medium sized tipi. We also had to borrow the tipi poles from the village. The only tipi poles we could borrow were tipi poles meant for a large tipi.
>
> We are from the Blackfoot tribe, it so happened there was a Blackfoot Elder at the encampment, so he asked us to set the tipi up at the west end of the encampment. We proceeded to set up the tipi. We thought it looked pretty good, even though the tipi poles were too long.
>
> One of the Elders from [another] tribe stood outside our tipi and hollered "*Hoho hwayyyy*!! Look at the Blackfoot tipi. The Blackfoot people claim they are experts in setting up tipis. Look at their tipi. The poles are so long that you could put another tipi upside down." It was the joke of that particular moment. In our tradition, we often make fun of what people do, but we never retaliate. We just laughed it off and waited for an opportune time to lash back. Unfortunately, our opportunity came sooner than we expected.
>
> After putting up our tipi, we drove to town to get groceries and other necessities we needed for the week-end celebration. While we were in town, a big gust of wind came, sending debris flying all over. We immediately drove back to the encampment to find out if our tipi survived the wind. To our amazement, the only tipi that was still standing was our little tipi with the long poles. All the other tipis blew down. Some of the older tipis tore. It was quite a mess. After assessing the damages, my husband stood in front of our tipi and hollered "*Hoho hwayyy*!!! Who says Blackfoot people don't know how to set up tipis. Whose tipi is still up after that *little* gust of wind blew by? Does that not prove that Blackfoot people are experts in setting up tipis?"

 | DOI: 10.1163/9789004415768_003

When Blackfoot Elder Mary Ruth shared this story during a workshop in 2007 in the traditional Blackfoot territory of Banff, Alberta, Canada, mathematicians were excited. As Mary Ruth recounts: "So, I told that story and Robert (a Sioux mathematician) right away got up and started drawing. I kept thinking, look at this geometry … and Robert was so excited about it, and I got excited the more he was drawing the geometry of the tipi." Mary Ruth and Robert were part of a collaborative workshop designed to bring mathematicians, Elders, and teachers in conversation to imagine alternatives for school mathematics curricula and pedagogy. This workshop, hosted by the Pacific Institute for the Mathematical Sciences (PIMS) and the Banff International Research Station (BIRS) at Banff in Alberta, was the second of three workshops. It and the other workshops were designed to address strategies for how school mathematics curricula and pedagogy might better recognize and honour the language and cultural resources Aboriginal students bring to the mathematics classroom. These workshop experiences are unique. They are week-long intensive retreats located high in the Rocky Mountains, where visual and performing artists, writers, and mathematical scientists enjoy the relative isolation and connection to the land to create, discuss and share innovative projects.

During the first workshop, participants described barriers to success for Aboriginal students and identified shortcomings of current approaches to mathematics curriculum and pedagogy for Aboriginal students. Most prominent among these were the cultural and social dissonance between school and students' Aboriginal communities; the inhospitable nature of public education systems for many Aboriginal students where Indigenous culture and history are ignored and students' potential is unrecognized; the absence of mathematics programs leading to success for all students; and the lack of professional development available to teachers to support their continued mathematical and cultural learning. These factors, when considered together, constitute insurmountable barriers to success in school mathematics for many students, and for Aboriginal students in particular.

The second workshop, a year later, focused on strategies to address these barriers. This workshop involved structured discussion with Aboriginal Elders and Aboriginal and non-Aboriginal mathematicians. The goal was to explore traditional ways of knowing, roles of language and culture in learning mathematics, and the cultural embeddedness of mathematical activities worldwide. Along with invited teachers, the group worked on mathematics problems together, shared solution strategies, and learned to appreciate others' ways of solving mathematical problems. It was at this workshop that Elder Mary Ruth first presented her tipi story to the group and challenged participants to explore the mathematics and science within it. In addition, relationships developed at this workshop resulted in invitations for the mathematicians to travel to various communities to work with teachers and Elders in schools.

Two years past the second workshop, a third was organized. Like the others, this workshop was a week in length and held in the same high-mountain institute. It brought together many of the same Elders, mathematicians, and teachers, and this time also included teacher educators. This particular workshop meeting involved nine Elders, 10 mathematicians, four mathematics teacher educators, and 19 teachers who worked in schools with a substantial proportion of Aboriginal students. During the meeting, an extensive amount of time was set aside for participants to work together in small groups that included a mathematician, an Elder, a teacher and a teacher educator, to design innovative and culturally based mathematics lessons. This chapter focuses on the experiences of one small group in this third workshop, which included Elder Mary Ruth (Blackfoot), Aboriginal mathematician Robert (Sioux), non-Aboriginal teacher educator Kathleen, non-Aboriginal teacher Linda, and workshop coordinator Melania. It focuses on what participants in this group said about their experiences while collaboratively developing mathematics lessons that are more accessible and responsive to Aboriginal culture and students. Although there is some research that articulates a process of curriculum development that is responsive to culture, community, and mathematics (e.g. Lipka, Yanez, Andrew-Ihrke, & Shehenaz, 2009), there is little research that explores the experiences of those who participate in its design. This research articulates what participants say about their experiences working together in a small group as well as in the larger group of workshop participants. The chapter begins with a discussion of culturally responsive mathematics education, then provides details of the study and its findings, and finally concludes with implications for developing future lessons and curricula that are culturally responsive. Much of the data for this chapter was collected as part of graduate research. This study and the workshops were developed based on a culturally responsive education approach.

CULTURALLY RESPONSIVE EDUCATION

Culturally responsive education (CRE) is a perspective of education that is based on students' culture, language, and traditions, validating who they are and at the same time offering a solid basis for further education. CRE supports the idea that the use of local knowledge and local ways of communicating and relating with each other will result in an improvement in academic performance (Burns, Keyes, & Kusimo, 2005; Demmert & Towner, 2003; Gay, 2002; Lipka et al., 2009). Inspired by Lipka et al. (2009), this research aims to create a curriculum based on students' Indigenous experiences, stories, and worldviews.

In CRE, students' culture is central to their education, in recognition that culture is a key element through which people develop their worldviews, and is also the way they understand life and relationships. Gutstein, Lipman, Hernandez, and de los Reyes (1997) define culture as "the ways in which a group of people make meaning of their experiences through language, beliefs, social practices, and the use and creation of material objects" (p. 712). That said, culture is defined constantly through social construction, and since "individual identities are constructed through the intersection of racial, ethnic, class, gender, and other experiences, it cannot be reduced to static characteristics or essences" (Gutstein et al., 1997, p. 712).

Culturally responsive mathematics offers the possibility of joining together two areas that are often considered unrelated; such were the goals of the workshops mentioned. In this perspective, mathematical and cultural knowledge are considered equally important and complement different knowledge systems and different worldviews. Culturally responsive teachers believe in the skills, abilities, and capacities of all students. They respect students' previous experiences and acknowledge that all learners possess cultural knowledge that can be a source of further connections and understandings (Villegas & Lucas, 2002). Culturally responsive teachers expect students to achieve high standards, and in order to do so they encourage and support students constantly while building upon their personal and cultural background (Delpit, 1995; Villegas & Lucas, 2002).

While narrowing CRE to the mathematical field, Ezeife (2002) mentions that mathematics is not typically related to students' culture and real life, which makes it difficult to engage with. However, Averill et al. (2009) mention that "Tate (1997) and others have argued that ignorance of the cultural diversity of students has contributed to the underachievement of some minority groups in mathematics" (p. 159). Mathematics is not a culture-free discipline (Averill et al., 2009). Indeed, research has shown that using language and cultural content in mathematics supports students' knowledge and academic success (Averill et al., 2009; Lipka, Sharp, Adams, & Sharp, 2007). This type of mathematics includes general math academic content and cultural knowledge (Adam, Alangui, & Barton, 2003).

Within culturally responsive education, teacher and student relationships are open, caring, and reciprocal. In this relationship, students and teacher can communicate what they are thinking and how they are feeling without being criticized or judged. At the same time the power is shared (Bishop, 2008), and in addition, both care for each other personally. But also, the teacher cares about students, which prompts high achievements in education (Gay, 2002b).

CRE teachers understand the different dynamics of identity and will try to bridge students' cultural systems with education (Gay, 2000). Teachers of CRE consider students' home cultures and languages as strengths which they can build upon (Gutstein & Peterson, 2006), so that "culturally relevant teachers utilize students' culture as a vehicle for learning" (Ladson-Billings, 1995, p. 161).

DEVELOPING CULTURALLY RESPONSIVE CURRICULUM

Gay (2002a) argues that students learn better when learning is based on familiar content:

> Common sense, professional experiences, and research findings tell us that students learn better content that is familiar, has high interest appeal, is

> challenging, and is presented in ways that are linked directly to their prior knowledge and ways of knowing. (p. 624)

Research has shown that culturally responsive curricula helps students to outperform academically. Therefore, culturally responsive curriculum is a path to use for culturally appropriate content and to overcome a Western perspective dominance in education. Gutstein and Peterson (2006) argue that rethinking the curriculum involves using cultural content that is based on students' experiences and knowledge. Their perspective involves teaching themes that are more accessible, familiar, and meaningful for students. As Delpit (1995) suggests:

> Appropriate education for poor children and children of color can only be devised in consultation with adults who share their culture. Black parents, teachers of color, and members of poor communities must be allowed to participate fully in the discussion of what kind of instruction is in their children's best interest. (p. 45)

Since the literature is not as extensive in curriculum as it is in teaching or in the relationship between students and teachers, I argue that knowledge gained in these related areas can be applied to curriculum development. Curriculum developers should pass through a process of self-reflection about the groups to which they belong, represent, or with whom they identify. Curriculum developers should also know who the students are, and if they are planning for a broader level of education (regional or national) they should leave space for the teacher to infuse local content and knowledge. Curriculum developers should also consider and include in the content key elements or rules for being part of the culture of power, which includes Western knowledge. Additionally, developers should create a respectful relationship among themselves and with the students, teachers, and community members of the particular context, so that they can all be involved in the planning process within a responsive approach.

Curriculum development should directly involve community members, Indigenous people in particular, and should include Indigenous knowledge (IK) systems in education. In their literature review on CRE, Castango and Brayboy (2008) state that one of the areas that has to be developed in CRE for Indigenous students is the involvement of Indigenous epistemologies into education.

According to Villebrun (2006):

> To decolonize education is to retrench and retrieve our traditions in the classrooms in our communities. We need to shore up the cultural and spiritual teachings of our peoples in urban centers where they have moved. Whenever possible we need to teach others about who we are as Indigenous Peoples, to share in all schools our stories of who we are in the lesson plans teachers follow so that our stories are told respectfully and responsibly. (p. 18)

RESEARCH METHODS

I, Maria Jose, am an educator who has had the opportunity to walk alongside Indigenous people on educational projects they have defined. Just recently, I began to identify myself as two-Spirit Mestiza, with Indigenous heritage. I felt honoured to have worked collaboratively and learned how to recreate a curriculum in order to make it culturally responsive. While working on culturally relevant curriculum development, I always wondered what others felt and learned from the experience, what we might learn from processes of developing culturally responsive curriculum (CRC) and others' experiences with CRC. This is the reason why I wanted to explore the experience of the stakeholders involved in developing lessons based on the tipi story for my master's thesis. I was born in Mexico City, and raised since I was two years old in the Cancun area. I have the blessing of speaking the Mayan language, of the territory I grew up in. But also, I have worked with and understand Ayuujk Jay, Pur'epecha, and Ñöñho, among others. I have been a visitor in Musqueam territory where I have had the blessing of studying for my MA and now PhD at UBC.

When designing my methodology, I, the researcher, thought carefully about how to work with Indigenous people in this research collaboratively, respectfully, and without imposition. Even though I, the author, have Native heritage, the colonialism process also impacts the heart and spirit of Indigenous people; thus, Latino American auto-identification with Indigenous backgrounds has a very complex and different process than in Canada. At the time this study was conducted, I did not identify myself as Indigenous. In addition, having the possibility of higher education and graduate studies is a privilege that only less than 5% of the Mexican population have. Therefore, I, for my master's thesis, questioned myself constantly on: How can a mixed race or a non-Indigenous person coming from educational privilege work with Indigenous people on a research project without Western-dominant approaches? More specifically, how can research be performed while acknowledging the colonizing damage caused to Native people? And finally, how to make this a collaborative and participatory project when only one author needs this study?

From Smith (1999), Steinhauer (2003), and Wilson (2001), I understand that beyond a research method, an attitude toward research and education must be developed, and within a perspective that respects people and their cultural background. According to Wilson (2001), it is all about relationships. Therefore, the relationships developed in this study between researcher and participants are the key element that guide the investigation.

Relationships are foundational to develop a confident and trustful environment to share and work collaboratively in any culturally responsive education project. Therefore, the relationship and ethics developed during this project were guided by the four R's of Kirkness and Barnhardt (1991), namely, respect, relevance, reciprocity, and responsibility.

In this study, respect was understood as regard with honour to the individual and to his/her cultural background. Respect was shown to who the person is, to his/

her experiences, knowledge, and ideas. Therefore, the semi-structured interviews were fashioned as a sort of dialogue. Accordingly, the participants shared their ideas around the topics the researcher wanted to explore, in such a way that participants followed their own interest in that topic, or those of their choosing. Respect was also shown in the relationship developed between the people involved, by listening and learning with the participant. I, the researcher, felt like a learner among these wise and committed people. Thus, the interviews or conversations were an opportunity to learn from each participant of the tipi group. Respect, then, was a foundational characteristic of the relationships built between those involved, and was expressed from the first contact with the participants, and throughout the duration of the study. A constant search for respectful space is needed for the creation of transformative research (Haig-Brown & Archibald, 1996).

Respect was also shown by sharing the research control and power. The participants, along with the researcher, gave meaning to the study and reviewed, re-created, and modified the analysis results. All participants of the study worked together in a shared-power relationship, and we were all respected as guides of the investigation. In this study, the researcher is not superior; on the contrary, the researcher is a learner and the participants the experts (Haig-Brown & Archibald, 1996), for they are the ones with the expertise and knowledge needed for the creation of lessons.

Relevance was understood as significant to this research and to the tipi lessons planning group. The research project is relevant and of interest to the people involved because we are working in areas of Indigenous education, and our findings and reflections can be used to inform our work. Likewise, the areas to be explored and analyzed are based on what we define as important, through the dialogue we had during the interviews and collaborative work. So, even though the analysis and the thesis was written by myself, Maria Jose, main investigator and author, the voices, interest, analysis, and findings where all reviewed and commented on with participants based on their interests.

In terms of reciprocity, the participants and the researcher worked together to develop lessons and create a community of learners. Through mutual reflection we constantly learned from each other. Reciprocity was developed by involving the participants in the research, sharing the power among equals (Maiter, Simich, Jacobson, & Wise, 2008), and creating a relationship similar to that of mutual collaboration. As indicated, the author and tipi workshop participants researched the experience as co-researchers, such that together we created understanding and knowledge, which led to "new levels of understanding for everyone" (Kirkness & Barnhardt, 1991, p. 7).

As researcher, I had the responsibility of using the data in the best interests of us all. There was an obligation to share, review, and modify the research. Furthermore, as part of responsibility, the researcher is committed to share with participants what was done with the study.

Beyond the four R's discussed thus far, I acknowledge that the study also drew upon participatory action research and case study. It focused on participants' perspectives during the process of developing culturally responsive math curriculum. Case study, like other qualitative research, has an intensive interest in personal perspectives and circumstances (Stake, 2005).

As indicated, in our case study, the research was explored through the voices of the participants, their opinions on the process of this curriculum development and culturally responsive education, and their further considerations and recommendations for these lessons. After developing the lessons together, and agreeing to collaborate on this research, which would explore our work in progress, we continued sharing our thoughts and conversations by phone. The information presented in this chapter and in my MA thesis was gathered through interviews in the form of open conversation by phone. There were semi-structured conversations with the curriculum developers and the organizer of the workshop. The semi-structured themes were guided by the topics that emerged within their first comments about their experience developing CRE math lessons, and from there all the areas mentioned were explored according to participants' interests.

This study and conversations occurred after the workshop's meeting. Participants agreed to be part of the study to learn more about this experience. The data was collected through an hour or hour-and-a-half conversation-interview in which the semi-structured topics of interest to the researcher were explored. The topics explored during the conversations were around their experience, working with the tipi group, processes of curriculum development, use of lessons in the classroom, culturally responsive math, and future curriculum development (see Appendix 1 for more details).

The conversations were transcribed, and each transcript was sent to each participant. Then another phone call or Skype conversation took place, in order to review the transcript together and to find out if the interviewee felt there was need to add any additional comments, or if the transcript showed what they intended to say. Also, during the following up phone call, if there was a theme not explored in the first conversation, then the researcher tried to find out if there was anything else the participant wanted to add around this area. Then, the researcher added the comments of the follow-up transcript review. And with that information, the researcher created narratives for each participant. With the overlapping topics and the research areas explored for the study, the analysis results were created. Finally, each participant's narrative and analysis results were sent to each person. And the last phone call took place in order to review with each one of them their narrative and the results, with their comments intertwined.

During the interviews, I, Maria Jose, kept track of the themes that emerged and wrote some notes that helped me follow the interviews and cover all topics. After reviewing the topics discussed previously, I created a mind map for each participant. As each topic emerged and was developed, a small mind map for it was created. If a participant addressed the same topic again, then I, Maria Jose, would continue enlarging the small mind map on that particular theme. I worked on one participant's comments at a time, and wrote the story of that experience after completing the individual mind map.

When asked, all but one participant agreed to use their actual names for this research. The participants are:

- Mary Ruth, an Elder who works in Indigenous education, is Blackfoot, and works on different projects with her own community as well as others, while collaborating with Elders and guiding diverse educational and governmental projects.
- Linda, a non-Indigenous teacher who has worked for 14 years within First Nations education, both on and off Indigenous reserves, and has worked with a high percentage of Indigenous students in both cases. She has worked in all levels from junior high to adult education.
- Robert Megginson, a Sioux mathematician, who works at the university level and participates in a range of different research projects that range from the geometry of Banach space to Indigenous American math education.
- Kathleen, a non-Indigenous math educator who teaches at the university level, but who also works in participatory action research projects with teachers who work with First Nations students in two communities in Alberta. Kathleen's extensive teaching experience comes from 17 years of teaching at the junior-high and elementary levels, as well as seven or eight years' experience teaching post-secondary education, which began during her post-graduate studies.
- Melania, the organizer of the Banff workshop, is one of the people who initiated and guided the workshop and from whom the idea of developing lessons emerged. She is a math educator interested in Indigenous education and culturally responsive education.

FINDINGS

In this next section, I offer each participant's "reduced" narrative and the results or conclusions for this chapter.[1]

Mary Ruth—Our Elder

Mary Ruth described her main concern and work: "My interest in helping our Native people on going to further education that is the big thing for me .... So we have to explore what it is that would encourage our students to be more interested in education, to be more interested in learning."

Mary Ruth's focus in the First Nations math workshops is on how to improve Native education and how to get Aboriginal students excited about math. For Mary Ruth, including Indigenous content into the actual curriculum is very important, and that is where the tipi project comes into play. Painting the curriculum with Native content will be beneficial to everyone. As she noted:

> [Banff] was a chance for Elders, mathematicians and Native teachers to get together to share some issues regarding First Nations students and to encode the Native culture into the existing English curriculum, the Alberta curriculum, or Canadian curriculum. Because I think in all the math workbooks and the other subject areas there is so much that could be taught from European or mainstream society, and if we have Native content in there our students would, hopefully, get more interested in the work because there is something they can identify with. And also, our students, our Native students, they will also learn the way we lived before the Europeans came.

As she explains, it will be favourable for First Nations students because bringing Aboriginal knowledge into the curriculum will spark interest in education, make them feel proud of their Aboriginal culture, and help them to better understand and identify with the official curriculum. At the same time, for her, this is valuable to non-Indigenous students because it will offer a more positive picture of Native people than the common stereotypes.

Mary Ruth's appreciation of First Nations knowledge is very important; she argues that Aboriginal people were developed in all areas of society, in all disciplines and fields. However, the way Western education describes Native knowledge seems limited and does not involve all the knowledge that such cultures had. For Mary Ruth, teaching all the areas in which Aboriginal cultures had development is important, because it positions Indigenous knowledge (IK) at the same level and value as Western knowledge (WK), even though Mary Ruth does not use the labels IK or WK. Presenting both knowledges with the same importance and the same value is key for Mary Ruth, and it guides her enthusiasm and her role within the tipi group. She also enjoyed working with a diverse group:

> I think we have a good distribution of participants: mathematicians, university mathematicians, the teachers, and the Elders. In fact, I am looking forward to the next meeting …. With the Elders, their stories will help our mathematicians and our teachers to develop for the younger kids story telling. It will help the younger kids learn so much. That is why I think this group is a very important group to get together and I hope we will continue to complete that package [of lessons] and have a chance to work at other things.

She and other Elders in her community have been working with students since the beginning of their schooling experience:

> So, we are very much involved in the school, in the education of our young people right from the day care … head start, early child hood development and the school. Then they have a youth council and we are working with them, it is really exciting.

She shared the tipi story thinking about the science in it, and how it created excitement among mathematicians:

> I thought you know the time that we set up the tipi, our tipi was long sided but long poles and small tipi, and I said why our tipi stood up and the others didn't? I was thinking of the science part of it, and that includes math, but I was thinking more on velocity and that and I thought that would be kind of exciting to explore and it was! … So I told that story and right away it created excitement, like for instance Robert right away he got up and started drawing. And I kept thinking, look at this geometry, and the velocity of the wind, why is it that our tipi didn't come down and the others did! It was sad but it did. Robert got so excited about it, and I got excited the more he was drawing those things.

Finally, Mary Ruth hopes that their work will expand to science, arts, language, and all topics: "I think in this particular thing with the math it can start and hopefully set a good foundation for our math teacher and then it can expand to other things."

Linda—Teacher

For Linda, working with the group of people was central, for as she said, "It is an exceptional opportunity that was provided, in having those people together already … it is nice to have a group to make lessons, rather than one person trying to make lessons by yourself." As a teacher, she usually plans lessons alone.

Linda considers it safe to have mathematicians with teachers, she believed Elders bring the cultural perspective and teachers spread the information:

> And every time you get mathematicians with teachers it seems to keep it safe, because they have a lot to share … I think it is good when the Elders have that focus and they want to be part of the process …. They have the cultural knowledge to put into the academic that is not readily accessible in a textbook, and it needs to be passed along. If you are passing along information to teachers, then you'll have, if there are 10 teachers in the room then the information is going to be transferred to the students in each of those rooms so you end up now with 300 students.

However, she felt it would be beneficial to include others (students, teacher candidates, beginning teachers) in such endeavours:

> It would be interesting to have students … possibly high school students or university students that are going into teaching or have their first year, second year of teaching under their belt, would have been productive to have in.

For Linda, the tipi story is a good story and it has the possibility to cross the curriculum: "It is a good story, it is a good concept. And it has a lot of learning potentials, you know? It can cross the curriculum in many different ways." The story could be used in math, but also science, history and so on:

> You can cross the curriculum and that is the basic idea that we got to, that you could do more than just the math and the science in it. You should do the language that is attached to the story and added in. The history of, going to the history of the location, that is a lot.

Overall it appeared that Linda, the teacher, had a practical concern about the lessons. She wanted them to happen, and many of her comments and statements are about how to improve the work as a team or design the lessons better and more quickly. She also gave suggestions on how to improve the discussion and communication within the tipi group and between the developers and the teachers implementing the lessons.

Robert—Mathematician

Robert, the mathematician, argues that the people who were working together were right, the mathematicians, Elders, and math educators.

> First of all, I thought that the group that was assembled was just right …. But it was so clear when I was working with people such as the mathematicians involved there from UBC and from other places, that mixture with some Native people like Ed and myself who are also mathematicians and also with the Elders and the mathematics educators. That was a great mix, it turned out that every group offered something that was very important.

He believed people worked very well together: "I just really think that this group has worked really well …. We had some magic." More specifically, he felt the people working at the workshop cared for students in different aspects:

> We all got together as people with a common goal of making it work for the students, and I think it was because there were people who cared, of students from different points of view. The educators cared about them as people to be educated, the mathematicians cared about students because they thought that everybody needed to know more mathematics in a younger age, and of course Native people care about their students because they are their people and want them to succeed.

Interestingly, Robert, the Indigenous mathematician, considers that Mary Ruth's (the Elder's) role was that of a facilitator who pushed the tipi group to continue the work and keep focus:

> Mary Ruth was often the facilitator, and she would keep us focused, pushing us, she would read people very well and she could see things that were interesting to people and let them go with this. She could see things were dead ends, and

> she could gently say let's go do this other thing, or why don't we think about that later, and that would work.

While arguing that the group was right, he indicated there is only one other group of people he would involve while developing lessons based on cultural content, namely administrators:

> I don't see another group that would be necessarily involved. Well, here is a group of people I would think of, and that is administrators … if you involve administrators right from the beginning, they already pre-solved; you don't have to convince them, they being at the meetings and they understand why something works, they already got the excitement.

Moreover, for Robert, there is respect between people:

> I think that there is a great deal of respect, and with each other …. I think that the respect has been the key. Of the fact that we all respect what each other are doing … the bonding through respect … everybody also understood that respect means to hear each other [and] that everybody was on the same page.

As for the tipi story, he believed it could be used in different topics; there are many interesting things that can be done with it, as he stated:

> There are some interesting physics experiments that can be done there. And experiments also with the force of the wind …. So again, there are lots of interesting things that can be done in collaboration with other teachers in the school. And then, what I will want to do is after the students have done all those explorations I will like to bring in the writing teacher to get the students to write about it, and write about what they learned. If there is a Native language program, write about it too.

Moreover, the lessons can be useful for all students, not only First Nations: "I think that the lesson plans are being developed here would be things that will be useful for students far beyond First Nations students."

Robert stated the peculiarity of having an open-ended curriculum development process guided by participants' decisions. However, he was not sure that this is a formula to follow or something that will always work. Robert argued that it needed different elements to come into play in order for it to work, but there is no formula or recipe to follow:

> In order to make it work, it has to be a combination; look at all the things that come together. First of all, the problem has to be rich, but we could have started with this very same rich problem and just have a very different group of people and might have not gone anywhere. So also requires a collection of people who can look at it from different viewpoints, the mathematicians that can look at things from a mathematical viewpoint and see the mathematical richness, the Elders they can see the cultural richness, the math educators they

> can see the pedagogical richness. It requires all of those people and if you can do that, then I think you have a chance to actually develop culturally relevant materials.

Kathleen—Mathematics Educator

For the math educator, Kathleen, the projects she is involved in are long-term, and so she and the teachers have a long-term relationship that they build up to create lessons for a particular classroom:

> I think, for me, the more powerful work happens when you know somebody well. And so developing a relationship with the teachers that I work with takes several years before we are actually able to trust each other to ask the hard questions when they sometimes come out …. And so I am really quite conscious of the need to develop long term relationship, for that what is developed in the curriculum is authentic and it needs something that can be useful for teachers … I think is pretty essential when developing a culturally responsive curriculum to have relationships that are long term.

She considers the need of having experts in different areas working together:

> The other thing that I think is really important is allowing everyone to have some expertise in the project. So that is where the setup of having Elders, mathematicians, and math educators, teachers, is really important because there is a variety of expertise that contributes then to the group.

Kathleen considers that it is very different to develop lessons and implement them with students, because you put a particular audience and context to it:

> It's different because we can then talk about what we saw, how the kids responded, what we thought worked well, what we need to do to prepare better. What kind of ideas the children had. With the work with PIMS we are doing it without children and that makes a huge difference …. And I think the kind of work that is defected necessarily involves the classroom context.

Kathleen stated that the work she is doing with teachers and the work the tipi group is developing is culturally relational, because it addresses the community's experience and interest (or the Elder's experience and interest). However, in her work with the teachers she is not using the tipi story because she wants to make it local, and because for her the tipi story is a starting point.

The story is a starting point to adapt to the local school context, as a living curriculum:

> So that again points to the idea that knowledge and the ideas from the stories are very unique and contextual and local. And how stories that are told are

> quite different between different communities. And so a story for all Aboriginal groups, that is not useful. And so it is a starting place and the stories are the starting place. But you work with those, it is a living curriculum, a working curriculum, a developing curriculum, an emerging curriculum, is not fixed.

Melania—Workshop Organizer

The workshop organizer, Melania, believes in the attendees; she indicates that the people invited to BIRS "really truly care about their community, and they want to do something for their community and it is not about themselves, and this is one of the main reasons why they were invited."

As an organizer, she invited the experts in culture and in math, and the facilitator:

> I knew that the voice of the Elder, the voice of the Elder is a voice that is respected in the communities and as well as their experience … they understand what is happening in the land, in the background and everything. So, it is important to hear what they have to say. It is also important to have mathematicians and math educators, but mainly mathematicians, because of the relation between culture and math. However, I wanted mathematicians that were very much involved in education, that is the reason why this particular group of mathematicians were also invited …. I thought that these conversations needed to happen. We got experts in math and experts in Aboriginal issues. They are the experts, they really are. And of course, you need someone to bring them together like the math educator and the facilitator. But they are the experts, the experts in the culture and the experts in the subject.

The groups needed to be balanced or to have an expert in each area:

> We wanted to create some groups that were balanced, we needed to have a mathematician, we needed to have an educator, we need to have an Elder, and we need to have a teacher. So, we built groups that were balanced in that sense. And in the case of Mary Ruth, she needed to work with Bob because they were very much into the tipi problem.

She described the important role of the teacher as the guide to change in schools:

> Teachers needed to be witnesses of this process, and they needed to see what and how it is happening. And they need it, because they are the ones that are able to change the culture in schools if they really want to do that, they are the harbingers of change. So they need to see that this is possible, that it is possible to have these two different groups come together and to come to consensus about the math, what is mathematics, what needs to be taught, and how these things should happen.

Melania was concerned with having only one culture reflected in the curriculum:

> When you teach math all the examples are really from the Anglo-Saxon culture, and I guess that that is the problem .... It will be a good idea to be more responsive to students, put a little bit of context from different cultures, including Aboriginal, then kids can see themselves reflected in the curriculum ... right now Aboriginal students don't see themselves reflected in the school materials, they don't see their life, they don't see much of their cultural reality in the curriculum, they don't see it in the math. That is the problem, I think that the Anglo-Saxon culture is very resistant to look at anything else.

As evidence in each of these "reduced narratives," the topics explored in each conversation revolved around these aspects: working collaboratively between mathematicians, Elders, teachers, and teacher educators, but in particular working with Elders and teachers; creating relationships among developers and between developers and implementers; and teachings for teachers wanting to develop CRE lessons. We now turn our attention to conclusions framed by these.

In essence, across the stories, the issues that were most important to the participants included the following: developing culturally responsive curriculum with participation from diverse stakeholders; paying particular attention to involving Elders and teachers; and understanding that a culturally responsive curriculum is a relational process.

DEVELOPING CULTURALLY RESPONSIVE CURRICULUM REQUIRES PARTICIPATION FROM DIVERSE STAKEHOLDERS

Co-development of lessons in a collaborative way offered one of the most important findings of this research. For every participant, the work done in cooperation was described with different positive phrases, such as magic, exceptional opportunity, bringing experts in dialogue, and so on. Evans, McDonald, and Nyce (1999) argue about the importance of collaboration when developing curriculum for Indigenous institutions. They mention that community participation and collaboration is one of the crucial aspects of their institutional curriculum development. Therefore, working with a group of people is foundational when developing lessons.

Participants' conversations reveal the need for collaboration and relationships among diverse participants while co-developing CRE lessons and curriculums. In Robert's terms, there have to be people caring about students from different perspectives: cultural, disciplinary, and pedagogical. As stated by Schwab (1978), representatives of students, the learner's milieu, teachers, the subject matter, and curriculum specialists have to be included in planning lessons.

CRE curriculum development needs to be a collaborative process, where people from different backgrounds and perspectives dialogue and work together.

According to the participants, the members assembled were key. Having a mathematician helped by including the field knowledge, as stated by Freire (1970), Parkay, Hass, and Anctil (2010), and Schwab (1978), and also as argued by Robert,

Mary Ruth, Melania, and Kathleen. The mathematicians were the experts in the discipline, but at the same time they were open to listen and learn with everyone.

The math educators brought together the field of education and the mathematical. At the same time, the math educators involved worked alongside teachers in particular schools and classrooms. It is noteworthy that there is a lack of emphasis in the literature on the importance of involving experts in teaching a particular subject, even though the math educators share the theoretical perspective on teaching the subject.

Different stakeholders should be involved as developers, and in particular, collaborate with Elders and teachers. Cajete (1999) primarily mentions the teacher, and includes community members such as parents. This research is aligned with Cajete's stakeholders, but is different because it suggests that Elders are not only sources of support, but need to be involved in development.

Involving Elders

This study, like most CRE work, includes community adults as developers (Delpit, 1995; Evans et al., 1999). As argued by Delpit (1995), involving adults with the students' cultural knowledge can bring relevance to any curriculum. Elders share knowledge that is often not available through textbooks, and at the same time offer guidance for including cultural content, and give the context and philosophical implications of the content (Katheleen, narrative). As stated by Freire (1970), community members are the ones from whom the curriculum should arise, and to whom it should come back. Thus, including community members is a fundamental aspect of Freire's model, as it has been for the tipi group participants. Hence, Elders are essential to restoring Indigenous education according to cultural knowledge, needs, and goals.

All participants considered it vitally important to include Elders in the project, especially for developing the curriculum and adding cultural perspective through all the process of curriculum development. Elders have more experience and offer feedback according to their struggles in school as Indigenous students (Melania). At the same time, Melania argued that the voice of Elders is respected in the community. According to Robert, working with Mary Ruth was exceptional, because she is good at explaining things, is involved in many cultural projects, provides different examples, and was a natural leader for the group.

Freire (1970) and Schwab (1978) agree on having community members as curriculum developers. According to Freire, from the first meeting community members should be part of identifying generative themes. The CRE approach articulates the need to include adults who share students' culture in the process of developing lessons (Delpit, 1995). For Indigenous' scholars the work done with Elders is praised, and everyone receives multiple teachings from it. For Hulan and Eigenbrod (2008), community participation is key in curriculum development. In addition, Elders spread Indigenous knowledge from one generation to the other.

For all participants, the role of Mary Ruth as Elder provided guidance, support, and leadership, but she also worked alongside each member of the group, offering an enriching experience to everyone. Therefore, the role of the Elder in CRE curriculum development is foundational.

Teachers

Teachers are needed when developing curricula, but also in the process of adapting and making lessons significant for the particular classroom and context, and for implementing and initiating the discussions between developers. Most CRE authors consider the classroom level as the scope for developing CRE lessons, and agree with this study that the final adaptations and the implementation of curriculum changes is done by teachers. However, as mentioned by participants, teachers should be part of the whole development process.

For Melania, teachers are the key group to involve for any change in education, since they are the ones who have contact with and impact on students, and are the ones implementing any curriculum. Professors can make any improvement a reality in school. Thus, teachers have to be part of any curriculum development. Kathleen, Mary Ruth, Melania, and Linda, after acknowledging the key role of teachers in curriculum development, are collaborating with teachers in endeavours of this kind.

Schwab (1978) has stated that teachers have the classroom and students' knowledge. CRE scholars such as Cajete (1999), Freire (1970), Ladson-Billings (1995), Peterson (2006), and Villegas and Lucas (2002) mention the key role of the teacher in curriculum development.

Teachers are the group of people with the practical knowledge for implementing any curriculum; they can make teaching an art, and turn theories into classroom reality. In this field, theory without practice is nonsense, while theories based on experience are solid enough to make an impact in school. Moreover, the teaching practice provides elements for any educational theory. Therefore, teachers offer experience, expertise, and knowledge that none of the other participants have.

In addition, teachers know how to develop lessons, read curricula, and interpret outcomes. Hence, as Kathleen argued, if the purpose is to develop lessons, there should be people involved that can do so. Thus, teachers are needed for developing lessons, and sharing practical knowledge about students and the classroom context.

CULTURALLY RESPONSIVE CURRICULUM IS A RELATIONAL PROCESS

One of the main aspects that participants mentioned during the interviews was around relationships. The key features, expressed by all participants, are developing CRE lessons with a group of people; developing trusting and long-lasting relationships among developers; and creating a relationship between developers and the teachers who implement the lessons. This relational aspect of curriculum development acquired major importance in this study, and was considered key to create respectful and trustful long-term relationships.

CRE curriculum development should have an initial period for constructing respect and trust. Only through a deep respect for the people assembled, their knowledge and experience, and the journey of Indigenous and non-Indigenous people walking side-by-side, can CRE development be possible (Silver et al., 2006). It takes time for respect and trust to be developed; however, it has to be the first goal of the developers. Building strong and confident relationships among developers will ignite respectful collaboration.

Likewise, relationships should be part of the whole process of curriculum development and be expanded to include the implementers. Implementers will provide the community/local knowledge and relationships with the community. Thus, CRE curricula require the commitment of the people engaged to be a long-term commitment in order to be more relevant, meaningful, and responsive curricula, and to assess and redesign the lessons planned.

According to Melania, Robert, Linda, and Mary Ruth, bringing together mathematicians, Elders, and teachers to co-develop lessons worked so well that they would do it again and again. However, they acknowledge that there is no perfect recipe to follow. Thus, there should be a combination of two main things: a group of people with different perspectives, and a mutual respect and excitement for working together and for making it happen. According to Robert, there was "magic" in this group, magic based on respect. The people in the group have to be excited about math and the cultural possibilities of the math problems that developers will be working on.

To work collaboratively, each participant must have an expertise, and have a perspective to share that no one else in the group has. While some of the participants mentioned that they could create lessons by themselves, they believed that to create CRE lessons they would have to take into consideration the perspective of Elders and teachers from Indigenous communities. As stated by Evans et al. (1999), when developing curricula in Indigenous contexts, collaboration is very important. Indeed, this study suggests that CRE lessons have to be developed collectively.

Evans et al. (1999) mention the challenges of giving voice to everyone, even when participants have different educational degrees and backgrounds. According to Melania, Robert, and Kathleen in this particular study, participants had expertise to share, and they all felt honoured to work with each other and the Elder. Thus, creating trustful and respectful long-term relationships is key for curriculum development work.

In addition, the relationship has to be extended between developers and teachers. Professors who will implement the curriculum must create a relationship with developers in order to give meaning to the lessons. According to Melania, Kathleen, and Linda, teachers using the curriculum need to create a relationship with developers. Such an extension of the relationship will offer support to the teachers, and will include them in the co-development process. Furthermore, the curriculum will be alive and emerging if developers and teachers create a relationship in which they redefine lessons, and possibly include relevant local knowledge.

Confident, respectful and trusting long-term relationships are key for developing CRE curriculum. Such development is a relational process that surpasses all stages of lesson design, going from the planning stage up to the evaluation. It is foundational that any CRE project devote a considerable amount of time and effort to relationships between co-developers, teachers and developers, and teachers and students. Delpit (1995), Gay (2000), Ladson-Billings (1995), and Villegas and Lucas (2002) argue for the relationship between teachers and students; nevertheless, according to this study, the relational aspect of CRE project goes beyond the classroom into the entire educational process.

CONCLUSIONS

Culture is central to learning; it shapes the thinking of individuals and groups (Ladson-Billings, 1994). As we share our different cultural backgrounds, what we learn from others can also become part of our culture, if we are open to it. All the members in the participating group learned something new from Mary Ruth's story and the conversations that followed, which has now become a part of them. This is one of the most important aftermaths of culturally responsive education: the expansion of individuals' and groups' cultural references.

According to the participants, the tipi story can be used for developing numerous math, science, cultural, arts, and history lessons. The story's potential is so vast that it can bring diverse possibilities for all subjects. Using cultural content that is rich and can be used in any discipline facilitates the holistic use of CRE lessons, and their exploration from different perspectives. Therefore, CRE lessons that can be used interchangeably in diverse fields are powerful, since the cultural content provides learning experience to any academic knowledge. Moreover, holism is one of Indigenous knowledge's characteristics, and brings that knowledge into a horizontal relationship with Western knowledge. Using a rich cultural content shows how powerful Indigenous knowledge and culture are.

Those developing CRE lessons should strive to find cultural content that can work as a catalyst for diverse themes, both within one discipline and among various fields. There is cultural content that can offer deep and challenging learning experiences across diverse grade levels and subjects. That content, with a teacher's and Elder's guidance, shows how profound and complex Indigenous knowledge can be, and at the same time how enriching it is; because of this, it can help students to respect and re-value Indigenous perspectives.

These are only some implications that emerged from this experience, however, they do not provide specific rules or steps to follow. As this summary implies, these are only recommendations that can be adapted to any CRE curriculum development; these areas will need further research. However, creating relationships founded on key elements including respect, trust, confidence, and dialogue is the basis, and would help any group find their own path when striving for solid CRE curriculum development.

NOTE

[1] If any reader would like to consult the full findings and narratives, my thesis is available online (https://circle.ubc.ca/bitstream/handle/2429/28889/ubc_2010_fall_athiemartinez_mariajose.pdf?sequence=1).

REFERENCES

Adam, S., Alangui, W., & Barton, B. (2003). A comment on: Rowlands & Carson "Where would formal, academic mathematics stand in a curriculum informed by ethnomathematics? A critical review." *Educational Studies in Mathematics, 52*(3), 327–335.

Atweh, B., Christensen, C., & Dornan, L. (1998). Students as action researchers: Partnerships for social justices. In B. Atweh, S. Kemmis, & P. Weeks (Eds.), *Action research in practice: Partnership for social justice in education* (pp. 114–138). New York, NY: Routledge.

Averill, R., Anderson, D., Easton, H., Te Maro, P., Smith, D., & Hynds, A. (2009). Culturally responsive teaching of mathematics: Three models from linked studies. *Journal for Research in Mathematics Education, 40*(2), 157–186.

Bishop, R. (2008). Te Kotahitanga: Kaupapa Maori in mainstream classrooms. In N. Denzin, Y. Lincoln, & L. T. Smith (Eds.), *Handbook of critical and Indigenous methodologies* (pp. 439–458). London: Sage.

Bishop, R., & Glynn, T. (1999). *Culture counts: Changing power relations in education.* Palmerston North: Dunmore Press.

Burns, R., Keyes, M., & Kusimo, P. (2005). *Closing achievement gaps by creating culturally responsive schools.* Charleston, WV: Edvantia.

Cajete, G. (1999). *Igniting the sparkle: An Indigenous science education model.* Skyland, NC: Kivakí Press.

Castango, A. E., & Brayboy, B. M. J. (2008). Culturally responsive schooling for Indigenous youth: A review of the literature. *Review of Educational Research, 78*(4), 941–993.

Delpit, L. (1995). *Other people's children: Cultural conflict in the classroom.* New York, NY: The New Press.

Demmert, W. G., & Towner, J. (2003). *A review of the research literature on the influences of culturally based education on the academic performance of Native American students.* Portland, OR: Northwest Regional Educational Laboratory.

Evans, M., McDonald, J., & Nyce, D. (1999). Acting across boundaries in Aboriginal curriculum development: Examples from Northern British Columbia. *Canadian Journal of Native Education, 23*(2), 190–205.

Ezeife, A. N. (2002). Mathematics and culture nexus: The interactions of culture and mathematics in an Aboriginal classroom. *International Education Journal, 3*(3), 176–187.

Freire, P. (1970). *Pedagogy of the oppressed.* New York, NY: Continuum.

Gay, G. (2000). *Culturally responsive teaching: Theory, research, and practice.* New York, NY: Teachers College Press.

Gay, G. (2002a). Culturally responsive teaching in special education for ethnically diverse students: Setting the stage. *Qualitative Studies in Education, 15*(6), 613–629.

Gay, G. (2002b). Preparing for culturally responsive teaching. *Journal of Teacher Education, 53*(2), 106–116.

Genat, B. (2009). Building emergent situated knowledges in participatory action research. *Action Research, 7*(1), 101–115.

Gutstein, E., Lipman, P, Hernandez, P., & de los Reyes, R. (1997). Culturally relevant mathematics teaching in a Mexican American context. *Journal of Research in Mathematics Education, 28*(6), 709–737.

Gutstein, E., & Peterson, B. (2006). *Rethinking mathematics: Teaching social justice by the numbers.* Milwaukee, WI: Rethinking Schools.

Haig-Brown, C., & Archibald, J. (1996). Transforming First Nations research. *Qualitative Studies in Education, 9*(3), 245–267.

Heshusius, L. (1994). Freeing ourselves from objectivity: Managing subjectivity or turning toward a participatory mode of consciousness? *Educational Researcher, 23*(3), 15–22.

Hulan, R., & Eigenbrod, R. (2008). Preface: Oral history and oral traditions. In R. Hulan & R. Eigenbrod (Eds.), *Aboriginal oral traditions: Theory, practice, ethics* (pp. 1–5). Halifax, NS: Fernwood Publishing.

Kirkness, V. J., & Barnhardt, R. (1991). First Nations and higher education: The four R's—respect, relevance, reciprocity, responsibility. *Journal of American Indian Education, 30*(3), 1–15.

Ladson-Billings, G. (1994). *The dreamkeepers.* San Francisco, CA: Jossey-Bass.

Ladson-Billings, G. (1995). But that's just good teaching: The case for culturally relevant pedagogy. *Theory into Practice, 34*(3), 159–165.

Lipka, J., Sharp, N., Adams, B., & Sharp, F. (2007). Creating a third space for authentic biculturalism: Examples from math in a cultural context. *Journal of American Indian Education, 46*(3), 94–113.

Lipka, J., Yanez, E., Andrew-Ihrke, D., & Shehenaz, A. (2009). A two-way process for developing effective culturally based math. In B. Greer, S. Mukhopadhyay, A. Powell, & S. Nelson-Barber (Eds.), *Culturally responsive mathematics education* (pp. 257–280). New York, NY: Routledge.

Maiter, S., Simich, L., Jacobson, N., & Wise, J. (2008). An ethic for community-based participatory action research. *Action Research, 6*(3), 305–325.

Parkay, F. W., Hass, G., & Anctil, E. J. (2010). *Curriculum leadership: Readings for developing quality education programs.* Boston, MA: Allyn & Bacon.

Peterson, B. (2006). Teaching math across the curriculum. In E. Gutstein & B. Peterson (Eds.), *Rethinking mathematics: Teaching social justice by the numbers* (pp. 9–13). Milwaukee, WI: Rethinking Schools.

Schwab, J. (1978). Testing and the curriculum. In I. Westbury & N.J. Wilkof (Eds.), *Science, curriculum and liberal education* (pp. 275–286). Chicago, IL: University of Chicago Press.

Silver, J., Hay, J., Klyne, D., Ghorayshi, P., Gorzen, P., Keeper, C., MacKenzie, M., & Simard, F. (2006). *In their own voices: Building urban Aboriginal communities.* Halifax, NS: Fernwood Publishing.

Smith, L. T. (1999). *Decolonizing methodologies: Research and Indigenous peoples.* London: Zed Books and University of Otago Press.

Stake, R. E. (2005). Qualitative case studies. In N. K. Denzin & Y. S. Lincoln (Eds.), *The Sage handbook of qualitative research* (3rd ed., pp. 443–466). Thousand Oaks, CA: Sage Publications.

Steinhauer, E. (2003). Thoughts on an Indigenous research methodology. *Canadian Journal of Native Education, 26*(2), 69–81.

Villebrun, N. (Dene National Chief). (2006). Athabaskan education: The case of Denendeh past, present, and future. In I. Abu-Saad & D. Champagne (Eds.), *Indigenous education and empowerment: International perspectives* (pp. 13–20). New York, NY: Alta Mira Press.

Villegas, A. M., & Lucas, T. (2002). *Educating culturally responsive teachers: A coherent approach.* New York, NY: SUNY Press.

Wilson, S. (2001). What is an Indigenous research methodology? *Canadian Journal of Native Education, 25*(2), 175–179.

APPENDIX 1

Conversation Guidelines

Experience

1. As I have mentioned, my interest is in the tipi lesson planning group at BIRS. Looking back at the workshop at Banff, could you tell me about your experience there?
   a. Why were you interested in participating? Since when you have participated in these workshops?
   b. How did you get involved in this workshop?

2. Could you tell me about your experience in developing the tipi lessons?
   a. Is there a particular aspect of the experience of developing the tipi lessons that stands out for you?
      Is there something important about the experience of developing the tipi lessons that stands out for you?
   b. How is this experience different or similar to how you usually design lessons?
   c. What are the challenges you experienced developing the tipi lessons?
   d. Would you want to do this again or not? Why?

Working with the group

3. What could you say about working with Elders, mathematicians, and educators co-developing lessons?
   a. How was the dynamic of the group?
   b. Do you think it is important to include all those groups of people in every lesson design?
   c. Is there any other group of individuals that you would include? Who and why?

Process of curriculum development

4. What is your opinion on the way we went about developing the tipi lessons?
   a. Do you remember how we went about choosing the tipi story?
      Could you elaborate on how we went about choosing the tipi story?
      How did the idea of using the tipi story come about?
   b. What would you keep the same about the tipi lesson planning?
   c. What would you change about the tipi lesson planning?

Use of the lessons in the classroom

5. How do you envision the lessons being used in the classroom?
   a. What do you feel the teacher needs to know or do in order to teach the lessons?
   b. Which students would most benefit from these lessons?

Culturally responsive math

6. The tipi lesson plan could be considered an example of culturally responsive math. Why do you think this is or is not?
   a. What does curriculum need in order to be considered culturally responsive?

Future curriculum development

7. Based on this experience, what suggestions would you give for the development of lessons based on cultural knowledge/culturally responsive math?

8. Could you elaborate on five things that stood out for you in this process?

Now that you know what the research is about and the interest I have in exploring our experience, is there anything I should have asked but I didn't?

Is there is anything you want to add that I haven't considered?

FLORENCE GLANFIELD AND GLADYS STERENBERG
(WITH DWAYNE DONALD)

4. UNDERSTANDING THE LANDSCAPE OF CULTURALLY RESPONSIVE EDUCATION WITHIN A COMMUNITY-DRIVEN MATHEMATICS EDUCATION RESEARCH PROJECT

INTRODUCTION

In this chapter, we share a conceptual framework on how we bring forth meaning about the notion of *culturally responsive education*[1] and how this phrase might be used to describe our understandings of research, mathematics, and culture. In the context of an invitation by the Band Council and school administration of Eagle Flight First Nation to develop a research program related to student performance in mathematics, we interrogate our responses by attending to the visible features of the landscape and the distinctive features of a sphere of activity.

According to the *Oxford English Dictionary*, there are two definitions of landscape: “all the visible features of an area of land” and “distinctive features of a sphere of activity” (“Landscape,” 2017). Thom and Glanfield (2018) write about how “expanding our focus beyond the visible features of landscapes involves recognition of how landscapes are constantly *co*emerging, *co*evolving, and *co*hering” (p. 1). In this chapter we want to explore the landscapes of our understanding of culturally responsive education and the “distinctive features of a sphere of activity” that we experienced when we were invited to work alongside Eagle Flight First Nation. Our observations of the landscape of culturally responsive education co-emerge and co-evolve as we participate in the sphere. We use our theoretical grounding in enactivism (Varela, Thompson, & Rosch, 1996; Maturana & Varela, 1987) to suggest that, as participants, we bring forth meaning of culturally responsive education as we observe the visible features and as we are “bringing forth the sphere of activity” with Eagle Flight First Nation.

BRINGING FORTH THE SPHERE OF ACTIVITY: VISIBLE AND DISTINCTIVE FEATURES

Enactivism, or enactive cognition, suggests that an individual brings forth meaning through the interaction of a physical world and the individual. Proulx (2008) writes,

 DOI: 10.1163/9789004415768_004

> The world of meaning is not in us, nor in the physical world, it is in the interaction of both in a mutually affective relationship. With my structure I make sense and give meaning to that physical world and bring forth a world of significance. It is a world of significance that is enabled by my structure, and also by the environment that I interact with. It is my structure that allows me to "see" or perceive things in the physical world, and so my structure allows me to give meaning to the attributes of the physical world. (p. 21)

In bringing forth a world of meaning, we are constantly interacting with the visible features of the environment. Rather than describing visible features as static objects, we conceptualize these as dynamic aspects of the landscape. Thom and Glanfield (2018) suggest that the visible features of the landscape are "the dynamic and collective expression of the interconnected patterns and recursive relations that give rise to and sustain it" (p. 3). They describe how air, wind, and precipitation continuously shape the visible features of the landscape and draw our attention to features that cannot be reduced to single entities. Indeed, they suggest that such features are not *only* "things" and claim that dynamic aspects of landscapes are in fact, "distinctive features of a sphere of activity" ("Landscape," 2017) that occasion the potential for new insights. Drawing on their research, we conceptualize a landscape as co-emerging and co-evolving, not some object that is acted upon. Indeed, through continuous interrelationships, "a landscape is inherently collective, constant, and recursive events of be(com)ing. Any and all landscapes are (ever) chang*ing*" (Thom & Glanfield, 2018). Such interrelationships and interactions are the spheres of activity that make up the landscape.

In our articulation of the landscape that we initially understand as culturally responsive education and through our descriptions of these visible features and the activities alongside Eagle Flight First Nation, we notice that the ways in which we talk about the visible features of the landscape are no longer static, but that the features are co-emerging and co-evolving. We observe how we collectively engage in this articulation and how we are bringing forth a sphere of activity. Thus, we explore the notion of the multiple landscapes that co-emerge within the sphere of activity as we interrogate the landscapes of research, mathematics, and culture. The spheres of activities that have informed our understanding of these three landscapes co-emerge in our understanding of culturally responsive education.

By viewing culturally responsive education from an enactivist perspective, we are prompted to consider how we bring our preconceived notions of culture to the act of responding. In this chapter, we articulate the ways in which our initial understanding of culturally responsive education could be described as a landscape where we only noticed the visible features. In the process of bringing forth the sphere of activity, we begin to notice the distinctive features of the sphere of activity, and the ways in which the features of the sphere of activity can bring forth, for us, new understandings of culturally responsive education.

A VISIBLE FEATURE OF THE RESEARCH LANDSCAPE: WHO MAKES THE DECISIONS?

My (Florence's) trip to Eagle Flight First Nation arose out of a telephone conversation with Samuel, a member of the Band Council whose portfolio included education for Eagle Flight First Nation. Samuel had graduated from our university and wanted someone from the university to be engaged in research at Eagle Flight School. Samuel shared with me that the Band Council and the school administration were very concerned about student performance on the Provincial Achievement Tests (PATs). Samuel also shared the history of the school with me. We agreed that I would go to visit.

According to Samuel, Eagle Flight First Nation has operated a school on the *Indian* Reserve for approximately 20 years. About three years ago, the Band Council decided that they would no longer financially support the transportation of students to provincially funded schools outside of the community. The Band Council wanted to ensure the success of the school in the community, and one way to ensure this success was for children to attend the community school. The Council was noticing that children were leaving the community to attend provincially funded schools in the town 40 kilometres away from the community.

Around the same time, the Band Council decided that one way to show that their school was "as good" as the provincially funded school was to have the children at Eagle Flight School participate in the Provincial Achievement Testing Program. The results over the three years that the school had participated in the program indicated that, overall, the performance of the children in the school was well below the provincial level of acceptable standards. Samuel expressed that student performance on the PATs in mathematics was a great concern for the Band Council with regards to the viability of the school.

The drive to the community took me through a portion of the province that I grew up in. The farm fields along the highway showed that the harvest was complete. The setting sun was reflecting the short stalks of the cut grain in some fields; in other fields, the freshly turned sod was evident, telling me that the farmers were preparing the field for next spring's planting. Further along in the trip, the deciduous trees that lined the road were bare. One could see the leaves on the ground at the base of the trees. Again, the setting sun was reflecting off the bare branches of the trees and the leaves on the ground, reminding me of the many years of seeing this while growing up on the ranger station not far from where I was on the highway. As I got closer to the community, I drove by a lake, the same lake that I had visited as a child. It was dusk now, yet I could still remember the outline of the cabin and it seemed that I was making that first visit to the lake all those years ago. I remember that we had arrived at dusk and I could see the shadow of the cabin in the lake. I looked over at the lake and saw the outline again. Of course, in reality, there were many new cabins at the lake, and the family cabin was no longer there. At the same time, the memories were so vivid.

The next morning I awoke and drove toward the community, a community situated in rolling hills. Driving into the community I noticed a small sign, "Eagle Flight First Nation" to indicate that I was now on the reserve. As I drove, it reminded me of a community where many of my family live. Homes and yards, some with fences and some without, dotted the sides of the road. A health centre, a recreation centre, a senior citizens' complex, band offices, a church, and two school buildings came into view.

The meeting that morning included Samuel, Stanley (Samuel's father, former chief and Band Councillor responsible for education), John (a school administrator), Colin (a high school mathematics teacher), Frank (a junior high mathematics teacher), and Larry (a consultant who had also been Samuel's and John's high school mathematics teacher in a provincially funded school in a neighbouring town). The first part of our conversation wove between our lived experiences and the reason we'd come together: that of discussing the possibility of engaging in research around performance on the mathematics PATs. During our conversations, I learned about the information that they'd already collected and analyzed, namely, student performance on the PATs, and teachers' perceptions of why the students performed in such a way. I learned that the community desperately wanted to have more of their young people eligible for employment with the oil and seismic companies in the area. The current situation was such that very few of their young people met the entry requirements for the post-secondary training programs required for these particular jobs.

It was then my turn to share who I was. I shared that I had grown up in strong relation with family in the northeastern part of the province. I spoke about my relationships with my parents and grandparents and how my father and grandfather taught me to be in relation to the land. I talked about visiting the area of the reserve several times as a child. I also told the group about my working life—that I'd taught high school mathematics in a northeastern community and had worked with Indigenous youth while teaching at a university in a different province. I shared that I would not be engaged in a research project where I would be involved in telling the community what they needed to do, but that I'd be willing to work in relation with the community. I commented that I was feeling frustrated by individuals prescribing what others needed to do and, therefore, did not want to be in a place where I was the one doing the prescribing.

After we talked about the possibility of working together, Stanley pulled out a magazine that he'd seen in the doctor's office the week prior to our meeting.

> He was very happy to see that the person interviewed for the magazine article (me) was the person that he was meeting with. Because he'd seen the article and learned about who I was throughout the meeting, he shared with me that he supported the idea that the community work with me. Stanley and I had developed a sense of a relationship, as he felt he'd been introduced to me through the article.
>
> By the end of the meeting, the group had decided that this could be a "community-based research project," one that engaged the Band Councillors, teachers, administrators, and children in the school in the process. I also talked about my two colleagues, Dwayne and Gladys, and mentioned that they might be interested in participating in this research. I—and Dwayne and Gladys if they were also interested—was invited to participate in a whole-school meeting, to be held six weeks later. (Glanfield, research journal entry, October 12, 2007)

BRINGING FORTH A DISTINCTIVE FEATURE OF A SPHERE OF ACTIVITY AROUND RESEARCH

To consider more fully the contextual complexities of this opening narrative, we wish to attend to the various discursive regimes that effectively delimit and circumscribe research projects initiated in partnership with Indigenous peoples and their communities. It is our view that the significance of colonialism as a social, cultural, historical, and educative force has not yet been meaningfully contemplated in Canadian educational contexts. For the most part, the average Canadian citizen comprehends colonialism as something that happened elsewhere, for example, in Africa or Asia, a long time ago.[2] This disposition is symptomatic of a deeply learned habit of disregarding the experiences and memories of Indigenous peoples.[3] This habitual disregard of Indigenous peoples stems from the colonial frontier experience and is perpetuated in the present educational context as a curricular and pedagogical logic of naturalized separation based on the assumption of stark, and ultimately irreconcilable, differences (Donald, 2009a, 2009b). These colonial frontier logics are predicated on those epistemological assumptions and presuppositions, derived from the colonial project of dividing the world according to racial and cultural categorizations, which serve to naturalize assumed divides and thus contribute to their social and institutional perpetuation. The overriding assumption at work in these logics is that Indigenous peoples and Canadians inhabit separate realities. The inherent intention is to deny relationality.

In the place we call Canada, where the history and memories of colonialism are frequently re-enacted, a huge, bureaucratic governmental institution called Indian Affairs was created to administer all things related to "Indians," and thus institutionalize the denial of relationality. To help with this effort, and to limit their clientele, the government decided to define *Indian* as such:

> *First.* Any male person of Indian blood reputed to belong to a particular band;
>
> *Secondly.* Any child of such a person;
>
> *Thirdly.* Any woman who is or was lawfully married to such a person. (Smith, 1975, p. 87)

A notable effect stemming from this definition of Indian in Canada has been to divide and disentitle individuals, families, and communities, and force conformity to interpretations of "Indianness" limited to the social, cultural, political, and legal interpretations of Indian endorsed by Canadians.

With these influences in mind, it is accurate to depict Indian as an abstract cultural concept rife with ambiguities. One aspect of Indian connotes the historical and ongoing relationship that First Nations have with both the British monarch and the Canadian government. This relationship recognizes the significance of family, clan, and tribe to Indigenous people through treaties, and implies an equitable partnership in which Indigenous peoples and Canadians will share land and resources, as well as the benefits that come from them. The other aspect of Indian was imposed by the Canadian political and judicial systems and personified in the form of the various Indian Acts. The Canadian government passed the Indian Acts of 1876 and 1884 with the intention of assimilating, disenfranchising and, eventually, facilitating the disappearance of Treaty Indians (Milloy, 1999, p. 21). Through these acts, Indians were displaced, and a series of draconian rules and regulations imposed on the day-to-day lives of the people and their communities (Richardson, 1993, pp. 95–106). This colonial regime has gone through different phases and seen various consequences. Indians have been considered obstacles to economic development and expansion; they have been treated as uncivilized children in need of Christianization and a proper education; they have been regarded as anachronistic residue left over from a dusky past that would eventually die off; they have been portrayed as tax burdens who want their Indian rights, but will give nothing in return (Tully, 2000, pp. 41–42).

Closely associated with this need to define *Indian* in these troubling ways is the need to relocate Indian to a place within the national landscape of the place now called Canada. The advent of the reserve system served this purpose. Indian reserves were created as a result of the treaty process and are basically patches of land that remain, *sui generis*, the ancestral territory of each First Nation that has traditional, legal, and constitutional sovereignty in that place. Strangely, though, and in curious parallel to the survival of the enigmatic label *Indian,* the reserve system has been maintained to the present day despite its obvious flaws. These two examples speak to the prevalence and resiliency of colonial frontier logics. The contemporary dilemma of the reserve is that it has become a homeland—a special place shared by relatives and friends—and location where distinct cultural, spiritual, and linguistic practices can be maintained, and paradoxically, a government-created and administered pocket of isolation often wracked by social and economic despair. Whether viewed as good or bad, or both, it seems clear that the "isolation of the reserve is the very essence of its being" (Buckley, 1992, p. 11). This problem of Indian isolation and exclusion, enacted through the Indian Acts and manifested through the reserve system, has become the primary conceptual means according to which Canadians understand and construct "Indianness" today. The reserve has also become the primary conceptual means by which First Nations people identify themselves and signify the places they call home.

Our research landscape is formed by the context of a reserve educational system. In the place we call Canada, First Nation reserves receive funding for education from the federal government through Indian and Northern Affairs Canada (INAC). First Nation communities decide whether or not to use the funding for their own schools or to support children who attend provincially funded schools. In this particular case, Eagle Flight First Nation made the decision to support their own community school and encourage as many children as possible from the reserve to attend the school. Since INAC funding is tied to student population, reserve schools need to attract and retain as many students as they can in order to maintain financial support for their educational system. One important way to ensure this is to enhance student achievement.

A VISIBLE FEATURE OF THE MATHEMATICS LANDSCAPE: IN WHAT WAYS ARE WE ABLE TO BE RESPONSIVE?

> In 2007, I (Gladys) was invited by Florence to work with her and Dwayne alongside a First Nation community. I was hesitant, unsure of my credentials. I had been working with a teacher at Big Rock First Nation School but had limited experience with Indigenous communities. I did not know the protocols, I did not have an Indigenous background. Indeed, I wondered why I was being invited to work at this particular research site.
>
> We began by meeting with the school community. We were asked to introduce ourselves by describing our background. Who was I? Where was I from? I was uncomfortable, as my sense of place seemed unfamiliar to me. Where was my place? I decided to share how I had found my place in mathematics education through my encounter with chaos theory. As a mathematics teacher, I encountered students' views of mathematics as solely arithmetic: certain, unchangeable, and absolute. Much of their previous experiences of mathematics had focused on training the mind through direct instruction of algorithms and mathematical procedures. However, I viewed mathematics in a more relational way. At this first meeting, I spoke about how my views of mathematics had shifted several years earlier when I read *Chaos* (Gleick, 1987). I remembered encountering the images: the Lorenz attractor, the Koch curve, the Mandelbrot set, fractal clusters. Drawn into the stories, I felt excited by these new mathematical possibilities for describing our world. Non-linearity, pluralism, and the dynamic nature of chaos resonated with my view

> of the nature of mathematics. Yet it was the computer-generated pictures that held my attention. I was enthralled by the colour and fluidity of shape. I began to play with my understanding of the universe. What if weather, population, and economic patterns were chaotic? I imagined tree branches, coastlines, and blood vessels as fractals and was fascinated by the harmonious arrangement of order and disorder occurring in natural contexts. This view of mathematics was relational: I was relating to the beauty, elegance, and imagery of mathematics. While I was introducing myself, I saw expanded possibilities of how I could relate to the teachers of this school community. As I related to mathematics, I found my place: I was an outsider on this research landscape but had inside knowledge of mathematics and of teaching. (Sterenberg, research reflection)

BRINGING FORTH A DISTINCTIVE FEATURE OF A SPHERE OF ACTIVITY AROUND MATHEMATICS

One possible response to the invitation by Eagle Flight First Nation is to focus on the achievement gap and the mathematical knowledge needed by both students and teachers to bridge this gap. This response might reflect preconceived notions that assert an authoritative status of mathematics and place mathematics educators in roles as outside experts who will provide "the way" to success. Indeed, Sterenberg's place within our research study might be primarily because of her role as a mathematics educator. On the surface, this might reflect assumptions based on a fixed and authoritarian view of mathematical knowledge. Swetz (2009) suggests that many mathematics programs are adopted by developing nations because of the mathematical authority of the curriculum developers. Describing how mathematics programs have been imposed through the United States Agency for International Development (USAID), he clearly demonstrates the challenges inherent in implementing a program that relies on authoritative mathematical knowledge. For example, he describes one African government's adoption of a mathematics program created by Zacharias, a physics professor from Massachusetts Institute of Technology (MIT):

> It seems that in the bland adoption of Western experimental curriculum into an African situation, no consideration was given to the cultural and socioeconomic impact. When USAID officials were questioned on Jerrold Zacharias's experience and expertise involving education in poor, developing, countries, they responded that his association with MIT was sufficient. Apparently, the aura of MIT's excellence in science and mathematics was omnipotent! This was an example of educational imperialism at its worst! Available research findings on mathematical teaching and learning in the developing world were ignored. (pp. 34–35)

In addition to asserting the mathematical knowledge of experts, this response assumes that one way to close the achievement gap in mathematics is through adherence to programs of study that reflect mathematical standards and systemic structures that promote accountability through high-stake testing. Indeed, current programs of studies across North America (e.g., Alberta Education, 2016; National Council of Teachers of Mathematics, 2000) emphasize goals, standards, expectations, general and specific outcomes, and achievement. The arrangement of mathematical concepts into a spiral curriculum of four organized strands suggests that student learning can be predetermined. Historically, the language used in the Alberta Program of Studies for elementary school mathematics (Alberta Learning, 1997) supported such a fixed and authoritative perspective. Knowledge was viewed as a commodity whereby students developed a "base of knowledge" and were expected to experience the "power and usefulness of mathematics" (p. 2). The content was stated as outcomes that were "measurable and identify what students are expected to know and do" (p. 13). Concepts were to be introduced with manipulatives and gradually developed linearly "from the concrete to the pictorial to the symbolic" (p. 13), and mathematical meaning was best developed "when learners encounter mathematical experiences that proceed from the simple to the complex" (p. 1). In this historic document, conformity to social, political, and economic goals was pursued. Schooling was functional as students were prepared for productive lives in a "rapidly advancing, technological society" (p. 2). The authoritative position of mathematics in society was rationalized, since proficiency in using mathematics "increases the opportunities available to individuals" (p. 2). Rather than seeking to change existing social structures, this program of studies reproduced social inequities by supporting an underlying ideology of authoritativeness, as "all students should receive a level of mathematics education appropriate to their needs and abilities" (p. 2).

Sterenberg's narrative points to the authoritativeness of mathematical knowledge in her description of students' views of mathematics as a certain and static discipline. Because mathematics is often viewed as absolute, a response focused on improving student achievement in mathematics might include imitating best practices of model programs, schools, and teachers who have experienced academic success in mathematics. Certainly, research about low-performing schools promotes the success of evidence-based models (e.g., Fleischman & Heppen, 2009), bilingual programs (e.g., Lindholm-Leary & Borsato, 2005), developmentally appropriate practices (e.g., Bredekamp & Copple, 1997), and culturally responsive professional development initiatives focused on process standards (e.g., Farmer, Hauk, & Neumann, 2005). Swetz (2009) problematizes the adoption of program models by drawing on the results of the 2003 Trends in International Mathematics and Science Study (TIMSS): "As a result of this assessment, several countries are now attempting to emulate Singapore by adopting its curricula and teaching methods. Beware!" (p. 36). His warning is that mathematical programs of study cannot be imported across cultural boundaries without considering the pedagogical implications and socio-economic complications of such an adoption. In reference to Singapore, he notes that student motivation and attitude might contribute more to high achievement in mathematics than programming does.

Programs of studies across North America reflect colonial logics whereby people with mathematical authority set the priorities. An underlying assumption that appears fixed is that not all students are capable of learning mathematics and that an authoritative person must decide on students' needs and abilities in order to implement the program at the students' levels. The stance is that students inhabit different realities. In addition, a focus on best practices suggests a unilateral predetermination of what is needed in all learning contexts and emphasizes a notion of mathematics as universal and culture-free. Mathematics appears to be fixed and universal because of the prevalence of absolutist philosophies of mathematics. These philosophies view mathematics as timeless because it builds on logics of deduction. Mathematical knowledge is "superhuman and ahistorical, for the history of mathematics is irrelevant to the nature and justification of mathematical knowledge, and it is pure, isolated, both value- and culture-free knowledge that happens to be useful because of its universal validity" (Ernest, 2009, p. 51). Mathematics is viewed as culture-free because "every trace of human effort and activity (and value-ladenness) is expunged from the final printed version" (p. 52). Moreover, by ignoring the historical development of mathematical knowledge in non-Greek places such as Africa, China, Middle East, India, Central and South America, the myth of mathematics as a European discipline is promoted and sustained. Colonial logics underpin this perspective as the European project of domination continues.

Gay (2009) challenges these fixed and static views by suggesting that mathematics is value-laden:

> [Teachers] continue to hold onto the mistaken notions that good teaching is culturally neutral, that "best practices" exist that are applicable to all learning circumstances and populations, and that some subject matter is "universally true," and therefore beyond the influence of particular cultures … [Best practices'] embedded meanings convey certain value preferences as well, including performance on measures of cognitive skills in standardized tests in high status subjects such as math and science. (p. 190)

Other researchers present the view of mathematics as a cultural practice. Davis and Hersh (1981) describe a variety of mathematical experiences and conclude:

> [The meaning of mathematics] is to be found in the shared understanding of human beings, not in an external nonhuman reality. In this respect, mathematics is similar to an ideology, a religion, or an art form; it deals with human meanings, and is intelligible only within the context of culture. In other words, mathematics is a humanistic study. It is one of the humanities. (p. 410)

The study of the history of mathematics supports the view of mathematics as a human endeavour sustained by cultural influences (Swetz, 2009). Lakoff and Núñez (2000) present mathematics as a "magnificent example of the beauty, richness, complexity, diversity, and importance of human ideas. It is a marvelous testament to what the ordinary embodied human mind is capable of—when multiplied by the creative efforts of millions over millennia" (p. 379). Davis (1996) argues that mathematics becomes an inhumanity when we ignore human experiences: "It is thus that we have created a system that values compliance over creativity, that spawns destructive behavior by destroying our experience, and that conditions learners to reach for the formulaic ahead of the imaginative" (p. 281).

Consistent with this perspective is a relational view of mathematics: mathematics is meant to be enjoyed as one relates to it. In the recent past, some programs of studies did not emphasize the relationality of mathematics and while expecting that students should "show *some* enjoyment" of mathematics by the end of the program (Alberta Learning, 1997, p. 2, emphasis added), they did not expect *all* students to develop such a relationship to mathematics. However, shifts in perspectives are becoming more evident:

> The new mathematics, as we shall see in detail, is one of relationships and patterns. It is qualitative rather than quantitative and thus embodies the shift of emphasis that is characteristic of systems thinking—from objects to relationships, from quantity to quality, from substance to pattern. (Capra, 1996, p. 113)

We believe that a distinct feature of the sphere of activity around mathematics is that learning and teaching mathematics is a relational and cultural practice. We wonder how culture is visible in educational settings.

A VISIBLE FEATURE OF THE CULTURAL LANDSCAPE: HOW SHALL WE ATTEND TO CULTURE?

> When Florence first shared the news that she had been invited to begin a research relationship with Eagle Flight School—and then invited Gladys and me (Dwayne) to join her on the project—I accepted without hesitation. Even though I was a beginning academic with an already lengthy list of research priorities and responsibilities, and even though I had very little experience working within the field of mathematics education, I was anxious to connect with the teachers and students at Eagle Flight School. The general busyness of academic life can sometimes make it difficult to get out of the city and visit First Nation communities and schools, and I was suffering the effects of this disconnect when Florence extended her invitation. Previous to moving to the University, I had been a teacher at Buffalo Runner First Nation School for a decade. I had accepted the offer to teach at the school even though I had little understanding of the complexities of life on a reserve. I was a generic urban *Indian* looking for a job and Buffalo Runner First Nation School seemed like a good place to start my teaching career. I figured that coming from the big city had its benefits and so thoughtlessly concluded that I would have much to offer the school and community. Ironically, I soon realized that I was the one who had the most to learn.

> The decade that I spent with the people of Buffalo Runner First Nation changed my life. During my time teaching there, I was gifted a beginner's education in wisdom traditions, spiritual practices, and philosophical foundations that continues on. The Elders, teacher colleagues, and community leaders whom I interacted with were generously supportive of the re-education process that I underwent while teaching at their community school. With their help, I became much more critically conscious of the ways in which colonial logics and structures have oppressed Indigenous peoples and their knowledge systems and continue to constrain their expression in the world today. I also participated in and witnessed ceremonial and spiritual traditions that resist colonial power, assert a place-based form of sovereignty, and thus enact Indigenous philosophies. This critical re-education in the Buffalo Runner First Nation context has been tremendously influential in my development as an educator and curriculum studies scholar.
>
> One of the more significant outcomes of this re-education process is a deep commitment to curriculum development projects that facilitate meaningful engagements with Indigenous philosophies and knowledge systems as these are understood and lived by Elders and community leaders. This is the type of engagement I had in mind when I eagerly accepted the invitation to begin a research partnership with teachers, students, and leadership associated with Eagle Flight School. However, after a few visits with teachers and students from the School, I realized that I was seeking to replicate the experiences that I had as a teacher and researcher at Buffalo Runner First Nation School. In a sense, I was elevating my experiences at Buffalo Runner First Nations School to "best practices" status and Eagle Flight School had become a place to implement these practices. The logic at play in the presumed transferability of "best practices" approaches is that the particularities of culture and context do not really matter very much.
>
> I had forgotten a key teaching learned from the people of Buffalo Runner First Nation School: it's all about relationship. The establishment of meaningful and trusting relationships with the teachers and students of Eagle Flight School would take time, energy, thoughtfulness, and an ethical commitment to carefully attend to the cultural and contextual particularities of their school and community. This is what would bring unity and spirit to our research partnership. (Donald, research reflection)

BRINGING FORTH A DISTINCTIVE FEATURE OF A SPHERE OF ACTIVITY AROUND CULTURE

Another possible response to the problem of improving Provincial Achievement Test scores in mathematics for students who attend Eagle Flight School is to follow an approach that emphasizes the development of culturally responsive education.

This response derives from Indigenous education initiatives around the world, which have promoted curriculum as a restorative tool that can help formerly colonized peoples heal and revitalize their communities through the identification, teaching, and learning of relevant notions of culture.

In the North American context, culture-based curricular and pedagogical approaches are similarly viewed by many educators and scholars as the key to changing the educational experiences of Indigenous peoples (Aikenhead, 2002; Au & Jordan, 1981; Battiste, 2000, 2002; Castagno & Brayboy, 2008; Ezeife, 2002; Gay, 2000; George, 2003; Ladson-Billings, 1994, 2009; Lipka & Stairs, 1994). To understand the intentions informing such calls for culturally responsive education for Indigenous youth, it is important to recall the difficult experiences that Indigenous peoples have had with formal schooling. In general, the colonialist solution to the perceived problem of bringing education to the "Natives" was to follow a coercive and abusive program of assimilation during which Indigenous children were subjected to a forced forgetting of the languages and knowledges learned from their families and communities. In the Canadian context, legislators repeatedly endorsed federal residential school policies that forcibly removed Indigenous children from their homes, housed them in harsh environments, unravelled their connections to their cultural values, identities, families, languages, and spiritual practices, and disrupted the functioning of family and cultural institutions (Miller, 1996; Milloy, 1999). At its height in Canada, the residential school program was an all-encompassing experiment in the resocialization and civilization of Indian children. To residential school administrators and teachers, who were usually also missionaries, the civilizing impetus necessitated a total remaking of the Indian child (Milloy, 1999). The response to develop culturally responsive education offers a way to disrupt colonial logics.

Castagno and Brayboy (2008) suggest that although the "plethora of writing on culturally responsive education that we review here is insightful, it has had little impact on what teachers do because it is too easily reduced to essentializations, meaningless generalizations, or trivial anecdotes—none of which result in systemic, institutional, or lasting changes to schools serving Indigenous youth" (p. 941). Their criticism with research on culturally responsive education reflects researchers' assumptions that culture is fixed. Castagno and Brayboy argue for more research to better understand the links between culture, community, curriculum pedagogy, learning, and academic achievement.

Verna St. Denis, a Cree/Métis scholar working at the University of Saskatchewan, confronts the problem of culturalism in provocative ways. St. Denis (2004) argues that the focus on cultural revitalization in Indigenous education in the wake of the residential school experience has led to the predominance of a vexing form of cultural fundamentalism in the field. "Adherence to cultural revitalization encourages the valorization of cultural authenticity and cultural purity among Aboriginal people and has helped to produce the notion and the structure of a cultural hierarchy. 'Authentic' cultural Aboriginal identity has become high currency" (p. 37). The author argues that cultural fundamentalism has created a particular problem for many Indigenous teachers and students who come to feel culturally inadequate as "Indians" because they cannot perform their culture or language in authoritatively authentic ways. Building on the significance of this analysis, St. Denis contends that the project of cultural revitalization frames Indigenous peoples as responsible for *losing* their culture and languages: "They are produced as reckless caretakers of their culture" (p. 43). Importantly, she forwards the view that this emphasis on cultural revitalization in Indigenous education distracts from and minimizes the historic and ongoing systemic racism and discrimination that has also so deeply affected Indigenous peoples:

> The popular notion that one has lost one's culture, as opposed to having one's culture stolen, places the responsibility for making appropriate cultural adjustments on those who for so long were the target of systemic and individual cultural change .... Describing Aboriginal youth as lost is a benign way to describe the effects of the discrimination, exclusion and sustained violence and aggression they face on a daily basis. (p. 43)

Writers and researchers working with culture-based approaches to curriculum and pedagogy in Indigenous education have struggled to present culture as something more complex than food, festivals, tipis, and legends. Indeed, generations ago Forbes (1973) wrote,

> We *must be sure that the essence, the deepest meaning, of Indian life-ways are dealt with.* All too often curriculum-builders have felt that by looking at a few baskets in the third grade, or by having children examine different housing-styles, they could adequately dispose of Native culture. It should be borne in mind, however, that the essence of Indianness has next to nothing to do with canoes, moccasins, or other material objects .... The meaning or significance of the Native American way of life basically revolves around *values* and it is this subject-area which should dominate curriculum-building. (p. 202, emphasis in original)

There are two general implementation models that apply to culture-based approaches to curriculum: exploration of specific cultural practices and development of culturally appropriate curriculum models. These two approaches are sometimes hard to distinguish because they often overlap. However, both generally involve the integration of specific cultural practices and teachings with standard classroom curricular orientations. A distinctive feature of a sphere of activity around culture is that Indigenous cultural teaching approaches offer opportunities for teachers and students to experience unique curricular and pedagogical engagements in the contemporary classroom. When teachers and students collaboratively explore forms of knowledge and ways of knowing not normally considered in classroom contexts, there is great potential for expanding understandings of human creativity, erudition, and learning. Our interrogation of a landscape that co-emerges and co-evolves informs our bringing forth meaning of culturally responsive education.

BRINGING FORTH MEANING OF CULTURALLY RESPONSIVE EDUCATION

By attending to our bringing forth the visible and distinctive features of the landscape of research, mathematics, and culture, we have deepened our understanding of culturally responsive education. The bringing forth of our understanding of culturally responsive education co-emerged as we were "bringing forth the sphere of activity" with Eagle Flight First Nation. The sphere of activity included the invitation by Eagle Flight First Nation to provide leadership in improving student mathematical performance on the PATs in mathematics. As outside experts, we could be expected to address the achievement gap and provide teachers and school staff with instructional tools and strategies that would result in increased student achievement. However, we interrupted this deficit stance by listening to Samuel (the Band Councillor responsible for education), his father Stanley (former chief and Band Councillor), the teachers, and the children. Our research has emerged in the moments of dialogical interaction, not in unidirectional responses. The unfolding of our work with Eagle Flight First Nation prompts us to deeply reflect on what it means to do research in an Indigenous context and the meaning of culturally responsive education. We grapple with our preconceived and static notions of cultural responsiveness. We wonder, what is meant by *culture*? It could mean that culture is static and in the case of mathematics education, that culture is a deficit. The construct of culture is an anthropological invention (St. Denis, 2009; Smith, 1999; Willinsky, 1999). McConaghy (2000) provides some insight:

> Culturalism, in brief, refers to the use of particular anthropological notions of "culture" by which "Indigenous culture" enters the field as "already read" ... culturalism incorporates the ideologies and discursive regimes of universalism, cultural racism, and cultural incompatibility in order to construct and perpetuate a "two race" binary. (p. xi)

Informed by this definition, the term *culture* can point to curricular and pedagogical positions that are founded on static notions of cultural authenticity and codified culture as unquestioned solutions to the problems faced by Indigenous peoples in the field of education. The word *responsive* implies that someone or something responds to another person or thing. We wonder, is this response unidirectional? We wonder, does this response involve taking turns, a mere back-and-forth interaction, where there are subsequent reactions to one another? The phrase *culturally responsive* implies that we are attentive to culture. In our bringing forth meaning of culturally responsive, we have come to think of culturally responsive education as *culturally relational*. What does this mean in the context of this particular research? It means that we relate, we are relational.

We were invited to engage in research, and we invited the community, as represented by the Band Councillors, to engage in research with us, and when we first met the school staff, we invited them to be co-researchers with us in this research. Our community-based research project investigates the ways in which community members, children in school, school staff, and school and community leadership come to develop a shared understanding of mathematics. In a sense, the three of us are relations of the individuals who work at Eagle Flight School, and we are relations of the whole community. Indeed, in bringing forth meaning around culturally responsive education, we have come to understand the ways in which we are shaped by this research and the ways in which it informs how we engage in this work. We wonder, what does it mean to be a relation and to work firmly in a relational context?

Through bringing forth meaning around research, mathematics, and culture, we have come to conceptualize culturally responsive education from an Indigenist research paradigm. Wilson (2007) uses the term *Indigenist* to label a research paradigm related to Indigenous perspectives. He chooses to use this term as he believes "that an Indigenist paradigm can be used by anyone who chooses to follow its tenets" (p. 193). Wilson also suggests that in order to describe and use an Indigenist research paradigm,

> researchers and authors need to place themselves and their work firmly in a relational context. We cannot be separated from our work, nor should our writing be separated from ourselves (i.e. we must write in the first person rather than the third). Our own relationships with our environment, families, ancestors, ideas, and the cosmos around us shape who we are and how we will conduct our research. Good Indigenist research begins by describing and building on these relationships. (p. 194)

What *culturally responsive education* has come to mean to us is that we live with the messiness of being in relation. This means that the data is collected by members of the school community and is shared with us. This means that we work with one another to develop an understanding of the ways in which the data is used. But, ultimately, this means that we are with the community in multiple ways. We work together or alongside one another to question the ways in which the notion of culture is typically taken up. We challenge our assumptions about what the culture says the students should be (e.g., Cree culture) and we honour the philosophies that underlie the culture.

Most significantly, when we live relationally we learn the value of listening. Listening is critical in being a relation. The Blackfoot concepts of *aoksisawaato'p* (visiting/renewal of relations), *aokakio'ssit* (be wisely aware; pay attention), and *aatsimaak'ssin* (responsibility to balance giving/taking reciprocity) inform our practice as culturally responsive. Reconsidering a culturally deficit stance, a perspective of the authoritativeness of mathematical knowledge, and strategies of best practices, we propose an emergent and generative relationality. We suggest that in our research, we bring forth a world of significance where knowing arises in action because of the highly relational nature of learning and social interaction as explicated by enactivists Maturana and Varela (1987): "This circularity, this connection between action and experience, this inseparability between a particular way of being and how the world appears to us, tells us that *every act of knowing brings forth a world*" (p. 26, emphasis in original). Knowing is doing. Knowing is enacted. Knowledge is relational (Davis, 1995). For us, culturally responsive education has a specific epistemological perspective:

> The *knowledge of knowledge compels.* It compels us to adopt an attitude of permanent vigilance against the temptation of certainty. It compels us to recognize that certainty is not a proof of truth. It compels us to realize that the world everyone sees is not *the* world but *a* world which we bring forth with others. It compels us to see that the world will be different only if we live differently. It compels us because, when we know that we know, we cannot deny (to ourselves or to others) that we know. (Maturana & Varela, 1987, p. 245, emphasis in original)

Moreover, culturally responsive education implies an ontological stance where

> we make the world we live in with one another. We're participants in the story as it unfolds. We aren't victims and we aren't outsiders. We are part of the universe, you and me, and the goldfish. We make our world with one another. (Waldrop, 1992, p. 321)

For us, culturally responsive education is profoundly dependent on the Cree concept of *wichihtowin,* a healing energy or medicine that is generated when we are actively together with the intention of honouring and respecting the relationships we are enmeshed within. The meaning of a culturally responsive education we bring forth draws on Davis's (1996) research. We believe that enactivist research focuses on the *now*—on emergent understandings, on immediate possibilities for action, and on the broader implications of such action:

> Knowledge, rather than being understood in objective or subjective terms—whereby persons and their understandings are regarded as essentially isolated and autonomous—is recast as those patterns of acting that allow our structures to be coupled, thus entangling us in one another's existence and implicating us in one another's knowing. (p. 235)

As educators, we adopt the ontological and epistemological stance of being in relation with Eagle Flight First Nation. As enactivist theorists, we "seek to reawaken our awareness of the active, negotiated, storied, and relational natures—that is, to give a dialogical rather than a monological account—of our selves" (Davis, 1996, p. 173). As co-researchers alongside the Councillors, staff, and children, we are seeking to honour meaningful engagements with Indigenous philosophies and knowledge systems as they are understood and lived by all in the relation.

NOTES

[1] In using this phrase, we acknowledge that the process of *education* is guided by curriculum and pedagogy, but also encompasses policies, philosophies, and communitarian partnerships that can often have a profound impact on the experiences of the students in the classroom. We use the phrase as a way *to* acknowledge the complexities that arise when these various influences intersect in the school setting.

[2] This generalization stems largely from work with pre-service and practising teachers in Alberta since 2003 and observations of Canadian cultural dispositions today. Evidence of this disposition was starkly revealed in a statement made by Canadian Prime Minister Stephen Harper at the G20 Summit in Pittsburgh on September 25, 2009, when he stated that Canada has "no history of colonialism" (Ljunggren, 2009). It is also informed by Mackey's (2002) notion of the *Canadian*-Canadian.

[3] See Barker (2009) for more on the persistence of imperial and colonial mentalities in Canada today.

REFERENCES

Aikenhead, G. (2002). Cross-cultural science teaching: Rekindling traditions for aboriginal students. *Canadian Journal of Science, Mathematics and Technology Education, 2*(3), 287–304.

Alberta Education. (2016). *The Alberta K-9 Mathematics: Program of studies with achievement indicators.* Edmonton: Alberta Education.

Alberta Learning. (1997). *Mathematics kindergarten to grade 6.* Edmonton: Alberta Learning.

Au, K. H., & Jordan, C. (1981). Teaching reading to Hawaiian children: Finding a culturally appropriate solution. In H. T. Trueba, G. P. Guthri, & K. H. Au (Eds.), *Culture and the bilingual classroom* (pp. 139–152). Rowley, MA: Newbury House.

Barker, A. (2009). The contemporary reality of Canadian imperialism: Settler colonialism and the hybrid colonial state. *American Indian Quarterly, 33*(3), 325–351.

Battiste, M. (2000). Maintaining Aboriginal identity, language, and culture in modern society. In M. Battiste (Ed.), *Reclaiming indigenous voice and vision* (pp. 192–208). Vancouver: UBC Press.

Battiste, M. (2002). *Indigenous knowledge and pedagogy in First Nations education: A literature review with recommendations.* Ottawa: Indian and Northern Affairs.

Bredekamp, S., & Copple, C. (1997). *Developmentally appropriate practice in early childhood programs.* Washington, DC: National Association for the Education of Young Children.

Buckley, H. (1992). *From wooden ploughs to welfare: Why Indian policy failed in the Prairie Provinces.* Montreal: McGill-Queen's University Press.

Capra, F. (1996). *The web of life.* New York, NY: Anchor Books.

Castagno, A., & Brayboy, B. (2008). Culturally responsive schooling for Indigenous youth: A review of the literature. *Review of Educational Research, 78*(4), 941–993.

Davis, B. (1995). Why teach mathematics? Mathematics education and enactivist theory. *For the Learning of Mathematics, 15*(2), 2–9.

Davis, B. (1996). *Teaching mathematics: Toward a sound alternative.* New York, NY: Garland.

Davis, P. J., & Hersh, R. (1981). *The mathematical experience.* Boston, MA: Houghton Mifflin.

Donald, D. (2009a). *The pedagogy of the fort: Curriculum, Aboriginal-Canadian relations, and Indigenous Métissage* (Unpublished doctoral dissertation). University of Alberta, Edmonton, AB.

Donald, D. (2009b). The curricular problem of Indigenousness: Colonial frontier logics, teacher resistances, and the acknowledgment of ethical space. In J. Nahachewsky & I. Johnston (Eds.), *Beyond presentism: Re-imagining the historical, personal, and social places of curriculum* (pp. 23–39). Rotterdam, The Netherlands: Sense Publishers.

Ernest, P. (2009). New philosophy of mathematics: Implications for mathematics education. In B. Greer, S. Mukhopadhyay, A. B. Powerll, & S. Nelson-Barber (Eds.), *Culturally responsive mathematics education* (pp. 43–64). New York, NY: Routledge.

Ezeife, A. N. (2002). Mathematics and culture nexus: The interactions of culture and mathematics in an Aboriginal classroom. *International Education Journal, 3*(2), 176–187.

Farmer, J., Hauk, S., & Neumann, A. M. (2005). Negotiating reform: Implementing process standards in culturally responsive professional development. *The High School Journal, 88*(4), 59–71.

Fleischman, S., & Heppen, J. (2009). Improving low-performing high schools: Searching for evidence of promise. *The Future of Children, 19*(1), 105–133.

Forbes, J. D. (1973). Teaching Native American values and cultures. In J. A. Banks (Ed.), *Teaching ethnic studies: Concepts and strategies* (pp. 201–225). Washington, DC: National Council for the Social Studies.

Gay, G. (2000). *Culturally responsive teaching: Theory, research, and practice*. New York, NY: Teachers College Press.

Gay, G. (2009). Preparing culturally responsive mathematics teachers. In B. Greer, S. Mukhopadhyay, A. B. Powell, & S. Nelson-Barber (Eds.), *Culturally responsive mathematics education* (pp. 189–206). New York, NY: Routledge.

George, P. (2003). The rainbow/holistic approach to Aboriginal literacy. *Canadian Journal of Native Education, 27*(1), 29–40.

Gleick, J. (1987). *Chaos: Making a new science*. New York, NY: Viking Press.

Ladson-Billings, G. (1994). *The dreamkeepers: Successful teachers of African American children*. San Francisco, CA: Jossey-Bass.

Ladson-Billings, G. (2009). *The dreamkeepers: Successful teachers of African American children* (2nd ed.). San Francisco, CA: John Wiley & Sons.

Lakoff, G., & Núñez, R. E. (2000). *Where mathematics comes from: How the embodied mind brings mathematics into being*. New York, NY: Basic Books.

Landscape. (2017). *English Oxford living dictionaries*. Retrieved from https://en.oxforddictionaries.com/definition/landscape

Lindholm-Leary, K., & Borsato, G. (2005). Hispanic high schoolers and mathematics: Follow-up of students who had participated in two-way bilingual elementary programs. *Bilingual Research Journal, 29*(3), 641–729.

Lipka, J., & Stairs, A. (1994). Negotiating the culture of Indigenous schools [Special issue]. *Peabody Journal of Education, 69*(2).

Ljunggren, D. (2009). *Every G20 nation wants to be a Canada insists PM*. Retrieved from https://www.reuters.com/article/columns-us-g20-canada-advantages/every-g20-nation-wants-to-be-canada-insists-pm-idUSTRE58P05Z20090926

Mackey, E. (2002). *The house of difference: Cultural politics and national identity in Canada*. Toronto: University of Toronto Press.

Maturana, H. R., & Varela, F. J. (1987). *The tree of knowledge: The biological roots of human understanding*. Boston, MA: Shambhala.

McConaghy, C. (2000). *Rethinking indigenous education: Culturalism, colonialism, and the politics of knowing*. Flaxton: Post Pressed.

Miller, J. R. (1996). *Shingwauk's vision: A history of native residential schools*. Toronto: University of Toronto Press.

Milloy, J. (1999). *A national crime: The Canadian government and the residential school system, 1879–1986*. Winnipeg: University of Manitoba Press.

National Council of Teachers of Mathematics. (2000). *Principles and standards for school mathematics*. Reston, VA: Author.

Proulx, J. (2008). Some differences between Maturana and Varela's theory of cognition and constructivism. *Complicity: An International Journal of Complexity and Education, 5*(1), 11–26.

Richardson, B. (1991). *Strangers devour the land*. Vancouver: Douglas & McIntyre.

Smith, D. (1975). *Canadian Indians and the law: Selected documents, 1663–1972*. Toronto: McClelland & Stewart.

Smith, L. (1999). *Decolonizing methodologies: Indigenous peoples and research*. London: Zed Books.

St. Denis, V. (2004). Real Indians: Cultural revitalization and fundamentalism in Aboriginal education. In C. Schick, J. Jaffe, & A. Watkinson (Eds.), *Contesting fundamentalisms* (pp. 35–47). Halifax, NS: Fernwood.

Swetz, F. J. (2009). Culture and the development of mathematics: An historical perspective. In B. Greer, S. Mukhopadhyay, A. B. Powell, & S. Nelson-Barber (Eds.), *Culturally responsive mathematics education* (pp. 11–42). New York, NY: Routledge.

Thom, J. S., & Glanfield, F. (2018). Live(d) topographies: The emergent and dynamical nature of ideas in secondary mathematics classes. In A. Kajander, J. Holm, & E. Chernoff (Eds.), *Teaching and learning secondary school mathematics – Canadian perspectives in an international context* (pp. 51–60). New York, NY: Springer.

Tully, J. (2000). A just relationship between Aboriginal and non-Aboriginal peoples of Canada. In C. Cook & J. Lindau (Eds.), *Aboriginal rights and self-government: The Canadian and Mexican experience in North American perspective* (pp. 39–71). Montreal: McGill-Queen's University Press.

Varela, F. J., Thompson, E., & Rosch, E. (1993). *The embodied mind.* Cambridge, MA: MIT Press.

Waldrop, M. M. (1992). *Complexity: The emerging science at the edge of order and chaos.* New York, NY: Simon & Schuster.

Willinsky, J. (1999). *Learning to divide the world: Education at empire's end.* Minneapolis, MN: University of Minnesota Press.

Wilson, S. (2007). Guest editorial: What is an Indigenist research paradigm? *Canadian Journal of Native Education, 30*(2), 193–195.

LISA LUNNEY BORDEN, DAVID WAGNER
AND NEWELL JOHNSON

5. SHOW ME YOUR MATH

Mi'kmaw Community Members Explore Mathematics

INTRODUCTION

Since 2007, in the spring of the year, Mi'kmaw students from across Nova Scotia and other parts of Atlantic Canada have been coming together at the annual Show Me Your Math (SMYM) event to share math fair projects created to demonstrate the results of their own ethnomathematical investigations. In the initial four years of the program, participation at the math fair has ranged from three to five schools, and starting in 2011 all nine schools under the Mi'kmaw Kina'matnewey (MK) system participated, and SMYM events have been developed in other jurisdictions including Wolastoqey communities in New Brunswick, and communities in Northern Ontario and Nunavut. Currently in MK schools alone, annual participation includes 200–300 students attending math fairs and thousands of students contributing projects. In this chapter we describe the SMYM event, its origins, and its impact on participating Mi'kmaw schools. We describe the transformative potential of such ethnomathematical activity for both students and their teachers and the enthusiasm of the Mi'kmaw students who, through this work, came to see that mathematical reasoning is part of their own cultural heritage and worldview.

Drawing on positioning theory as conceptualized by the social psychology work of Harré and van Langenhove (1999) and its development in the context of mathematics education by Wagner and Herbel-Eisenmann (2009), we describe how this project is challenging traditional storylines in classrooms. We argue that SMYM is contributing to a greater sense of equity by addressing critical questions of identity and power and enabling community voices to be seen as a source of authority.

Stories from the communities and the classroom exemplify how this project is lived out in schools and communities. These stories illustrate some of the many benefits of this project and give examples of the impacts on both the students and the communities. We will conclude with strategies and suggestions for how such an event might be enacted in another community.

 | DOI: 10.1163/9789004415768_005

Context

The Mi'kmaq are Indigenous inhabitants of Atlantic Canada. These communities have a unique jurisdictional agreement with the Government of Canada, known as the Mi'kmaw Kina'matnewey (MK) agreement, which gives the MK communities control over their education system and collective bargaining power. The MK communities have a stated goal of decolonizing education by incorporating Indigenous knowledge, culture, and values in their curricular and pedagogical practices. Such a decolonized approach to education, which allows for the inclusion of Indigenous worldviews, has been advocated as a necessity to meet the needs of Mi'kmaw students (e.g., Battiste, 1998; Battiste, 2000; Orr, Paul, & Paul, 2002). Yet these communities are also responsible under the agreement to offer provincially comparable curriculum. More recently, in response to a call from the Department of Indian and Northern Affairs Canada (INAC), the communities are also expected to implement accountability measures. Thus, they have begun a "school success planning" initiative. As such, participating schools are now required to write provincial assessments in mathematics and literacy. There are significant tensions for teachers who must find ways to demonstrate success related to the values embodied in provincial curriculum while centring on community knowledge, language, and values.

The marginalization of Mi'kmaw youth from mathematics is a long-standing concern in Mi'kmaw communities. While it is difficult to gather accurate statistics on the number of Mi'kmaw students pursuing educational paths involving mathematics and sciences, community leaders recognize and articulate concern about the disengagement of their students from these subjects. Similarly, interested parties within Indigenous contexts across Canada have expressed concerns about low participation in STEM fields. The government's national working group on education (Indian and Northern Affairs Canada, 2002) has said that a key area to be addressed in Indigenous education in Canada is the development of culturally relevant curricula and resources in areas of mathematics and science where there is currently an identified weakness. Although not specific to Canada, this argument is supported by a National Council of Teachers of Mathematics publication that stated that Indigenous people in North America have the lowest participation rates of all cultural groups in advanced levels of mathematics (Secada, Hankes, & Fast, 2002).

Ezeife (2003), Secada, Hankes, and Fast (2002), and others have identified a key reason for the disengagement of Indigenous youth from mathematics and science—the discrepancy between their own cultures and the cultural values embedded in school-based mathematics programs. Cajete (1994) found that when science is taught from a Western cultural perspective, it acts in opposition to the values of traditional culture for Indigenous students, which affects their performance in mathematics and science because it simply is not connected to their daily lives. Similarly, Lunney Borden (2010) has shown that the lack of attention to value differences and the use of inappropriate pedagogical strategies are among the factors that disconnect school-based mathematics from Mi'kmaw ways of reasoning mathematically.

As a result of value differences, many children choose to opt out of mathematics because the cost of participation is too high, demanding that they deny their own worldview in order to participate in the dominant view of mathematics. Doolittle (2006) has elaborated on this cost of participation for Indigenous students, and Gutiérrez (2012) has raised the same issue for minority groups in general. The incidence of conflicting worldviews has led many Indigenous students to either ignore the possibility of studying science or mathematics or to struggle within these disciplines. This marginalization is a serious issue for Indigenous communities that look to younger generations to acquire the skills and knowledge needed to move their communities closer to the realities of self-government in this modern age.

CONCEPTUALIZING CULTURALLY RESPONSIVE MATHEMATICS

Before sharing stories from SMYM, we situate this work within the broader field of culturally responsive mathematics education. We believe that SMYM presents a decolonized approach to mathematics learning and embraces many aspects of culturally responsive education advocated for by researchers in areas of Indigenous education, ethnomathematics, and equity. These we will be outlined in this section.

Inequities and marginalization in mathematics education can be addressed through a decolonized approach. A dominant theme in the literature is that Indigenous education should seek "to heal and transcend the effects of colonization" (Cajete, 2000, p. 181). Hampton (1995) argued that Indigenous education cannot ignore the reality of colonization but rather must address the issue directly. Indigenous education needs to move toward decolonization, which can be seen as a process of "deconstruction and reconstruction" (Battiste, 2004, p. 10) that "engages with imperialism and colonialism at multiple levels" (Smith, 1999, p. 20). This demands the critical examination of the hegemonic structures of mainstream education that continue to perpetuate the values of colonialism (Battiste, 2004).

The quest for greater cultural relevance in the teaching of mathematics has resulted in many projects that have attempted to document the cultural mathematical knowledge within Indigenous communities of North America (Barta et al., 2001; Lipka, 2002; Lipka, Wildfeuer, Wahlberg, George, & Ezran, 2001; Poirier, 2007). However, this documentation of cultural mathematical knowledge (known as ethnomathematics) is more extensive in Brazil and Africa (Powell & Frankenstein, 1997). While such scholarship often describes Indigenous understandings, it only occasionally provides substantive help toward designing innovative and sustainable curricula. Furthermore, it often fails to guide teachers' pedagogical practice in response to the impacts of colonizing curricula.

Some applied research has treated Indigenous knowledge of mathematics as add-ons to non-Indigenous curricula, as in basket and beadwork patterning (see Secada et al., 2002). Doolittle (2006) has argued that this can lead to an oversimplification of complex ideas, which he referred to as the "cone on the range"—the complexity of a tipi is reduced to an example of a cone. Such trivialization does not honour the knowledge or learning needs of Indigenous students nor does it promote substantive change in teaching practice.

The *Math in a Cultural Context* series published by the University of Alaska, Fairbanks stands out as the result of an exceptional project that has provided both learning resources and teacher supports. This Alaskan initiative, ongoing for over 25 years (Lipka, 1994; Lipka & Adams, 2004; Lipka, Mohatt, & The Ciulistet Group, 1998; Lipka et al., 2001) has brought together Yup'ik teachers, Elders, and university faculty as a research group to develop culturally based mathematics. The project originally began with The Ciulistet Group, which was a voluntary group of Elders and teachers, many of whom travelled significant distances for meetings, which Lipka has claimed was a testament to their desire to improve education for Yup'ik children. Although this group is no longer involved in the research, the work of transforming mathematics education in these Yup'ik communities has continued.

The university researchers worked collaboratively with community members, relying heavily on the Elders, to deepen their understanding of Yup'ik mathematical and scientific knowledge. The group chose the research focus, and the knowledge they gathered allowed them to develop lesson units that could be implemented in schools. The process of translating this Indigenous knowledge into formal curriculum was difficult and required extensive support from the Elders. The group developed units and took them to the Elders, who then debated the details and gave feedback. The group then recreated the units and brought them back to the Elders. The process began again until a satisfactory product was produced. Lipka (1994) stated:

> The premise behind developing a Yup'ik mathematics is threefold: (1) to show students that mathematics is socially constructed; (2) to engage students in a process of constructing a system of mathematics based on their cultural knowledge; and (3) to connect students' knowledge of "their mathematics" through comparisons and bridges to other Indigenous and Western systems. (p. 25)

This threefold approach is central to Lipka's "both/and" notion of transforming curriculum. The MCC experience was very influential in the evolution of SMYM. We believe that substantive change in mathematics education will happen when Indigenous knowledge and community worldviews are given a position of privilege alongside traditional school-based views of mathematics. Decolonization is not about valuing one way of knowing over another, but rather about valuing multiple worldviews for their contributions to each other. As we show below, SMYM is an event that provides a space for such valuing, inviting Mi'kmaw students and their teachers to explore and value the mathematical ways of knowing within their own community context.

We believe the structure of SMYM also incorporates many aspects of culturally based pedagogy for Indigenous students. Cappon (2008) argued that Indigenous education holds the view that learning is holistic, lifelong, experiential, rooted in Indigenous languages and cultures, and spiritually oriented. He claimed that it is a communal activity with all community members playing a role and that it integrates both Western and Indigenous knowledge. Orr, Paul, and Paul (2002) have shown how this notion of bridging Indigenous knowledge with mainstream curriculum has been worked at effectively by Mi'kmaw teachers in one of the participant communities. These teachers provided a quality education through the incorporation of cultural practical knowledge along with Indigenous pedagogical, relational, and political consciousness.

Lunney Borden (2010) examined calls for transforming mathematics education with two schools participating in SMYM. Through ongoing discussions in the form of *mawikinutimatimk* (coming together to learn together), the participants in her study highlighted areas of tension that, she argues, must be addressed in order to transform mathematics education for Mi'kmaw learners. These areas include: (1) the need to learn from Mi'kmaw language; (2) the importance of attending to value differences between Mi'kmaw concepts of mathematics and school-based mathematics; (3) the importance of attending to ways of learning and knowing; and (4) the significance of making ethnomathematical connections for students. Within each of these categories, teachers identified conflicts that arise when worldviews collide and identified potential strategies to address these tensions. SMYM was often pointed to by participants as an approach that addresses many of these areas.

Through a decolonized approach that is responsive to the calls for transformation from Indigenous communities, which are increasingly reflected by settler neighbours (e.g., uptake on the calls to action from the Truth and Reconciliation Commission of Canada [2015]), SMYM brings about a greater sense of equity for Mi'kmaw students. Although equity may be interpreted differently in different contexts, it is often focused on addressing disparity among various groups by improving access to quality mathematics and improving the achievement of students who have historically been marginalized by school mathematics. Our work embraces a broader conception of equity, as outlined by Gutiérrez (2012) who argued that equity should be defined in four dimensions: access, achievement, identity, and power. While access and achievement are important components of equity, the more critical concepts of identity and power must also be addressed to ensure true equity is enacted.

Gutiérrez (2012) has argued that many students have been marginalized by mathematics because they are asked to deny their identity in order to participate in the dominant view of mathematics. For Gutiérrez, "Students should be able to become better persons in their own eyes, not just in the eyes of others" (p. 19). She argued that ethnomathematics and culturally relevant mathematics provide the opportunity to address issues of identity within mathematics education. She described identity through a window/mirror metaphor, claiming that "students need to have opportunities to see themselves in the curriculum (mirror) as well as have a view onto a broader world (window)" (pp. 19–20).

Gutiérrez (2012) has argued that attention to identity does not have a goal of replacing "traditional" mathematics with a predefined "culturally relevant mathematics," but rather to strike a balance between the number of windows and mirrors provided to any given student in his/her math career" (p. 20). However, she also argued,

> Even if students have access to quality mathematics, achieve a high standard of academic outcomes as defined by the status quo, and have opportunities to "be themselves and better themselves" while doing mathematics, it is not enough to call it equity if mathematics as a field and/or our relationships on this planet do not change. As such, a final piece of equity involves *Power*. The Power dimension takes up issues of social transformation at many levels. (p. 20)

The power dimension involves not only questioning who has power in the classroom in terms of who participates, who talks, and so on, but also as it pertains to using mathematics to question power relationships in the world and seeing mathematics as a human endeavour.

In this project, we consider power in the context of authority. Where are the voices of authority in mathematics education? Typically, a student might see the textbook, the teacher, and the field of mathematics as being the authorities: those who define what mathematics should be learned and what counts as mathematical reasoning. There are other sources of authority at work for students, though many of them may be marginalized. Amit and Fried (2005) referred to the multiple sources of authority as "networks of authority." However, Herbel-Eisenmann and Wagner (2010) showed how mathematics classrooms generally authorize a single authority (the discipline of mathematics presented and mediated by the teacher), not the multiple sources that would be possible. We will demonstrate that SMYM challenges dominant, discipline-focused conceptions of authority by revitalizing old storylines in which Elders, community leaders, parents, grandparents, crafts people, and so on are positioned as authorities, now in the context of developing mathematical understandings.

RESPECTFUL POSITIONING

The application of positioning theory was instrumental in the development of our research choices in Mi'kmaw communities. This theoretical perspective on positioning follows from an edited book (Harré & van Langenhove, 1999) and an elaboration of the theory in the context of mathematics education (Wagner & Herbel-Eisenmann, 2009). In general, *positioning* is taken to refer to the way people use action and speech to arrange social structures. Words and actions evoke images of known storylines and positions within those stories. When one person engages a storyline, others may comply with this positioning or resist it.

Davies and Harré (1999) explained how positioning theory draws attention to "immanent" practices, as opposed to "transcendent" discourse structures (e.g., a student's relationship to the transcendent discipline of mathematics), which Wagner and Herbel-Eisenmann (2009) showed to be a common scholarly focus in mathematics education (e.g., Evans, 2000). With positioning theory's attention to immanent relationships, aspects of human interaction become more noticeable, because the larger discourses are ignored, even characterized by Davies and Harré as myths. Wagner and Herbel-Eisenmann, however, maintained the promotion of attention to immanent practice without relegating larger discourses as inconsequential. This is because people take discourses as being real in their own right and thus powerful in immanent interactions. Wagner and Herbel-Eisenmann argued against stripping mathematics of its power (demythologizing it) in favour of inviting new mathematical narratives that recognize mathematics in places that it has often been ignored or marginalized (remythologizing mathematics). This informed our expressed desire to seek strategies to privilege Mi'kmaw mathematical knowledge and bring it alongside school-based mathematics.

THE EMERGENCE OF SHOW ME YOUR MATH

The research began with Lunney Borden and Wagner interviewing Elders in Mi'kmaw communities to identify some of their everyday practices (both traditional and current) that could be deemed mathematical. This typical approach to ethnomathematics research (cf. Powell & Frankenstein, 1997) relies on Bishop's (1988) articulation of activities that are potentially mathematical (practices that involve counting, measuring, locating, designing, playing, or explaining) and on the assumption that any mathematics is an artifact of a particular culture.

As we will describe in this chapter, the SMYM event emerged from these initial conversations. SMYM was certainly part of the research project, which began with the conversations with Elders, but this event belonged to the community more than it did to the project. Data in the project includes recordings (and transcriptions) of the interviews with Elders and of the conversations in which SMYM was organized. The SMYM event generated publicly accessible student work, which gives us insight into the impact of the SMYM event. We also have listened to accounts from teachers and participant students about their involvement, but most of these were not recorded, because they emerged in unexpected situations.

The larger project was initiated by David Wagner, who was joined by Lisa Lunney Borden virtually at the beginning. She had had a 10-year teaching career in one of the participant schools before beginning her doctoral studies with Wagner as her supervisor. Newell Johnson is a teacher in a different participant school, had a history of collaboration with Lunney Borden, was one of the teachers who participated in the initial meeting about SMYM, and has been a leader in its success. In this chapter, we refer to individual authors in the third person at times to clarify unique involvement or voice that is not shared among the three of us. In particular, Johnson's voice is important, as she shares her personal experiences with the SMYM event and associated classroom experiences, and it is necessary to clarify aspects of the SMYM story that involved Lunney Borden and Wagner but not Johnson.

Lunney Borden and Wagner were excited to hear the stories the Elders were sharing with them in the early stages of the research project. However, in reflection, they saw themselves as mediators, interfering with the intended process of connecting students to the mathematics in their communities. As researchers, they were careful to orient their conversations with participants around mutual respect, yet still found connections with colonialist storylines as they observed that some participants were trying to be helpful by telling them what they wanted to know. While this spirit of cooperation seemed to be generative for the planned research, it was disturbing to have participants ask if they were telling Lunney Borden and Wagner what they wanted to hear. As researchers, they did not want to be seen as the ultimate audience, but they often felt positioned in that way. It is important to recognize that Lunney Borden and Wagner were complicit in this positioning; it was not a matter of the Elders and knowledge keepers choosing this relationship for them against their will. Lunney Borden and Wagner also worried about authenticity, because participants seemed to be subjecting themselves to their agenda. They wanted to hear more about the participants' agendas.

The interpersonal relationships appeared to be similar to the diagram in Figure 5.1. The community experts were responding to researchers' requests for information to pass on to teachers, who in turn would pass it on to the children. The children would then do something called "work" for the teachers. The path of communication is quite linear, passing through people less central to the communities. Lunney Borden and Wagner, as researchers, had some involvement in the community (and Lunney Borden had further involvement unrelated to research). Some of the teachers lived outside the community, but even the teachers who lived in the communities and were Mi'kmaq (including Johnson) were connected to and associated with curricular demands that came from outside the community. This is why we have the researchers and teachers straddling the boundary of the community in Figure 5.1.

New storylines were necessary. Attending to positioning opened up new opportunities. Lunney Borden and Wagner realized that the conversations would be more authentic if the children themselves talked with Elders and others to find

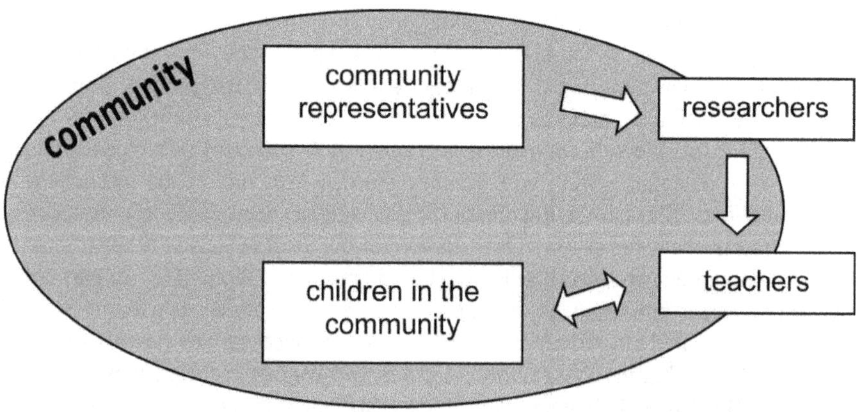

Figure 5.1. Interpersonal positioning in the initial ethnomathematical work

mathematics in traditional and modern community practices. They, as researchers, needed to remove themselves and the teachers from the position of mediums. Also, reflecting on Morgan's (1998) research that underscores the importance of audience in students' mathematical writing, they realized that positioning the children as the ultimate audience in the ethnomathematics conversation does not afford them opportunities to address an audience other than their teacher, nor an imperative to engage in real problems/issues faced by their community.

As a result, Lunney Borden and Wagner initiated a new series of conversations structured to prompt community members to talk and listen to each other about everyday mathematics in traditional and modern community settings. From their dissatisfaction emerged the Show Me Your Math (SMYM) event that changed the interpersonal positioning in substantive ways.

Lunney Borden and Wagner approached some schools with the idea for a SMYM contest and invited teachers (one of whom was Johnson) to come to a meeting to discuss this possibility. They imagined an event in which students would be invited to do projects exploring the mathematics in their everyday lives. They planned to have students submit their work to a website that would be hosted by the Atlantic Canada First Nation Help Desk, an existing infrastructure that supports communication among communities via the Internet. Such "contests" are commonly used in this region to promote community-based education and to develop cultural resources for teachers and students. While this plan was well received, Johnson and the other teachers in the initial meeting wanted more than a website. They suggested that schools host local mathematics fairs and send selected students to a regional fair where they could share their work with others.

To substantiate the break from the school tradition of students doing work for teachers as audience, it was suggested that instructions for the contest be given in a video. The teachers in the workshop contributed to the structuring of the video, but the researchers produced it. It featured Indigenous people, including an Elder, a middle-aged teacher, and children, all asking the viewer (the student) to show their math (their community's math). We all felt that the form of this video would help students see the community as their audience, instead of their teachers. In response to this prompt, schoolchildren interviewed Elders, experts in crafts, and others to explore mathematics done in their communities in historic and modern times. They published their work on an Internet site hosted by the Atlantic Canada First Nation Help Desk and also presented their work to the region's communities in a math fair. (Selected student projects are now available at http://showmeyourmath.ca.)

The interpersonal relationships structured with the SMYM event are more complex than the relationships in the initial research configuration. Figure 5.2 illustrates an attempt to diagram these relationships. The three of us (Lunney Borden and Wagner as researchers and Johnson as a lead organizer) were and continue to be in conversation with other teachers (and some other community representatives, including Elders). These conversations initiate student investigations that involve students in conversation with experts in the community. The students report back to

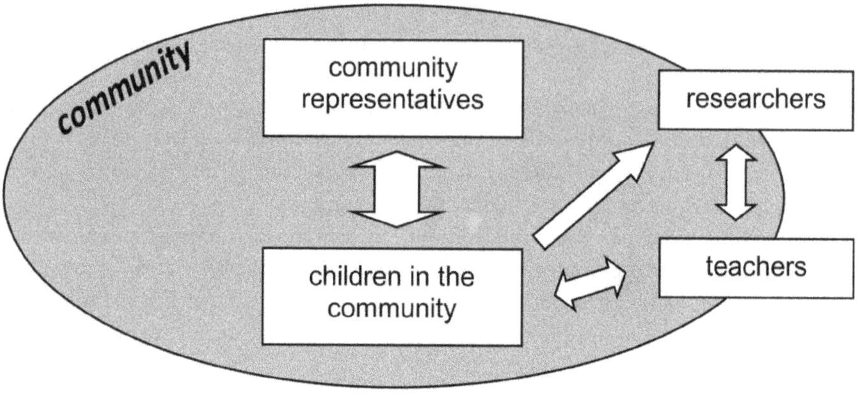

Figure 5.2. Interpersonal positioning with the "Show Me Your Math" event

the community, to their teachers and to the researchers at mathematics fairs and also to the outside world, because some of their work is made available online. In our view, the most important difference between this set of relationships and the earlier one is the communication between children and others in the community, illustrated by the big bidirectional arrow in the middle of Figure 5.2.

Wagner and Lunney Borden (2012) elaborated on these shifting positioning structures and associated storylines, and on the power of positioning theory for interrogating relational practice. This chapter has a different focus. We draw upon examples of student work and stories from participants in the SMYM event to demonstrate how this project is transforming mathematics education in participant schools and contributing to a more culturally responsive mathematics program. With the community positioned as the source of mathematical authority and the students positioned as researchers and disseminators of mathematical knowledge, this project has created an opportunity to realize Wagner and Herbel-Eisenmann's (2009) imperative. It has opened up the opportunity for students, teachers, and community members to "identif[y] with storylines that are not traditionally a part of mathematics classroom discourse" (p. 13). These new storylines have challenged traditional conceptions of authority and transformed the relationship between students and mathematics as a field of study. In the next section, we will share stories that illustrate how SMYM has achieved greater equity through attention to these questions.

DEVELOPING NEW STORYLINES: SEEING THE COMMUNITY AS A SOURCE OF AUTHORITY

The nature of this project was that children were encouraged to learn in relationship with community members, family members, Elders, and peers. Such communal activity invites all community members into the conversation about mathematics. This kind of learning promotes respect for community knowledge among students.

It helps them to see that mathematical reasoning is a part of their own cultural identity. It also positions the community as a source of mathematical knowledge, thus dispelling the myth that mathematical knowledge comes from teachers and textbooks. In this section, we share stories from the project that exemplify how students have learned from the community.

A common experience for students involved learning from Elders, in particular parents and grandparents. The following story comes from Johnson's observations of her nephew engaging with his grandparents (Johnson's parents) as he was working on his project. This exemplifies the kind of family involvement that has been prompted by the Show Me Your Math event:

> Sitting around the kitchen table at my parents' place, a place that I visit every day after work, I was listening to a conversation between my father and my eight-year-old nephew. My nephew had just informed my father that he had to do a project for Show Me Your Math, and that he knew what he wanted to do.
>
> He reminded my father about the hide that my parents had prepared over the summer. After recounting those memories, my nephew asked my father if he did anything with the hide. My father told him about the beautiful drum that was created from that piece. My nephew then asked him if he had to use any math to figure out the measurements, and my father's eyes opened wide. If my father were any younger, he would have been jumping up and down with excitement. My father explained to him that he had to cut out circles from the hide, bigger than the drum top and he further explained to him why they had to be measured that way.
>
> My nephew told my father that he wanted to do a project on the hide and he wanted my father to show him how he could measure pieces to make the drums. My father asked him if he was serious, and even tempted him to do a project on something else, something else that my father thought the teachers would appreciate. But my nephew was adamant in his decision. He insisted that he wanted to do this project and he wanted to see how he could measure the hide to make drums.
>
> As I was listening, I was really surprised, and my mother was becoming more and more excited at the thought of my nephew taking such a keen interest. After some time into the conversation, my father turned to me and asked if I had anything to do with this, and as I smiled I said "No." He gave me that look, that look of knowing that I was really enjoying the events as they were unfolding.
>
> My nephew was brainstorming on how he could make circles on his hide that was now drawn on paper, and he began hunting for objects that were circular. He found some milk covers and some larger juice covers. In all, he found three different sizes and he began to trace them on his paper. He noted that he needed

> two of each, one for the top of the drum and one for the bottom of the drum. He made some measurements of the circles and drew some circles on the side of his poster. He was able to count how many different size drums that this piece of hide would make. He had arbitrarily given the covers a diameter measure of 12, 24, and 36 inches.
>
> After completing this task, he remembered that there was a picture on the computer of my parents standing beside the hide while working on it. He went to the computer and found the picture, printed it off and glued it onto his poster. He was so proud of his project. Being a math teacher, I was the proudest auntie on that day. My parents were also proud and were amused by the fact that I was in my glory and was so thrilled at the enthusiasm, creativity and determination that my nephew had brought into the completion of his Show Me Your Math project.

We see the shift from the traditional expectation that mathematics is learned in mathematics class. In this case, the mathematical discussions are occurring between a boy and his grandfather and are stemming from an out-of-school experience. We see that this project is one that is of interest to the child and stems from his own curiosity. His problem is not one that has been assigned by a teacher or a textbook, but rather has emerged from his own desire to understand one of his experiences. We note the grandfather's resistance to this shifting storyline in his suggestion that the child do something that the teachers might prefer. This exemplifies the deep-rooted nature of the belief that mathematics is of the school and not of the community. This is reminiscent of earlier ethnomathematical conversations with Elders in this project who drew distinctions between school-based mathematics and community mathematical reasoning (this distinction is elaborated on in Wagner & Lunney Borden, 2015). These Elders, like the grandfather in this story, did not see their knowledge as something that would be valued by the school; they did, however, make the argument that learning community ways of reasoning would benefit students' common sense.

We also see our plans for changing the sense of audience being played out in this episode. The boy seems well aware that he will share this knowledge with peers, teachers, and the community. His school holds a local math fair in advance of the regional math fair. His attention to detail, and the inclusion of the picture of his grandparents next to the hide, demonstrate his desire to celebrate his grandparents' knowledge and to share this with his peers. This also demonstrates how the community knowledge is repositioned and given high priority in this project. This knowledge is not an add-on or a tangential connection to the privileged mathematics; rather, it is given a central focus with the mathematics emerging alongside the community knowledge.

This second example comprises Johnson's reflections on a teaching experience. It exemplifies how students' needs and interests are at the centre of SMYM projects. In her classroom, participation in the SMYM project is something that the students know is an expectation for the course. It addresses the major outcomes for an independent research unit in the Grade 11 mathematics curriculum prescribed by the province and used by the community schools. The story relates to a project done by two sisters who wanted to honour the memory of their mother.

> There were these sisters that were in my class, and they had just recently lost their mother to a fatal heart attack. Their mother was taking a course in Ontario over the summer, and prior to her coming home nearing the completion of her course, she had suffered a massive heart attack. This mother was younger than me and the community was in shock. She left behind a husband and five beautiful children. At the beginning of the school year when I announced in class that students will be doing this project and explained to them that it would have to be something related to our culture, the girls casually called me to their desk, and asked if they could do a project on moccasins. I was really excited and anticipated their creative thinking. I asked if this is something that they knew how to do, and they looked to me with sadness. I sat down beside them and asked them what was wrong. They explained to me that their mother was doing a course in Ontario where she was learning to make moccasins. She was training to be a leather artisan and moccasins were her specialty. A few weeks before her course completion, she had died alone in her dorm, being away from her family for almost the entire summer.
>
> They wanted to know what their mother's passion was and they were inspired by the thought of having the opportunity to bring this into the classroom. They completed their project, bringing in some of their mother's tools and actual moccasins that their own mother had created. They stated at the end of the project that their mother had to know a lot of math in order to be able to create moccasins. I was very happy that the girls had the opportunity to explore their mother's work and to bring pride and understanding to her passion.

This story highlights how participation in SMYM can allow students a space to explore something that is important and meaningful to them. These students saw this project as an opportunity to celebrate their mother and to learn more about her craft. Also, in sharing their learning with peers, teachers, and the community, these young women had an opportunity to celebrate their mother's talents and her life. This story and other experiences like it remind Johnson that her role as a teacher is constantly changing, including the need to be a provider of information, a facilitator, a guide, and a learner.

This recognition of the teacher's role as learner is juxtaposed with a typical mathematics class interaction structure, which involves teachers asking questions to which they already know the answers. With SMYM, the students are not completing tasks that have a clearly defined process and solution; rather, they are using their understanding of mathematical concepts and their own community knowledge to create new mathematical knowledge. These students are now positioned as the

Figure 5.3. Eight-point Mi'kmaw star

experts, the researchers who have discovered something that can be shared. The teacher can then be seen as a legitimate audience, as she does not already know about the content of the projects. As such, the students are engaged in authentic mathematical investigations with an expectation that they will have an authentic audience who will see them as experts in their chosen project.

Student interests are often the focus of mathematics learning that occurs as a part of SMYM. The "Mi'kmaw star" is a well-known design among Mi'kmaw people (see Figure 5.3). The shape is often coloured, using the four traditional colours of red, black, white and yellow, each having spiritual significance. This star shape has been the subject of several SMYM investigations.

In the first year of SMYM, one high school student (Johnson's daughter) chose to explore aspects of paper folding and cutting that would enable her to quickly create an eight-point star. Through a series of explorations, she developed an approach that involved three folds and one cut, creating a perfect star. This project came from her own unique explorations. Having seen that paper folding and cutting can be useful in geometric reasoning, she chose to apply this strategy to something of interest to her. During the math fair, she shared with those in attendance her stories of discovery and also invited visitors to try it out for themselves with her guidance. Figure 5.4 shows her doing this folding.

During a more recent SMYM fair, two Grade 12 students explored the geometry involved in making a somewhat more complicated eight-point star design often used in beadwork patterns and quilting patterns. They used their school-based geometry to create the design with intricately detailed borders, but they also spoke about Elders in their community who were skilled at using these designs for beadwork and

Figure 5.4. Folding the eight-point star

quilts. These students concluded that using the mathematics they learned in school was far more complicated than the Elders' way of making a similar design. One of the girls explained:

> We were just saying how it's cool how our ancestors and our Elders can make these stars without using geometry. It's all proportioned correctly without geometry. They don't need all these steps, they just do it on their own. … A lot of our Elders use them still, in quilts and stuff like that. To them it's normal because they already know.

In this example we see that students are responding to and thus transforming their relationship with school-based mathematics by making connections with community practices. The students recognized that the textbook approach provided one way of looking at geometric construction, but equally valued their Elders' knowledge. In this example, the degree of interaction between the students and the Elders does not seem to be fully evident; the student seemed to refer to the Elders symbolically as representatives of the whole community. As far as we know, the girl quoted above did not interact with Elders in her project work, but rather reflected on things she saw in her community. This resembles a form of learning traditional to Mi'kmaw people—observing in silence.

The students referred to the Elders as possessing this knowledge and not needing to do all the mathematics. However, when prompted as to how they thought these Elders did this work, the students replied that they did not know and admitted that they "should have asked them that." This is not intended as a critique of the students' project but points to a way in which this experience could have been enhanced for these students, and perhaps for the Elders as well. That being said, we see again that SMYM investigations encourage students to think critically about a textbook's authority and to examine how community knowledge can be equally privileged.

These few examples of student work and interactions give a snapshot of how traditional storylines in mathematics classes are changing as a result of SMYM. In each example, we see how community knowledge is celebrated and given a place of privilege alongside of traditional school knowledge, thus enabling students to see how mathematical reasoning is a part of their own cultural heritage. We also see the important role of authentic engagement in mathematical investigations that address students' interests and needs. In each case, students were able to choose a project that was meaningful to them, to conduct an investigation where the results were not already known by the teacher, and to be positioned as the researcher and expert in this area.

The role of audience is significant in each project. Students were aware that they would be the experts in relation to their chosen projects; they also were aware that there would be an audience that included their classmates and peers as well as other students, teachers, and community members. Being asked to share their work with a wider audience gives greater purpose to the task. Each of these pieces contributes to transformation as students enter into a new relationship with the field of mathematics. In the next section, we will describe some of the impacts such transformation is having in classrooms.

TRANSFORMING CLASSROOMS

Many teachers who have had students participate in SMYM have remarked on the benefits of such participation. Lunney Borden (2010) heard many positive comments about this project during her research discussions with teachers in two schools, with one teacher remarking that this type of learning was what was needed more often to help students make more meaningful connections to mathematics on a daily basis. The SMYM math fair is a highly regarded and much celebrated event each year within the Mi'kmaw schools, and each year more participants contribute projects. Many teachers have shared with us how great the impact has been on their schools. We have ourselves often been overwhelmed by the enthusiasm and excitement in students' voices as they share with others the ideas they have learned. In this section, we share Johnson's stories of how this project is impacting her classroom. We feel her experiences are representative of the transformative potential of SMYM.

Even before the SMYM event started, many teachers in Mi'kmaw Kina'matnewey (MK) schools, like Johnson and Lunney Borden prior to 2005, searched for ways of representing and explaining mathematics to make it more comprehensible for their students and more in line with students' ways of thinking, especially as this relates to the concepts, preconceptions, and misconceptions students bring to the learning of mathematics. Many MK teachers, in particular those like Johnson who grew up in the communities, often looked to their own cultural practical knowledge and searched for any information that could be used to further enhance methods of representing and explaining concepts. Johnson sees the SMYM event as one clear way to invite students to look at their culture and derive mathematical concepts and representations. The students' investigations help their teachers expand their repertoires for connecting mathematics with cultural practices. Johnson and other teachers have remarked that they find it interesting to see what the students bring to the learning and how their mathematical learning is evolving and emerging. There have been many Show Me Your Math fairs and numerous projects done by students in the early grades, and parents and teachers will still continue to remark that they are amazed by the mathematical content knowledge that the students bring into their own learning.

Johnson has noted that in her own mathematics classroom, she is constantly looking for engaging activities that also develop students' mathematical concepts. She has often developed activities where students are invited to learn outside of the classroom, yet this has been met with criticism. For example, other educators have told Johnson she is giving students free time and that they are not engaged in learning. Johnson has observed that, on the contrary, students are often more engaged than if they were to learn from a textbook using conventional methods. She has found that Mi'kmaw students are more inclined to be motivated if they are given the opportunity to create, explain, and investigate on their own terms, in their own language, and in their own realm of knowing and constructing.

It is often difficult to advocate this type of learning in the high school, where there are specific curriculum outcomes that students must achieve, exams to prepare for, and assignments that require completion. However, Johnson continues to create activities where all these domains are addressed through activities designed to foster more culturally relevant learning environments. She notes that this is quite challenging; however, as the years pass, she is finding it easier to draw from her own cultural experiences and those in the students' SMYM projects to challenge the taken-for-granted assumptions in the curriculum. This does not mean ignoring the curriculum guide, but rather finding activities that are more project-based and culturally relevant, like those done for SMYM.

She has seen the benefits of such activities during her peer mathematics tutoring sessions. The peer tutors often create math centres and activities to engage their learners. Johnson finds it amazing to watch students take so much interest in attempting to reach struggling learners, and the motivation that drives them is inspiring. She often draws from the experiences with this group, inspired to create engaging, culturally relevant activities that the tutors can pilot during these peer-tutoring sessions.

Johnson believes that when her students are involved in ethnomathematics, and learning mathematics related to their culture, it cultivates their cultural identity. The SMYM projects allow the students to develop their own learning, to select topics that interest them, and to present their projects in meaningful ways that address their own learning styles. This type of activity allows students to arrive at a solution in more ways than one, where a hands-on approach to learning is often the method of choice. This form of project-based learning has proven to be an effective method of teaching for her. Each year, all of her students submit projects for both semesters, something seen as uncommon at the high school level. It has been surprising to the staff and principal. The principal once asked Johnson, "How do you get them to do their projects, to hand them in on time, and to get all the students to do one?" She had no real answer at the time; however, after reflection on this question, she felt that the students took ownership of their own learning. They chose topics that interested them, and used their own culture as the basis for their projects.

Johnson also notes that her repertoire of activities has increased since she initiated the SMYM projects as a part of her classroom routine. She has accumulated many projects over the years, and with the students' permission, she includes their projects in her lessons whenever possible. One example of this involved a lesson on quadratics. There was a project that had been submitted by Grade 11 students in the previous semester on this very topic. She brought out the project and displayed it in the classroom. She told the students to have a look at it and to see what connections they could make from the project and what they were learning. As such, Johnson noticed a connection with Tompkins (2002), who wrote, "I see my role as a facilitator as one of trying to set up structures that will acknowledge the differential power in the room and create maximum opportunities for the marginalized voices to be heard" (p. 418).

Johnson reflects that, like many teachers, she often wonders, "How do I get my students to engage in critical thinking and in what context?" She has noted that SMYM has been one of the ways that she has been able to deal with these questions. She believes her fundamental obligation as an educator is to encourage students to acquire knowledge with their own eyes and not through the eyes of others, and to hear with their own ears and not through the ears of others. The best way to do this is to take on projects where students investigate their own truth.

Bishop (1990) demonstrated this concept, stating that Western mathematics is one of the most powerful weapons in the imposition of Western culture. He argued that Western ideas of length, area, volume, weight, time, and money would have been imposed on Indigenous societies. These concepts were imposed because there was a need to assimilate cultures in order to advance European trade. Bishop (1990) also stated:

> [To] decontextualise, in order to be able to generalize, is at the heart of western mathematics and science; but if your culture encourages you to believe, instead, that everything belongs and exists in its relationship with everything else, then removing it from its context makes it literally meaningless. (p. 57)

Removing the context of mathematics and teaching abstracted concepts in isolation is meaningless for students. The SMYM projects that the students are encouraged to do allow them to study mathematics in its context.

We believe that the purpose of education is to activate potential, rather than just retrieve and dispense facts. Students need to be given opportunities where they can explore and develop their own thinking, their own beliefs, and their own understandings. They need the opportunity to make meaningful connections between their learning and their own cultural identity and experiences. The SMYM projects have allowed Johnson to understand where her students are coming from and what they can bring into the classroom, and have created a space where their voices are heard. We believe that the implementation of this type of program in the Mi'kmaw education system will continue to foster a learning environment that will allow students to reach their maximum potential.

CHALLENGES AND NEXT STEPS

While SMYM has brought about positive impacts on schools, we acknowledge that there are areas where these projects could be improved and enhanced. One key area of ongoing investigation in this project involves exploring strategies to strengthen connections between SMYM projects and prescribed curriculum outcomes. In particular, we would like to bring together Elders, teachers, mathematics education researchers, and perhaps mathematicians to make the mathematical connections more explicit. Johnson expressed her own challenges with connecting project ideas to the required curriculum, and although she has found more ways to make these connections more authentic with each year of SMYM projects, she and other teachers have stated that it would be beneficial to be supported in this area.

In an effort to address this concern, Lunney Borden has worked with many MK schools and public schools to do inquiry projects for SMYM (Lunney Borden & Wiseman, 2016) and these projects have been shared in a new section of the website called Mawkina'masultinej, meaning "Let's learn together!"[1] Projects have focused on birch bark biting (Lunney Borden, 2015), paddle making, maple syrup making, eels, drum making, baskets, beadwork, and so on with each project beginning in a community context and allowing mathematics to emerge. Teachers have remarked that this approach is helping them to better address curriculum pressures. One Grade 5/6 teacher remarked, "We can meet more outcomes than you can possibly imagine" (MK Teacher, personal communication, June 2015), and explained how much easier it was to address curriculum outcomes by teaching in this way than by trying to fit outcomes into isolated courses. This approach shows great promise for addressing curriculum concerns, though there is an ongoing need for financial resources and professional learning to continue this work.

Additionally, we have the goal of eventually using student projects and the inquiry projects to develop mathematics modules that could be used to support mathematics learning in Mi'kmaw classrooms. We see this as an important next step that may also address the need for more explicit connections to curriculum. Such modules would also enable the spirit of SMYM to impact classrooms on a more regular basis, rather than just at an annual affair.

RECOMMENDATIONS

We believe that the structure of SMYM is something that may be of benefit to any community that seeks to challenge the dominant storylines in their mathematics program and to allow students to remythologize mathematics from their own community perspective. There were several factors that we believe have contributed to the success of this project. First, it was important that there was buy-in from the schools. The teachers themselves determined that there should be a math fair to showcase student projects. The teachers worked together to make it come to fruition. While the specific format of the fair varies from year to year and from location to location, teachers see great value in having students share their work with an audience of peers, parents, grandparents, community members, and others who attend.

Second, the video prompt that was developed to invite students to conduct the ethnomathematical research featured community members, Elders, and children inviting students to show their math. The ideas for the content of this video were developed collaboratively with teachers. We feel this video was a key component in establishing a sense of audience and in challenging traditional storylines that often have a textbook or teacher making requests of what mathematics should be done. We suggest that another community might want to create their own video prompt, drawing on examples from their own community.[2]

Third, schools and the entire Mi'kmaw Kina'matnewey (MK) system have embraced SMYM and have celebrated the student work. Many of the participant schools host their own local math fairs, inviting parents and community members in to view the students' projects. Walls of many schools are covered with SMYM projects. Schools commit time and funds to ensure that students can participate in the regional math fair. Some schools also provide small honoraria to Elders and craftspeople to come into schools to work with students in their classrooms to conduct whole-class projects and investigations (especially at the younger levels). Web space is provided to house a site displaying students' projects, and MK cites SMYM as one of the major events of the academic year, often providing additional funds to support the project. SMYM is a relatively simple project to organize, but it can have a significant impact on increasing culturally responsive pedagogical practices in schools.

ACKNOWLEDGEMENTS

This research was supported by a grant from the National Science and Engineering Research Council of Canada, entitled Atlantic Regional CRYSTAL (ARC): Enhancing the Culture of Science, Technology, and Mathematics across the Atlantic Provinces.

NOTES

1 See http://showmeyourmath.ca/mawkinamasultinej-lets-learn-together/
2 The original SMYM video prompt can be viewed at http://showmeyourmath.ca

REFERENCES

Amit, M., & Fried, M. N. (2005). Authority and authority relations in mathematics education: A view from an 8th grade classroom. *Educational Studies in Mathematics, 58*, 145–168.

Barta, J., Abeyta, A., Gould, D., Galindo, E., Matt, G., Seaman, D., & Voggessor, G. (2001). The mathematical ecology of the Shoshoni and implications for elementary mathematics education and the young learner. *Journal of American Indian Education, 40*(2), 1–27.

Battiste, M. (1998). Enabling the autumn seed: Toward a decolonized approach to Aboriginal knowledge, language, and education. *Canadian Journal of Native Education, 22*, 16–27.

Battiste, M. (2000). *Reclaiming indigenous voice and vision*. Vancouver: UBC Press.

Battiste, M. (2004, May 29). *Animating sites of postcolonial education: Indigenous knowledge and the humanities*. Plenary address to CSSE, Winnipeg, Manitoba.

Bishop, A. (1988). *Mathematical enculturation: A cultural perspective in mathematics education*. Dordrecht: D. Reidel.

Bishop, A. (1990). Western mathematics: The secret weapon of cultural imperialism. *Race and Class, 32*(2), 51–64.

Cajete, G. (1994). *Look to the mountain: An ecology of indigenous education*. Durango, CO: Kivaki Press.

Cajete, G. (2000). Indigenous knowledge: The Pueblo metaphor of indigenous education. In M. Battiste (Ed.), *Reclaiming indigenous voice and vision* (pp. 181–191). Vancouver: UBC Press.

Cappon, P. (2008, May 1). Measuring success in First Nations, Inuit, and Métis learning. *Policy Options*, 60–66.

Davies, B., & Harré, R. (1999). Positioning and personhood. In R. Harré & L. van Langenhove (Eds.), *Positioning theory: Moral contexts of intentional action* (pp. 32–51). Oxford: Blackwell

Doolittle, E. (2006). Mathematics as Medicine. In *Proceedings of the Canadian Mathematics Education Study Group Conference* (pp.17–25). Calgary, Alberta: Canadian Mathematics Education Study Group.

Evans, J. (2000). *Adults' mathematical thinking and emotions: A study of numerate practice*. London: Routledge/Falmer.

Ezeife, A. (2003). The pervading influence of cultural border crossing and collateral learning on the learner of science and mathematics. *Canadian Journal of Native Education, 27*(2), 179–194.

Gutiérrez, R. (2012). Context matters: How should we conceptualize equity in mathematics education? In B. Herbel-Eisenmann, J. Choppin, D. Pimm, & D. Wagner (Eds.), *Equity in discourse for mathematics education: Theories, practices, and policies* (pp. 17–33). New York, NY: Springer.

Hampton, E. (1995). Towards a redefinition of Indian education. In M. Battiste & J. Barman (Eds.), *First Nations education in Canada: The circle unfolds* (pp. 5–46). Vancouver: UBC Press.

Harré, R., & van Langenhove, L. (Eds.). (1999). *Positioning theory: Moral contexts of intentional action.* Oxford: Blackwell.

Herbel-Eisenmann, B., & Wagner, D. (2010). Appraising lexical bundles in mathematics classroom discourse: Obligation and choice. *Educational Studies in Mathematics, 75*(1), 43–63.

Indian and Northern Affairs Canada. (2002, December). *Our children – Keepers of our sacred knowledge. The final report of the minister's national working group on education.* Ottawa: Minister's Working Group on Education. Retrieved July 14, 2009, from http://dsp-psd.pwgsc.gc.ca/Collection/R41-9-2002E.pdf

Lipka, J. (1994). Culturally negotiated schooling: Toward a Yup'ik mathematics. *Journal of American Indian Education, 33*(3), 14–30.

Lipka, J. (2002). *Schooling for self-determination: Research on the effects of including native language and culture in the schools* (ERIC Identifier ED459989).

Lipka, J., & Adams, B. (2004). *Culturally based math education as a way to improve Alaska native students' math performance* (Appalachian Collaborative Centre for Learning, Working Paper No. 20). Athens, OH: ACCLAIM Research Initiative, Ohio University.

Lipka, J., Mohatt, G., & The Ciulistet Group. (1998). *Transforming the culture of schools: Yup'ik Eskimo examples.* Mahwah, NJ: Lawrence Erlbaum Associates.

Lipka, J., Wildfeuer, S., Wahlberg, N., George, M., & Ezran, D. R. (2001). Elastic geometry and storyknifing: A Yup'ik Eskimo example. *Teaching Children Mathematics, 7*(6), 337–343.

Lunney Borden, L. (2010). *Transforming mathematics education for Mi'kmaw students through mawikinutimatimk* (Doctoral dissertation). University of New Brunswick, Fredericton.

Lunney Borden, L. (2015, June 23). Learning mathematics through birch bark biting: Affirming Indigenous identity. In S. Mukhopadhyay & B. Greer (Eds.), *Proceedings of the 8th International Mathematics Education and Society* (Vol. 3, pp. 756–768). Portland, OR: MES8.s

Lunney Borden, L., & Wiseman, D. (2016). Considerations from places where indigenous and western ways of knowing, being, and doing circulate together: STEM as artifact of teaching and learning. *Canadian Journal of Science, Mathematics and Technology Education, 16*(2), 140–152.

Morgan, C. (1998). *Writing mathematically: The discourse of investigation.* London: Falmer.

Orr, J., Paul, J., & Paul, S. (2002). Decolonizing Mi'kmaw education through cultural practical knowledge. *McGill Journal of Education, 37*(3), 331–354.

Poirier, L. (2007). Teaching mathematics and the Inuit community. *Canadian Journal of Science, Mathematics and Technology Education, 7*(1), 53–67.

Powell, A., & Frankenstein, M. (Eds.). (1997). *Ethnomathematics: Challenging Eurocentrism in mathematics education.* Albany, NY: SUNY Press.

Secada, W., Hankes, J., & Fast, G. (Eds.). (2002). *Changing the faces of mathematics: Perspectives of Indigenous people of North America.* Reston, VA: National Council of Teachers of Mathematics.

Smith, L. T. (1999). *Decolonizing methodologies.* London: Zed Books.

Tompkins, J. (2002). Learning to see what they can't: Decolonizing perspectives on Indigenous education in the racial context of rural Nova Scotia. *McGill Journal of Education, 37*(3), 405–422.

Wagner, D., & Herbel-Eisenmann, B. (2009). Re-mythologizing mathematics through attention to classroom positioning. *Educational Studies in Mathematics, 72*(1), 1–15.

Wagner, D., & Lunney Borden, L. (2012). Aiming for equity in (ethno)mathematics research. In B. Herbel-Eisenmann, J. Choppin, D. Pimm, & D. Wagner (Eds.), *Equity in discourse for mathematics education: Theories, practices, and policies* (pp. 69–88). New York, NY: Springer.

Wagner, D., & Lunney Borden, L. (2015). Common sense and necessity in (ethno)mathematics. In K. Sullenger & S. Turner (Eds.), *New ground: pushing the boundaries of studying informal learning in science, mathematics, and technology* (pp. 113–128). Rotterdam, The Netherlands: Sense Publishers.

ROBIN AVERILL, PANIA TE MARO, DAYLE ANDERSON, HEREWINI EASTON AND MARAMA TAIWHATI

6. BICULTURAL MATHEMATICS TEACHER EDUCATION AND RESEARCH

Supports and Challenges

Kia mau ki te kaupapa: Uphold the reason and purpose
Kia tika te mahi: Do it correctly
Whakamahia: Just do it

INTRODUCTION AND BACKGROUND

Addressing diversity and equity are crucial issues for all involved in education. This is particularly so for contexts enriched with Indigenous communities, cultures, and languages, so many of which have been severely damaged by the effects of colonization and globalization. Mathematics educators work in an area often portrayed or seen as "culture-free," and therefore may face greater challenges than those in other curriculum areas to address these issues. Research into teaching practices that are culturally responsive to Indigenous groups is important in order to aid development of equitable education processes and outcomes, and to acknowledge, reflect, and nurture these cultures.

This chapter shares the authors' reflections on factors that have enabled and challenged their teaching and research with respect to bicultural practices within their pre-service mathematics teacher education courses, over 10 years of working together. These are framed with descriptions of the New Zealand context, the wider literature, and contextual factors that have impacted their work and their teaching and research.

Consistent with Ladson-Billings's (1994) aim for culturally relevant teaching, empowering "students intellectually, socially, emotionally, and politically by using cultural referents to impart knowledge, skills, and attitudes" (p. 18), culturally responsive education is defined in this chapter as that which deliberately incorporates historical and current cultural referents in the areas of values, contexts for learning experiences, behaviours, and ways of being. The team's work draws from two broad cultures—Indigenous Māori and New Zealand European—and hence the term "bicultural" is used. The research team includes Māori and New Zealand European members, who have all planned, taught, and researched together within

 DOI: 10.1163/9789004415768_006

their pre-service mathematics education courses. Alongside preparing student teachers to teach mathematics in elementary schools, they aim in their teaching and research to provide tools for student teachers and teachers that facilitate bicultural emphasis in mathematics education programmes. Specifically, the authors have endeavoured to provide ways to incorporate a range of aspects of the Māori culture (e.g., pedagogies, protocols, language, values, learning contexts, and worldviews) in order to enhance the mathematics learning opportunities of Māori students, and increase understanding of the Māori world for all students. The frameworks were designed to facilitate safe and practical steps student teachers and teachers can make toward incorporating bicultural approaches in their teaching. The team hopes that their work will contribute toward ensuring that in New Zealand schools, Māori students can succeed mathematically while seeing themselves as Māori and having their cultural identities enriched, and that all students can develop deeper knowledge and understanding of, and can delight in, aspects of New Zealand's Māori heritages. They see bicultural teaching as a step, appropriate to the nation's history, toward acknowledging and addressing the cultural diversity of New Zealand classrooms.

Throughout this article, brief translations of Māori terms used are given in parentheses. More detailed translations are not provided where Māori terms are generally not easily matched in English. Hence, the translations should be read as indicative rather than complete.

The New Zealand Context

New Zealand has a population of nearly 4.5 million. The main school types are elementary (for children in Years 1–8 or 1–6), intermediate (Years 7–8), and secondary (Years 9–13). There are roughly 2,500 schools; most use English as the main language of instruction, and 70 are Māori immersion schools. Approximately 18% of New Zealanders are of Māori ethnicity and 68% New Zealand European. Ethnicities of the combined elementary and secondary school student group are roughly 21% Māori, 57% New Zealand European, 9% Pasifika,[1] and 8% Asian, with the remaining 5% made up of students from many other ethnic heritages. The proportion of students of Māori ethnicity is increasing relative to most other groups.

As a group, Māori are very diverse, with differences across and within iwi (tribes). Māori culture and language have suffered greatly through the effects of colonization. Around 60% of Māori have English as their first and only language; few are native Māori speakers. Ministry of Education policy states that "success in education for Māori relies on them achieving that success as Māori," and that it is critical "that Māori culture is acknowledged and included in the learning process," and emphasizes the importance of the curriculum being "responsive and relevant to Māori identity" (Ministry of Education, 2008, p. 5). Often used in teacher appraisal, the New Zealand Teacher Professional Standards (e.g., Ministry of Education, 1998) include reference to teachers' responsiveness to the principles of the Treaty of Waitangi, New Zealand's founding document signed in 1840 promising partnership between Māori and the

British crown and protection of Māori *taonga* (treasures). This inclusion is a further signal of policy intentions that teachers respond to the needs of all students, and of their Māori students in particular, in ways that are consistent with, and acknowledge, the sentiments and promises made in the Treaty. Furthermore, the current national curriculum (Ministry of Education, 2007) includes responsiveness to the Treaty of Waitangi as one of eight principles to be reflected in curriculum delivery. Such policy was not emphasized when many New Zealand teachers learned mathematics, hence many teachers lack confidence in knowing how to put the policy in place in their own teaching, and bicultural practices are not yet widespread.

Recent New Zealand studies with children and student teachers, including Māori (Baskerville & Bondy, 2010; Bevan-Brown, 2006; Ritchie, 2003, 2015) illustrate how bicultural perspectives can be reflected in special and early childhood education, and in teaching drama and technology. Strategies include:

- recruitment of Māori, cultural support and mentoring, greater involvement of *whānau* (extended family), and bicultural pre- and in-service teacher education (Bevan-Brown, 2006);
- development of student teachers' concepts of partnership and respectful relationships, commitment to knowledge of *te reo* and *tikanga Māori* (Māori language and customs), interactive teaching, paralleling Māori knowledge with Western (New Zealand European) knowledge, and exploring cultural values (Ritchie, 2003); and
- storytelling and drawing teaching content and approaches from customary Indigenous approaches (Baskerville & Bondy, 2010; Kisker et al., 2012; Nicol, Archibald, & Baker, 2013; Trinick, Meaney, & Fairhall, 2015).

The authors of these studies have the following goals in common: providing ways to help ensure New Zealand educators hold sincere commitments to Māori and other learners; advancing biculturalism; and incorporating suitable bicultural resources and approaches in their teaching.

Culturally Responsive Education

Indigenous and culturally responsive and relevant education have attracted increasing international attention over recent years (Castagno & Brayboy, 2008; Gay, 2010; Greer, Mukhopadhyay, Powell, & Nelson-Barber, 2009; Ladson-Billings, 1994; Ministry of Education, 2008, 2013; Ogbu, 1992; Rodriguez & Kitchen, 2005; Villegas & Lucas, 2002). Initial teacher education programmes are challenged to be responsive to the contemporary demands of the education systems within which they sit. In many countries, changes in societal beliefs and expectations and stakeholder conceptions of knowledge, teaching, and learning are causing those involved in initial teacher education to examine and modify their practices (Bishop, 2012; Conner, 2009; Fickel, Macfarlane, & Macfarlane, 2017; Korteweg & Nicol, 2009; Mills, 2008; Villegas & Lucas, 2002). Increasingly, emphasis is being placed on examining ways to promote equity and opportunity for all students and to encourage in them respect and empathy for others' cultures (Darling-Hammond & Bransford, 2005; King et al., 2008). Korthagen, Loughran, and Russell (2006) are among many who emphasize the importance of reflection on practice for helping student teachers develop sound pedagogical practices. Aspects of Korthagen et al.'s (2006) principles for change in teacher education programmes provide very useful ideas toward examining research into culturally responsive teacher education. For example, "learning about teaching requires a shift in focus from the curriculum to the learner" (p. 1029), and "learning about teaching is enhanced when the teaching and learning approaches advocated in the programme are modelled by the teacher educators in their own practice" (p. 1036) are particularly pertinent.

Indigenous people being able to see themselves within mathematics learning and contributing to Western society are very important goals. The tensions between the gate-keeping roles of mathematics in society, school mathematics, and loss of Indigenous mathematical power are important considerations for developments in mathematics education, developments which should be carried out with, rather than about, Indigenous people (Doolittle & Glanfield, 2007). Examples of involvement of Indigenous groups with mathematics curriculum development and implementation are given in the work of Lipka and his team (Kisker et al., 2012; Lipka et al., 2005), Nicol and associates (Baker, Nicol, & Archibald, 2009; Nicol et al., 2007), and others (Perry & Howard, 2008). These studies all indicate that substantial and appropriate cultural (Indigenous) links are possible within mathematics instruction, and that such development is achievable when consultation and partnership are in place between committed researchers and community members.

Culturally responsive teaching can be limited by several factors. For example, teachers may lack knowledge of Indigenous culture/s and view mathematics and heritage cultures as separate entities. Harrison (2010) reported on difficulties for Australian teachers and their students caused by the requirement that teachers include Indigenous perspectives in their teaching of all Australian children. Harrison's view was that most Australian teachers lack knowledge of Australia's Indigenous peoples and hence are susceptible to presenting pedagogy that "fantasises about the lives of indigenous people before the British invasion rather than confronting the historical and political controversies of the present" (p. 99).

Exploring social justice within mathematics instruction (Gutstein, 2006), ethnomathematics (Barton, 1996; D'Ambrosio, 1999), and multicultural education (Banks, 2006; Sleeter & Grant, 2007) are valuable areas of research offering much to enhance the mathematics learning of students in Indigenous, minority, and marginalized groups. Although overlaps do exist, the authors of this chapter position their work separately from these fields in that their work targets reflecting an Indigenous culture in mathematics programmes in a holistic sense.

Thus, both the national and international literature convey strong calls for culturally responsive education practice and research, and offer useful ways forward. The work to date provides guidance toward curriculum development (Lipka et al.,

2005; Nicol et al., 2007; Perry & Howard, 2008), teacher education (Villegas & Lucas, 2002), reflecting Indigenous practices in instruction (Demmert & Towner, 2003; Doolittle & Glansfield, 2007; McMurchy-Pilkington, 1996), bicultural practice (Bevan-Brown, 2006; Ritchie, 2003), and considerations for teachers working with students of ethnicities other than their own (Rodriguez & Kitchen, 2005). Less often explored are the supports and challenges that research teams investigating bicultural teaching face when carrying out their work.

Understanding factors that enable and constrain teaching and research in relation to culturally responsive education in general, and bicultural educational practices in particular, is important in order to help the progress of these challenging areas of endeavour. The authors believe that similarities and differences may exist between factors impacting on culturally responsive research involving Indigenous communities across research teams working within different research settings. An examination of the factors that enable and hinder research into bicultural teacher education and teaching will contribute to the literature by providing a lens through which teams can anticipate and consider how to maximize the use of potential supports and minimize the effects of likely challenges to their work. Here, one team's reflections on their work are presented to stimulate and inform wider understanding of the supports and challenges to culturally responsive education involving Indigenous people, to help equip teams and individuals for the work still to be done.

Next, before presenting the team's reflections, a description of the team and their teaching and research is provided.

THE TEAM'S TEACHING AND RESEARCH

The team's pre-service teaching and research into their teaching has explored the use of teaching strategies that include elements from both Māori and New Zealand European paradigms. In their courses they have tried to model ways that student teachers can include bicultural perspectives within mathematics teaching, and have used a series of research studies to examine the effectiveness of their efforts. The overall goals of their teaching and research include:

- equity of access to mathematics achievement;
- acknowledgment and incorporation of te reo and tikanga Māori within mathematics instruction; and
- ensuring that New Zealand children (and in particular Māori children) have opportunities to learn about being Māori, whether their teacher is Māori or non-Māori.

The five authors of this chapter have worked together in pre-service mathematics teacher education and related research projects over many years. Three are Māori (two female, one male) based within Te Kura Māori (our faculty's School of Māori and Pacific Education studies), and two are New Zealand Europeans (both female) based within another school of the faculty. Experienced pre- and in-service mathematics teacher educators, collectively the team have also taught in Māori-medium, bilingual (te reo Māori/English), and English-medium primary and secondary schools.

The initial teacher education programs within which the team has worked together include a three-year degree, a four-and-a-half-year conjoint degree, and a one-year graduate diploma. The mathematics education courses taught by the team have all been compulsory courses within these programs. The longer programs enabled exploration of different bicultural approaches in different years with the same groups of students. The student teachers involved in the courses have been generally favourably inclined toward bicultural teaching approaches, but have reported that they noticed little exposure to such approaches in the courses of curriculum areas other than mathematics education.

Shared teaching is usual team practice, and shared decision-making regarding planning, teaching, and research has been achieved using discussions and consensus in regular meetings focused on course delivery and/or research. In most cases, course delivery has included a combination of lectures for large student teacher groups (entire cohorts) and workshops with groups of up to 30 student teachers. Different combinations of the team worked in each course and research project, with involvement determined by availability and the staffing of the courses within which particular studies took place. All team members involved in each course or research project contributed ideas in relation to course design, course content, and research. In addition, members of the team were involved in their own individual and/or other combined research projects, many of which were related to bicultural mathematics education (Te Maro, Higgins, & Averill, 2008), involvement which has also informed the team's combined work.

The research methodologies used across the research projects include elements consistent with Māori-centred research (Cunningham & Durie, 1998), kaupapa Māori research (Bishop, 2005; Smith, 2005), and culturally sensitive research in which culture is positioned as central to the research process. Researchers are committed to "maintaining the cultural integrity of the participants and other members of the community" (Tillman, 2002, p. 6) and considering "the extent of their own cultural knowledge, cross-race and same-race perspectives … related to the research process" (p. 6). Literature regarding the use of models (Brown, 2007), curriculum tools (Remillard, 2005), and teacher learning (Davis & Krajcik, 2005) have also informed our work.

Student teachers were participants for all but one of the studies. For the remaining study, the participants were three of our graduates, then in their first year of teaching (Table 6.1). Exploratory and consultative approaches, empirical data from questionnaires, interviews, focus groups, and lesson observations were used alongside data gathered through less formal but equally rigorous culturally responsive approaches such as informal conversations, noticing body language, and learning and working in Indigenous settings.

As a brief summary, the team found a combination of holistic and specific school-based, personal, and pedagogical elements to be necessary for bicultural mathematics

Table 6.1. Research projects relating to framework development

Investigation focus	*Data collection methods*	*Participants*
Bicultural components of an initial teacher education course (first framework)	Questionnaire	Student teachers
Use of cultural metaphor (second framework)	Focus groups, questionnaires, written work	Student teachers
Reflection of Treaty principles in mathematics teaching program (third framework)	Interviews	Three early career teachers
Bicultural model (fourth framework)	Questionnaire, written work	Student teachers

teaching. They used their study data alongside international and New Zealand literature to explore four frameworks for planning, describing, implementing, and analyzing bicultural mathematics teaching strategies (Averill, Anderson, et al., 2009; Averill, Te Maro, Taiwhati, & Anderson, 2009) (Table 6.1). The frameworks, three new and one pre-existing, were explored in relation to their applicability to teacher education and school settings.

The earliest framework, useful for course planning and analysis, allows for examination of bicultural perspectives in courses within two broad areas: course structure (protocols, pedagogies, assessment, and personnel); and course content (language, readings, context, learning activities, and planning). The second, the holistic model, is useful for course design. For the second model, a rich cultural activity becomes a course theme and a metaphor for the learning within it. Throughout the course, student teachers work on a modified version of the cultural activity, in the case of the study that of making a modified *tukutuku* panel (traditional woven panel), which served as a means of weaving together the cultural and mathematical educational aspects. The third framework, a pre-existing model increasingly used in New Zealand educational settings to discuss cultural responsiveness (Baskerville & Bondy, 2010), uses three principles often associated with our nation's founding document, the Treaty of Waitangi, partnership, protection, and participation (Graham, 1997), to examine practice. Finally, the latest framework, designed to be used within pre- and in-service teacher education, includes as themes: knowing and understanding each other as people and as learners (Anderson, Averill, Te Maro, Taiwhati, & Higgins, 2010), knowing and understanding each other's cultures (Averill, Taiwhati, & Te Maro, 2010), and enhancing feelings of cultural identity.

Elements of the frameworks have been used formally and informally in the team's own mathematics education pre-service teacher education courses, and the frameworks are now used in one of their faculty's teacher education professional studies pre-service courses for teachers of all subject areas at early childhood, elementary, and secondary school levels. Within New Zealand the team's work offers teaching practices consistent with the principles of the Treaty of Waitangi and Ministry of Education policy (Ministry of Education, 2008, 2013). Looking internationally, the work offers a range of pedagogical, practical, and structural ideas for implementing culturally responsive teaching with/in Indigenous and multi-ethnic contexts.

Few examples of bicultural mathematics education teaching and research teams, including Indigenous members, exist internationally, but hopefully such teams are increasing in number. In this team's work, each person has done their best throughout to be culturally sensitive to others in the team and to the student teachers. Through informal discussions among the team about their work, it became apparent that examining more formally how the work and factors impacting on it had been perceived by individual team members would be worthwhile, to inform our own teaching and research and that of other bicultural teams. The next section of the chapter shares the team's reflections on supports and challenges to their shared bicultural teaching and research work.

TEAM REFLECTIONS

The purpose of gathering and analyzing team members' reflections on the supports and challenges experienced in implementing a bicultural focus in mathematics teacher education courses and carrying out our associated research was to inform the team's own work and that of others involved in similar work. The research question was: What supports and challenges have affected the team's bicultural pre-service mathematics education teaching practice and research?

The five authors' reflections were gathered after more than a year had passed following the completion of the fourth framework. The research team's perceptions of the supports and challenges impacting on their work were gathered individually in written form, using first open and then directed questions. Written reflections were used to allow time for consideration and refinement of opinions and the way in which they were expressed, enabling deeper reflection than would be afforded by interviews or completing a survey. Open questions were used in the first instance in an attempt not to restrict researchers' thinking and responses. In the open section of the data gathering, team members were asked to write individually about the supports for, and challenges to, bicultural teaching practice and research that they had experienced.

All were then invited to provide further individual written reflections in relation to a list of potential supports and challenges put together by the two lead authors of this chapter, the two who had been most heavily involved across the teaching and research projects. Drawn from the experiences of the two researchers, the list was compiled collaboratively before responses to the open questions were collected. The potential supports and challenges listed included: personal engagement and commitment over time; appropriateness (e.g., of intent, personnel, processes, research design, and tools); educational and institutional policy; responding to the diversity of Māori cultures and views; institution-based change; enculturation of student teachers into school communities; time pressure from multiple and increasing calls on time and expertise; effects of historical and political contexts; and community perceptions of cultural diversity and mathematics. Participants were asked not to feel limited to responding to these ideas alone, but to include any other perceptions that these may have triggered. The collected data showed that the list of ideas enabled more detail in the responses to be collected than was achieved through the open questions alone.

Time demands on participants required by a range of projects outside of the combined research led to a departure from the team's usual practice of involving all members in sorting and analyzing the data. Instead, the ideas from the written descriptions were grouped thematically by the lead author using "investigative" or "inductive" analysis (Onwuegbuzie & Teddlie, 2003, p. 370; Patton, 2002, p. 453). All ideas from participants were included. The entire team was then consulted regarding the themes, subthemes, selected quotes, and initial and subsequent drafts of the accompanying discussion.

SUPPORTS AND CHALLENGES: A COLLECTION OF REFLECTIONS

Factors affecting the team's bicultural mathematics education teaching and research fell within three main areas: responsiveness to educational perspectives; responsiveness to culture; and the team's working together. Elements of each will be discussed in turn using team members' own words to allow inclusion of the detail and emphasis present in their reflections. The supports discussed regarding bicultural teaching and research included personal engagement and shared team commitment, the transformative results of earlier work, and policy and curriculum levers. The identified challenges included time constraints and doing justice to the diverse articulations and manifestations of Māori concepts according to *tīpuna* (ancestors), *whānau* (extended family/ies), *hapū* (subtribe/s), and *iwi*. Further elements were felt at times to enable and at other times to hinder bicultural mathematics education teaching and research (e.g., personal cultural knowledge and enculturation of graduate teachers into schools).

Responsiveness to Educational Perspectives

The team reflections indicated the importance for the team of being responsive to educational and institutional policies, their students' and New Zealand school students' needs, and feedback on their work. These areas provided both supports and challenges to the work. Over the 10-year timeframe of the work, the Māori language and culture have become increasingly visible within New Zealand society and educational policies. Such changes have impacted the team's work:

> When we first began incorporating bicultural perspectives into our programs in a substantial way, we needed to preface using them by explaining our pedagogical decisions and teaching some basic Māori words and pronunciation in order for them to be accepted. Now, student teachers are increasingly expecting and asking for ideas for incorporating Māori language and culture into their teaching, so the justifications are no longer needed. (Robin)

The team has used three areas of policy and curriculum levers in particular to inform their teaching and research and to facilitate others' use of their work: the professional standard requiring teachers to address the principles of the Treaty of Waitangi in their teaching (Ministry of Education, 1998); the Ministry of Education's (2008, 2013) policy for Māori education; and team members' involvement in related Ministry of Education publications, research, and resource writing projects (Averill, Hindle, et al., 2014; Hynds, Averill, et al., 2015; Hynds, Meyer, et al., 2013). Responsiveness to the Treaty of Waitangi is emphasized in the guiding documentation of many New Zealand educational institutions and in the teacher professional standards. The team's study with recently graduated teachers aligned data analysis closely to the three principles often used to interpret the meaning of the Treaty in educational settings—protection, participation, and partnership. In one researcher's view:

> Utilizing the Treaty principles within mathematics education can help ensure that mathematical processes and content are able to be explained using Māori cultural tikanga and *kawa* [customary procedures, practices, and protocols] with integrity. (Herewini)

In 2007, the Ministry of Education distributed a draft version of a policy document on Māori education for consultation. The team used the draft and the final versions to guide the direction of, and promote, their work:

> *Ka Hikitia* and other ministry publications are useful because we can talk to their expectations for our student teachers—[the ideas we are putting forward] aren't just *our* "crazy" ideas. (Pania)

The team's work has also been enhanced through consideration of ways bicultural perspectives have been incorporated in other Ministry of Education publications and projects (Bishop, Berryman, Tiakiwai, & Richardson, 2003; Macfarlane, 2004), and through their efforts toward implementation of these and similar ideas within their own programs:

> Pāngarau i roto o te marautanga o Aotearoa [the mathematics curriculum in the Māori language, Ministry of Education, 1996] has opened the door for us to use Māori cultural contexts in education programmes. (Herewini)

Members of the team have been involved in a range of research and resource development projects funded by the Ministry of Education (Ministry of Education, n.d.; Te Maro, Averill, Higgins, & Tweed, 2009; Te Maro et al., 2008), involvement which has supported the team's work:

> The fact that our vision and purpose for bicultural teaching has been relevant to the Ministry's and our own institution's goals has helped our work. The Ministry of Education's involvement of members of our team has informed our team's research and vice versa. (Dayle)

However, despite recent progress being made with regard to inclusion of bicultural perspectives in educational developments nationally, questions remain among team members regarding the extent to which education policy recognizes and portrays Māori worldviews, indicating their concern that the policies and associated documents (e.g., curriculum) themselves may limit bicultural teaching.

Factors affecting the team's research that relate to their own institution have been both positive (e.g., policy providing impetus, faculty professional development focus) and negative (e.g., institutional and course structures, multiple demands on lecturers, and reduction of face-to-face time in initial teacher education courses). The policy directions of the team's institution most clearly related to their work include responsiveness to the Treaty of Waitangi, enhancing the educational opportunities of Māori students, and including bicultural perspectives within the expected learning outcomes of pre-service teacher education courses. The ideals associated with these directions provide policy-based justifications for their work. However, the policy directions are not always translated into practice, and team members' energy, morale, and work are affected by the comparatively low levels of bicultural practice adopted by some other teams:

> How do you ask for space in courses when the objectives state that *te reo me ōna tikanga* will be used/considered "where appropriate"? When is te reo me ōna tikanga *not* appropriate? When is weaving a Māori worldview through courses *not* appropriate? How do you convince people that they *can* do it at whatever level they are, and by doing so there is something they can offer *and* they can grow, particularly when they fall back on the same old barriers? Who does this asking? How do they do it? So, if the institution's structures are not bicultural, how do we really expect that student teachers will go out with robust knowledge under their belts? When it is only a handful of staff that pay true respect to bicultural and culturally responsive methodologies, students notice it and ask why the others don't do it, or ask why they should when it is only the Māori and some of their friends delivering it. How do you ask others to be as awesome as this team, that is, dedicated to authentically and with integrity being bicultural practitioners? (Pania)

Lack of consistency across the authors' faculty in addressing bicultural perspectives in courses led to a professional development focus for staff involved in teacher education of developing understanding and use of bicultural perspectives. This enabled the team to share their research across the faculty and facilitated learning about further bicultural approaches from colleagues. However, members of the team believe that the effects of the professional development have been somewhat undermined by subsequent institutional changes that have included increased separation of staff roles, many program and course revisions, and reduced face-to-face lecturer-student time, factors that have also negatively affected the team's bicultural teaching and research:

> The increased division of the faculty's schools has made it more difficult for us to work as a team and involve staff from other schools in our teaching. Having Te Kura Māori staff members as part of our curriculum team no longer happens. Reduced teaching time has put pressure on bicultural aspects of my own teaching practice: I do less now (sadly). (Dayle)

> How do you ask people who are designing courses and programs to include bicultural perspectives when they are already worried about the contact time they are losing? (Pania)

There is no doubt that the program changes mentioned above, and increasing calls on research team members' time and energy, have reduced the team's focus on their shared work. All have during the time of the research been working toward their own qualifications, contributed to funded research projects outside the team's work, and provided advice to external organisations, and most now hold positions of curriculum or administrative responsibility. Team members moving out of the faculty and awareness of each other's time demands have affected how much they call on one another and how often they can teach together and even meet. The team characteristic of commitment to the *kaupapa* (purpose, theme) has now become a challenge to continuing the team's work because of the increased calls on team members:

> If the same people keep being asked to [assist other lecturers and teachers] they will burn out. If you then keep asking them and get annoyed because they are really busy with other people's work they have been asked to do, does it occur to anyone that while there is one person knocking on their door, 10 others have been there before and 10 more might come behind? Then it is their own people who need them. Trying to work for staff, the institution, and the students here, then other institutions who need their expertise, then other organisations, then … the list goes on and it is hard to say no. (Pania)

> Time pressures have been a constraint, for example brought about by individuals' other work commitments and requirements with regard to personal study for more advanced qualifications and other research. (Dayle)

However, seeing the effects of their exploration into bicultural practices on their students' thinking and teaching and on others has motivated team members:

> Our own knowledge base has advanced as the learning took hold of students and lecturers. Working with students and teachers in schools and noting changes in attitude and practice over time was the ultimate high. (Herewini)

> Most rewarding for me is seeing people outside our team now choosing to begin their presentations with proverbs, songs, or Māori greetings, and looking for such emphasis in others' work, or using our research to inform their own initial teacher education courses or teacher professional development workshops. (Robin)

Next we move to a second theme that held both supports and challenges for the research—factors relating to cultural appropriateness and risk-taking, and the enculturation of beginning teachers into school communities.

Responsiveness to Culture

Being true to Māori perspectives and the diversity of Māori views are vital considerations for the team's work. Ways in which the group has set about responding to these considerations include: having Māori team members who are based within Te Kura Māori; consideration of Māori team members' comfort regarding the shared teaching and research; and consultation within and outside of the team.

The work was enabled through having Māori team members willing to share their knowledge and experience and supporting non-Māori in the team, and non-Māori taking responsibility for developing their own personal knowledge of te reo and tikanga Māori and using this knowledge:

> One of the key supports is having Māori members in the team with cultural expertise who are approachable and look for ways to build deeper understanding of their culture in others. They encourage Māori cultural "independence" in non-Māori group members, whilst still being willing and available to advise and guide. (Dayle)

> As much as possible I work on developing my own personal knowledge of *te Ao Māori* (the Māori world) and in my teaching I tend to reference where I have seen practices and ideas and use phrases such as "as I understand it…" and "in my experience." (Robin)

The Māori team members reported team processes and their personal background in te Ao Māori as supports:

> Being able to use my language and have my worldview acknowledged in the work that I do is a definite bonus; I don't feel alien in my own country! (Pania)

> Fortunately, the research team has been extraordinarily sensitive, supportive, encouraging, and respectful to the work that has happened. The research design has always been in consultation and often the framework development was guided by us [Māori members]; however, it has been inclusive. (Marama)

> The greatest supports for me came from colleagues, literature, and from my own tribal ontology, epistemologies, and vernacular, extracting tikanga and

> kawa [customary procedures, practices, and protocols] from the Māori words to enhance understanding of mathematical processes, content, and practice. My motivation and feelings of integrity remained intact throughout the journey. (Herewini)

However, they also discussed tensions they have experienced in relation to drawing learning experiences from aspects of the Māori language, culture, and worldview. Challenges regarding culturally based tensions experienced by Māori team members in the research included the time needed to think through and enunciate cultural issues and personal standpoints, maintaining cultural integrity within the ways taonga are adopted, and the risk of false impressions being given through the ways in which our work is expressed (e.g., through its being conveyed in English using Western writing and teaching paradigms):

> When you're so used to doing and saying things because they just occur as the way you are, trying to articulate what you're meaning to say takes time and thought. One of the major cautions for me in our work is ensuring that the integrity of the Māori language that is being used is not hindered, watered down, or simplified at the expense of something else, [such as helping] readers have an opportunity to gain a possible view. However [this work] is a beginning point; it only touches the surface in terms of the depth and scope of the Māori language. There is certainly more that can be explored. (Marama)

> One of our frameworks drew heavily on tukutuku. Maintaining tribal and cultural integrity of tukutuku and its representation was and is of concern. How can we take a "traditional" framework and use it with *mana* [integrity] without it losing its original mana and *wairua* [spirituality]? We have to reassure ourselves that our tīpuna wouldn't mind us using tukutuku for this purpose, as it advances knowledge and its essence is intact. Is it tokenism? Is it misappropriation? What will whānau think? These questions were real; however, the responses from students, colleagues and whānau outweighed the doubts. (Herewini)

> I guess that creating a framework that may appear to be linear can give a false picture of what we expect to happen, which is that everything will come together holistically. (Pania)

Despite these reservations (or tensions) regarding personal cultural comfort, Māori team members were committed to the purpose of the work:

> I'll be straight about this; we will never get it perfect. What we must ensure is that the integrity of ourselves as Māori is never compromised. It is never a perfect situation, but that's all that should be said about it. It is not only about us as Māori, but ensuring our colleagues and readers are moving along a journey or pathway too. (Marama)

Informal consultation within and outside the team, and in particular that carried out by Māori team members with their Te Kura Māori colleagues outside the research team, has been used to help ensure the work has been carried out in culturally appropriate ways, providing further team confidence in team decisions:

> Within my own school [Te Kura Māori], I was able to go back and discuss certain aspects of tikanga Māori or get clarity around my views to ensure that what I thought was not only personal to me, but important for Māori. (Marama)

> Because of the levels of consultation that occur in any of the work that we do I have felt comfortable that aspects of appropriateness are fine. (Pania)

For non-Māori team members, there have been challenges relating to cultural responsiveness, their personal knowledge of te Ao Māori, cultural risk-taking, and taking care not to over-consult with Māori colleagues:

> We face challenges constantly in working out what is possible, safe, comfortable, suitable, and appropriate to do for our team as ethnically located people and as lecturers, and for student teachers and their teaching. Personally, developing sufficient cultural knowledge to find a path and tread it carefully and soundly, knowing there are discomforts and many views in and amongst Māori as well as in and amongst our staff, student teachers, and society, has always been challenging. (Robin)

> We know we are always open to being challenged about our work, both from Māori and non-Māori. For me, I would defend my attempts at bicultural teaching with the thought that modelling it was more important than was the danger of implications of doing it wrongly. Better to be seen to be giving it a go than not doing so because I was scared of doing the wrong thing. This attitude has been encouraged by our Māori colleagues. Modelling ways of finding out more about te Ao Māori is also important. (Dayle)

> I have faced dilemmas regarding when to check with whom, about what. I always want to use suitable teaching practices but do not want to burden my Māori colleagues by constantly checking with them or positioning them as cultural experts and gate-keepers. In our research writing things feel safer than in my teaching, as everything is checked by all team members. (Robin)

A common theme expressed across the research team was a concern that many of our student and graduate teachers may not experience school communities philosophically aligned with or rich in the use of bicultural perspectives. Team members had mixed views regarding the likelihood that the bicultural perspectives used in the pre-service courses would be transferred to teachers' own practice, and worried that the bicultural practices promoted may be lost or diminished as early career teachers come to grips with the demands of their jobs and are enculturated into school environments:

> In my experience, early career teachers have been responsive if they have something to work with as a starting point. If the ideas, course design, or framework has been developed prior to teaching then it sets up learners for the ideas and expectations of what will be happening. (Marama)

> Student teachers may enthusiastically take on board the want, the passion, and the understanding of being bicultural teachers; however, what happens to them in the schools that they go to is not something we can do much about. If those schools do not see a need for such an emphasis, it would be difficult for an early career teacher to make much of a difference. (Pania)

However, one student teacher's written work regarding their teaching practicum shows that some schools do have environments well suited to early career teachers being able to capitalize on existing bicultural practices:

> [I knew] I would have to fit in with existing school practices and there may not be opportunity to [work the way] I might want. However, it was immediately clear that my school has a strong bicultural approach and promoted the Treaty of Waitangi principles of protection, partnership, and participation. The principal said "we believe celebrating diversity is crucial to sound education in Aotearoa New Zealand and becoming a global citizen." Te reo is used throughout the school. The day starts with *waiata* [song/s] and *karakia* [prayer/s] and the children constantly hear and use te reo. In my first week I saw active participation and partnership between school and the community when whānau were invited to an evening to discuss the achievements of the Māori students. This all meant that I would be able to build on what already existed in the school, which meant I had more chance of being successful in my [bicultural teaching] goals. (Excerpt from student teacher's work)

The potentially heavy demands of schools' calls on our graduates' cultural knowledge and expertise were also of concern to the research team:

> What we also need to be wary of and caution our students about is becoming the "Māori expert" teacher, the one who takes responsibility for "doing" the Māori stuff in the school, which is actually a heavy load and responsibility particularly if you have to respond to any sort of community questions or backlash. We need to assist our student teachers to know how to approach the leadership and management of schools for support. (Pania)

A Team Working Together

The size and constitution of the pre- and in-service mathematics teacher education group at our institution has varied over time, as have the work-related responsibilities of individual team members. The research team comprises roughly half of the wider group. The way the wider mathematics education group works together has been important for progress in the bicultural teaching and research we have done. Although a variety of beliefs and opinions regarding bicultural teaching exist amongst the wider group, there is an expected way of doing things in relation to practice in general and bicultural practice in particular. Decision-making is shared and all team members have equal status in planning and teaching. Making te Ao Māori visible in teacher education and working in ways culturally appropriate for Māori (e.g., attending to suitable protocols and tikanga) are believed important, and the group is open to new ideas. The group constantly refines and builds on previous practices using what we are learning from our student teachers, our research, and the research of others.

Factors relating to personnel that have been particularly important for our bicultural teaching and research include: having team members rich in Māori heritage; knowing each other well; the team's shared commitment to the kaupapa; and the personal characteristics of team members.

We have been fortunate to have had research team members (Māori and non-Māori) working together within course design, preparation, and implementation of many of our initial teacher education mathematics courses and within in-service mathematics teacher professional development for most of the past 10 years. Lecturers have been present in one another's lectures and, at times, lectures have been shared. The close collaboration has allowed us to know one another well as individuals, as culturally located people, and as mathematics educators. Ongoing consultation across team members, shared understandings of our programs, teaching methods, and resources, and shared priorities for subsequent developments have enabled our work.

Several researchers described the commitment of their colleagues to the goals of our bicultural research, despite its challenges, as a key affordance:

> It is inspiring having people who invest of themselves to "push" a kaupapa that they know will be challenged along the way. These colleagues are committed to finding out why facing the challenges is worthwhile, and passionate about finding ways to make it easier for student teachers to engage and be passionate about being biculturally aware and responsive. (Pania)

> For me the supports to this work include being part of a group who share the same vision and concerns: mutual support and sharing of ideas for new ways of representing Māori culture in our pre- and in-service teaching. (Dayle)

Ways of working as a team that engendered feelings of collective ownership were seen as important for developing strong relationships involving mutual respect, negotiation, initiative, open-mindedness, and support for one another and one another's ideas, all of which contributed to the research:

> The *manaakitanga* [care, respect, ways of looking after] and *aroha* [love, friendship] displayed by the group I have been working with comes in terms of generosity of time, mentoring, assisting, growing others like me. (Pania)

> Te *taha whānau* [the family perspective and involvement] is inherent in the relationships created through the phenomenon of *ako* [reciprocity of teaching and learning] in our work. (Herewini)

> We are all contributors, experts, and learners in everything we do. The group is humble and generous about who controls and holds power in the work. (Pania)

A final challenge to the team's work relates to limited opportunities to explore the use of the frameworks in schools. Team experience in using the bicultural perspectives in school settings could lead to further refinements and increased team confidence in promoting them for teachers' use.

DISCUSSION AND CONCLUSION

To summarize the team's reflections, team attributes and processes both contributed to the development of bicultural teaching and research. A shared sense of purpose and commitment, knowing each other well, and personal attributes such as cultural expertise were identified as supports. Processes such as alternating leadership for different projects, respecting each other's expertise and commitments, and consultation within and outside the team helped to develop trusting relationships between team members and their confidence in the appropriateness of collective and personal actions related to bicultural teaching. Time was a precious commodity. The team worked to the goals set and was flexible, for example, in terms of members opting in or out depending on their personal comfort with the goals at hand and other demands on their time and expertise. The team processes built mutual respect and a sense of collective ownership. In line with the team's fourth framework (Averill, Te Maro, et al., 2009), the team worked well together by knowing who was in the *waka* (canoe), "developing ways to paddle in time with one another and in the same direction and using and acknowledging everyone's strength" (p. 35) and having someone willing to steer the course. The institutional changes that reduced opportunities for such team interaction and development were seen as constraining the work.

The team's shared focus was preserved over time because the group enjoyed and valued working together, they shared responsibility for the teaching and research, and the task at hand remained, as bicultural practice in teacher education and schools has not yet been fully realized. The work was approached in such a way that the team reached consensus through discussion and seeking the opinions of others; views and possible apprehensions were voiced, heard and respected. The perspectives, wisdom, and sensitivities of the Māori team members were held paramount, but other members' apprehensions were also valued and respected. Such working reflects the principles of partnership, protection, and participation inherent in the Treaty of Waitangi, principles that could provide a useful guide for others working in this area.

While shared views and goals were considered a support to their practice and research, awareness of the limitations of this situation would also have proven useful.

For example, an examination of whether others in the wider mathematics education group felt excluded or unable to contribute might have enabled further developments. To a certain extent, the team's work developed in response to teachers' views that differed from the team's in that many pre- and in-service teachers, when the work began, would not have recognized the need for culturally responsive mathematics education or may have viewed such teaching as tokenistic.

Other conclusions from the results presented in this chapter include the affordances provided by identifying and using commonalities between the team's goals and government and institutional goals and initiatives. The team's reflections regarding the vital importance of Indigenous people's involvement within their team's work are consistent with others' views regarding carrying out developments in partnership with, rather than about, Indigenous people (Doolittle & Glanfield, 2007). Maintaining integrity while drawing from cultural values and identity provided both assurance and tensions for Māori team members, highlighting these as important aspects to consider when working in culturally mixed teams. Non-Māori experienced tensions regarding their personal knowledge of te Ao Māori and making decisions about utilizing this in ways comfortable for themselves, Māori, and non-Māori. Although team members experienced such tensions, none were identified between team members, perhaps due to strong inter-team relationships and shared commitment to kaupapa. In reflecting on the team's work, Marama said "you know, colour didn't ever come into it, we all just got on with it and did the job together."

The chapter illustrates ways that can assist Indigenous people's seeing themselves within the curriculum by providing an example of one team's attempts to focus student teachers' attention on learners and to model advocated learning approaches (Korthagen et al., 2006) through positioning culture as central to the development (Tillman, 2002). This work adds to the literature that promotes attending to issues of equity and inter-cultural knowledge and respect in relation to Indigenous and culturally responsive education by highlighting possible impacting factors on cross-cultural teams and their work and providing a model for those involved in such work.

Incorporating and privileging the worlds of Indigenous groups in education is vital for acknowledging, celebrating, and maintaining cultural diversity, for cultural development, and enhancing student motivation and achievement. There is no time to lose. The authors encourage cross-cultural research and teaching teams, particularly those including Indigenous members, to analyze the supports, challenges, and tensions within their own work to understand more clearly the factors impacting on it and on each other, and to maximize its effectiveness.

Returning to the opening thoughts of this chapter, *kia mau ki te kaupapa*—keep the main purpose of the work (incorporating a wide range of Indigenous elements into classroom interactions and instruction) at the forefront of thinking. *Kia tika te mah*—with partnership, respect, and trust, draw from those with cultural and curriculum expertise to ensure that what is done is done as appropriately as possible. Now, the challenge, *whakamahia*—just do it:

> We didn't really know what to expect, but by giving it a go we now have "real" evidence to advance practice. Just because it isn't clear, this isn't an excuse not to try. (Herewini)

NOTE

[1] Students with Pacific Nations heritage.

REFERENCES

Anderson, D., Averill, R., Te Maro, P., Taiwhati, M., & Higgins, J. (2010). Knowing each other as learners: Māori students learning mathematics. In V. Green & S. Cherrington (Eds.), *Delving into diversity: An international exploration of issues of diversity in education* (pp. 45–56). New York, NY: Nova Science.

Averill, R., Anderson, D., Easton, H., Te Maro, P., Smith, D., & Hynds, A. (2009). Culturally responsive teaching: Three models from linked studies. *Journal for Research in Mathematics Education, 40*(2), 157–186.

Averill, R., Hindle, R., Hynds, A., Meyer, L., Penetito, W., Taiwhati, M., Hodis, F., & Faircloth, S. C. (2014). "It means everything doesn't it?" Interpretations of Māori students achieving and enjoying educational success "as Māori." *Set: Research Information for Teachers, 2*, 33–40.

Averill, R., Taiwahti, M., & Te Maro, P. (2010). Knowing and understanding each other's cultures. In R. Averill & R. Harvey (Eds.), *Teaching primary school mathematics and statistics: Evidence based practice* (pp. 167–180). Wellington: New Zealand Council for Educational Research.

Averill, R., Te Maro, P., Taiwahti, M., & Anderson, D. (2009). Culturally responsive mathematics teaching: A bicultural model. In R. Averill & R. Harvey (Eds.), *Teaching secondary school mathematics and statistics: Evidence based practice* (Vol. 2, pp. 27–46). Wellington: New Zealand Council for Educational Research.

Baker, J., Nicol, C., & Archibald, J. with Athie, M. J., Novakowski, J., & Rayner, A. (2009, April). *Conceptualizing culturally responsive education.* Paper presented at the American Educational Research Association (AERA) Conference, San Diego, CA.

Banks, J. A. (2006). *Cultural diversity and education: Foundations, curriculum, and teaching* (5th ed.). New York, NY: Pearson/Allyn & Bacon.

Barton, B. (1996). Making sense of ethnomathematics: Ethnomathematics is making sense. *Educational Studies in Mathematics, 31*(1–2), 201–233.

Baskerville, D., & Bondy, A. (2010). A bicultural approach to teaching and learning for diversity. In V. Green & S. Cherrington (Eds.), *Delving into diversity: An international exploration of issues of diversity in education* (pp. 57–69). New York, NY: Nova Science.

Bevan-Brown, J. (2006). Beyond policy and good intentions. *International Journal of Inclusive Education, 10*(2–3), 221–234.

Bishop, R. (2005). Freeing ourselves from neocolonial domination in research: A kaupapa Māori approach to creating knowledge. In N. K. Denzin & Y. S. Lincoln (Eds.), *The Sage handbook of qualitative research* (3rd ed., pp. 109–138). Thousand Oaks, CA: Sage.

Bishop, R. (2012). Pretty difficult: Implementing kaupapa Māori theory in English medium secondary schools. *New Zealand Journal of Educational Studies, 47*(2), 38–50.

Bishop, R., Berryman, M., Tiakiwai, S., & Richardson, C. (2003). *Te Kotahitanga: The experiences of year 9 and 10 Māori students in mainstream classrooms.* Hamilton, New Zealand: Māori Education Research Institute (MERI), School of Education, University of Waikato.

Brown, M. (2007, April). *Toward a theory of curriculum design and use: Understanding the teacher-tool relationship.* Paper presented at the American Educational Research Association (AERA) Conference, Chicago, IL.

Castagno, A., & Brayboy, B. M. J. (2008). Culturally responsive schooling for indigenous youth: A review of the literature. *Review of Educational Research, 78*(4), 941–993.

Conner, L. (2009). Initial teacher education programmes in transition: Tensions and possibilities. *Curriculum Matters, 5*, 121–137.

Cunningham, C. W., & Durie, M. H. (1998). *A taxonomy, and a framework for outcomes and strategic research goals, for Māori research and development.* Palmerston North, New Zealand: Massey University School of Māori Studies.

D'Ambrosio, U. (1999). Mathematics, history, ethnomathematics and education: A comprehensive program. *The Mathematics Educator, 9*(2), 34–36.

Darling-Hammond, L., & Bransford, J. (Eds.). (2005). *Preparing teachers for a changing world: What teachers should learn and be able to do.* San Francisco, CA: Jossey-Bass.

Davis, E. A., & Krajcik, J. S. (2005). Designing educative curriculum materials to promote teacher learning. *Educational Researcher, 34*(3), 3–14.

Demmert, W. J., Jr., & Towner, J. C. (2003). *A review of the research literature on the influences of culturally based education on the academic performance of Native American students.* Portland, OR: Northwest Regional Education Laboratory.

Doolittle, E., & Glanfield, F. (2007). Balancing equations and culture: Indigenous educators reflect on mathematics education. *For the Learning of Mathematics, 27*(3), 27–30.

Fickel, L. H., Macfarlane, S., & Macfarlane, A. H. (2017). Culturally responsive practice for indigenous contexts: Provenance to potential. In C. Reid & J Major, (Eds.), *Global teaching: Southern perspectives on teachers working with diversity* (pp. 101–127). New York, NY: Palgrave Macmillan.

Gay, G. (2010). *Culturally responsive teaching: Theory, research, and practice* (2nd ed.). New York, NY: Teachers College Press.

Graham, D. (1997). *Trick or treaty?* Wellington, New Zealand: Institute of Political Studies, Victoria University of Wellington.

Greer, B., Mukhopadhyay, S., Powell, A., & Nelson-Barber, S. (2009). *Culturally responsive mathematics education.* New York, NY: Routledge.

Gutstein, E. (2006). *Reading and writing the world with mathematics: Toward a pedagogy for social justice.* New York, NY: Routledge.

Harrison, N. (2010). Teaching indigenous perspectives in the 21st century classroom: An exploration of quality pedagogy in Australian schools. In V. Green & S. Cherrington (Eds.), *Delving into diversity: An international exploration of issues of diversity in education* (pp. 99–107). New York, NY: Nova Science.

Hynds, A., Averill, R., Penetitio, W., Meyer, L., Hindle, R., & Faircloth, S. F. (2015). Examining the impediments to indigenous strategy and approaches in mainstream secondary schools. *International Journal of Leadership in Education: Theory and Practice*, 1–23. doi:10.1080/13603124.2015.1051130

Hynds, A., Meyer, L., Penetito, W., Averill, R., Hindle, R., Taiwhati, M., & Hodis, F., with Faircloth, S. (2013). *Evaluation of He Kākano professional development for leaders in secondary schools: Final report.* Wellington, New Zealand: Ministry of Education.

King, J., Adams, R., Kim, Y., Belcastro, A., Thorpe, S., & Greene, W. (2008). Supporting purpose-driven teaching in Southern Oregon, USA: A Teacher Education for the Future project report. *Pacific-Asian Education, 20*(1), 8–18.

Kisker, E. E., Lipka, J., Adams, B. L., Rickard, A., Andrew-Ihrke, D., Yanez, E. E., & Millard, A. (2012). The potential of a culturally based supplemental mathematics curriculum to improve the mathematics performance of Alaska Native and other students. *Journal for Research in Mathematics Education, 43*(1), 75–113. doi:10.5951/jresematheduc.43.1.0075

Korteweg, L., & Nicol, C. (2009, April). *Culturally responsive teacher education for non-Aboriginal urban teachers.* Paper presented at the American Educational Research Association (AERA) Conference, San Diego, CA.

Korthagen, F., Loughran, J., & Russell, T. (2006). Developing fundamental principles for teacher education programmes and practices. *Teaching and Teacher Education, 22*(8), 1020–1041.

Ladson-Billings, G. (1994). *The dream keepers: Successful teachers of African American children.* San Francisco, CA: Jossey-Bass.

Lipka, J., Hogan, M. P., Parker Webster, J., Yanez, E., Adams, B., Clark, S., & Lacy, D. (2005). Math in a cultural context: Two case studies of a successful culturally based math project. *Anthropology and Education Quarterly, 36*(4), 367–385.

Macfarlane, A. H. (2004). *Kia hiwa ra! Listen to culture: Māori students' plea to educators.* Wellington, New Zealand: New Zealand Council for Educational Research.

McMurchy-Pilkington, C. (1996). Ina te mahi he rangatira: Māori women as rangatira in marae kitchens. In B. Weber (Ed.), *He paepae korero: Research perspectives in Māori education* (pp. 19–34). Wellington, New Zealand: New Zealand Council for Educational Research.

Mills, C. (2008). Making a difference: Moving beyond the superficial treatment of diversity. *Asia-Pacific Journal of Teacher Education, 36*(4), 261–275.

Ministry of Education. (1996). *Pāngarau i roto i te marautanga o Aotearoa.* Wellington, New Zealand: Learning Media.

Ministry of Education. (1998). *Interim professional standards for primary school deputy/assistant principals and teachers.* Wellington, New Zealand: Ministry of Education.

Ministry of Education. (2007). *The New Zealand curriculum for English-medium teaching and learning in years 1–13.* Wellington, New Zealand: Ministry of Education.

Ministry of Education. (2008). *Ka Hikitia – Managing for success: Māori education strategy 2008–2012.* Wellington, New Zealand: Ministry of Education.

Ministry of Education. (2013). *Ka Hikitia – Accelerating success 2013–2017: The Māori education strategy.* Wellington, New Zealand: Ministry of Education. Retrieved from http://www.education.govt.nz/ministry-of-education/overall-strategies-and-policies/the-maori-education-strategy-ka-hikitia-accelerating-success-20132017/the-maori-education-strategy-ka-hikitia-accelerating-success-2013-2017/

Ministry of Education. (n.d.). *Figure it out series.* Retrieved from https://nzmaths.co.nz/figure-it-out

Nicol, C., Archibald, J., & Baker, J. (2013). Designing a model of culturally responsive mathematics education: Place, relationships and storywork. *Mathematics Education Research Journal, 25,* 73–89.

Nicol, C., Archibald, J., Kelleher, H., Brown, L., Hutchingson, V., Nielsen, W., & Owuro, J. (2007, April). *Culturally responsive approaches to mathematics teaching and learning in Aboriginal communities.* Paper presented at the American Educational Research Association (AERA) Conference, Chicago, IL.

Ogbu, J. U. (1992). Understanding cultural diversity and learning. *Educational Researcher, 21*(8), 5–14, 24.

Onwuegbuzie, A. J., & Teddlie, C. (2003). A framework for analysing data in mixed methods research. In A. Tashakkori & C. Teddlie (Eds.), *Handbook of mixed methods in social and behavioral research* (pp. 351–383). Thousand Oaks, CA: Sage.

Patton, M. Q. (2002). *Qualitative research and evaluative methods* (3rd ed.). Thousand Oaks, CA: Sage.

Perry, B., & Howard, P. (2008). Mathematics in indigenous contexts. *Australian Primary Mathematics Classroom, 13*(4), 4–10.

Remillard, J. T. (2005). Examining key concepts in research on teachers' use of mathematics curricula. *Review of Educational Research, 75*(2), 211–216.

Ritchie, J. (2003). Bicultural development within an early childhood teacher education programme. *International Journal of Early Years Education, 11*(1), 43–56.

Ritchie, J. (2015). Disentangling? Re-entanglement? Tackling the pervasiveness of colonialism in early childhood (teacher) education. In V. Pacini-Ketchabaw & A. Taylor (Eds.), *Unsettling the colonial places and spaces of early childhood education* (pp. 147–161). New York, NY: Routledge.

Rodriguez, A. J., & Kitchen, R. S. (Eds.). (2005). *Preparing mathematics and science teachers for diverse classrooms: Promising strategies for transformative pedagogy.* Mahwah, NJ: Lawrence Erlbaum.

Sleeter, C. E., & Grant, C. A. (2007). *Making choices for multicultural education: Five approaches to race, class, and gender.* Hoboken, NJ: John Wiley.

Smith, L. T. (2005). On tricky ground: Researching the native in the age of uncertainty. In N. K. Denzin & Y. S. Lincoln (Eds.), *The Sage handbook of qualitative research* (3rd ed., pp. 109–138). Thousand Oaks, CA: Sage.

Te Maro, P., Averill, R., Higgins, J., & Tweed, B. (2009). Fostering the growth of teacher networks within professional development: Pāngarau teachers working in wharekura. In *Te Poutama Tau evaluation report 2008: Research findings in pāngarau for years 1–10.* Wellington, New Zealand: Ministry of Education.

Te Maro, P., Higgins, J., & Averill, R. (2008). Creating strong achievement gains for Māori children in mathematics English-medium classrooms. In *Findings from the New Zealand numeracy development projects 2007* (pp. 37–49). Wellington, New Zealand: Ministry of Education.

Tillman, L. C. (2002). Culturally sensitive research approaches: An African-American perspective. *Educational Researcher, 31*(9), 3–12.

Trinick, T., Meaney, T., & Fairhall, U. (2015). E haere ana koe ke hea? Spatial thinking: Traditional Māori systems and the language of spatial orientation. In R. Averill (Ed.), *Mathematics and statistics in the middle years: Evidence and practice* (pp. 146–161). Wellington, New Zealand: New Zealand Council for Educational Research.

Villegas, A. M., & Lucas, T. (2002). *Educating culturally responsive teachers: A coherent approach.* Albany, NY: SUNY Press.

ROBERTA HUNTER, JODIE HUNTER AND TREVOR BILLS

7. ENACTING CULTURALLY RESPONSIVE OR SOCIALLY RESPONSE-ABLE MATHEMATICS EDUCATION

INTRODUCTION

Mathematics is often described as a gatekeeper, and yet the importance of all students being given opportunities to develop mathematics competencies for life and work is widely agreed upon. The benefits that accrue have significant impact nationally both in economic (Hanushek & Woessmann, 2012) and in personal ways within many aspects of an individual's life. The benefits include such factors as enhanced employment opportunities, health and well-being choices, and even life expectancy (Tate, 2013). Nevertheless, within Aotearoa, New Zealand, as in many other countries, there are ethnic groups of students who have a long history of low achievement as a result of the structural inequities and institutionalized hegemonic practices they encounter in mathematics classrooms (Hunter & Hunter, in press). The group we are addressing in this chapter are students of Pāsifika ethnicity who historically have as a group achieved well below their Pākeha[1] and Asian counterparts (New Zealand Qualifications Authority, 2013) in Aotearoa, New Zealand mathematics classrooms. In this chapter, our goal is to illustrate how the outcomes for this group of students are reversed when teachers enact high-quality and inclusive pedagogy that incorporates the students' language, identity, and culture. We will draw on student, teacher, and *fānau* (extended family and community) voices to inform, to examine, and to explore the actions and positive impacts when mathematics education builds from, in, and with the culture of Pāsifika learners.

PĀSIFIKA STUDENTS IN AOTEAROA, NEW ZEALAND

In this chapter, we use the term Pāsifika peoples as an umbrella term to include all the many different terms and spellings used to describe people of Pācific origin and ethnicity. This includes such terms as Pācific, Pāsifika, Pācifica, Pācific Islander, and Polynesian. The term Pāsifika peoples is a collective term, constructed in New Zealand and used predominantly by policy-makers, government agencies, and education institutions to describe people of Pācific descent who reside in New Zealand and who, as Schuster (2008) states, "identify themselves with their indigenous Pacific countries of origin because of ancestry or heritage, family and

 | DOI: 10.1163/9789004415768_007

cultural connections with Samoa, Cook Islands, Tonga, Niue, Tokelau, Fiji, Solomon Islands, Tuvalu and other Pacific countries" (p. 12).

In this chapter, we also use the term Pāsifika peoples but at the same time remain cognizant of the significant and unique differences in cultural practices and beliefs and cultural and ethnic complexities which relate to the different groups of peoples within the broader term (Manu'atu & Kepa, 2002; Mara, 1999; Mara, Foliaki, & Coxon, 1994; Nakhid, 2002). In using the umbrella term Pāsifika peoples, we are not suggesting that this is a homogenous group, or that Pāsifika peoples who fall within this broad descriptor would apply it to themselves rather than using their island nation name. For example, Schuster (2008) describes how a member of the Samoan community stated, "I didn't realise I was a Pāsifika person or an Islander until I arrived at the airport in Mangere, before that I was a Samoan!" (p. 11). However, the term is used widely by policy-makers, and national educational data for this group of students is most often collated, analyzed, and reported under the generic term of Pāsifika peoples, and so we too use it in a similar utilitarian way.

Up to the present time, research has most often taken a deficit theorizing approach to exploring the educational underachievement of such children from ethnic minority/low-income families (Gorinski & Fraser, 2006), like those we write about in this chapter. Although there is evidence of a growing interest in Pāsifika research taken from a strength-based or success perspective (Bills & Hunter, 2015; Evans, 2011; Hunter, 2013; Hunter & Hunter, in press; Lipene, 2010; Passi, 2011; Theobald, 2013), a predominant focus of previous research has been placed on why Pāsifika students might be failing in comparison to other groups of learners in schools. In this chapter, we aim to present a strength-based mathematics approach that has been developed for Pāsifika students in Aotearoa, New Zealand, termed Developing Mathematical Inquiry Communities (DMIC). This mathematics program was designed within ambitious teaching (Kazemi, Franke, & Lampert, 2009) while also drawing on what Gaye (2010) described as "culturally responsive" and Atweh and Ala'I (2012) termed "socially response-able mathematics."

DEVELOPING MATHEMATICAL INQUIRY COMMUNITIES

DMIC builds on best international research (Kazemi, Franke, & Lampert, 2009; Wood, Williams, & McNeal, 2006) around pedagogical practices that have engaged students in inquiry, reform, and ambitious mathematics teaching and learning. DMIC focuses on building classroom communities of mathematical inquiry. Central to student induction into DMIC classrooms are their participation and communication patterns that support them to gradually become proficient in engaging in and using reasoned mathematical practices (Hunter, 2008).

Mathematical practices develop through socially constructed mathematical interactions (Ball & Bass, 2003). Mathematical practices extend beyond the use of mathematical knowledge and the development of strategies or solutions; they are the practices that proficient problem solvers use and include providing mathematical explanations, representing and "justifying claims, using symbolic notation efficiently, defining terms precisely, and making generalizations [or] the way in which skilled mathematics users are able to model a situation to make it easier to understand and to solve problems related to it" (RAND, 2003, p. xviii). At the heart of the development and use of mathematical practices are ways of talking and reasoning, and ways of asking questions, challenging others, and engaging in mathematical argumentations.

The role of the teacher is central to developing such learning communities and, given that proficiency in the use of mathematical practices is closely aligned to construction of a positive mathematical disposition, this highlights the importance of *all* students being provided with opportunities to participate in mathematical practices. To ensure that *all* students are provided with equitable access to participate in mathematical practices substantially increases the demand on teachers. In this chapter, we will illustrate why it is critical that teachers understand the culture and values of their students and the importance of drawing on these to engage Pāsifika learners in order for them to also retain a strong cultural identity.

In the next section, we will examine and explore the many strength-based attributes which schools in Aotearoa, New Zealand, have available to build from, in, and with the culture of Pāsifika learners when Pāsifika values, culture, and languages form the foundations for learning in mathematics classrooms. We will provide examples of how these have been embedded within DMIC to provide wider and deeper accessibility to the mathematics inquiry and reasoning while at the same time ensuring that the students do so in culturally appropriate ways.

EMBEDDING CULTURALLY RESPONSIVE OR SOCIALLY RESPONSE-ABLE MATHEMATICS AND PĀSIFIKA CULTURE, LANGUAGES, AND VALUES

Although Pāsifika peoples are a diverse group of people, the thread of commonality which binds them and the Indigenous Māori of Aotearoa, New Zealand, together in a rich tapestry of shared relationships, heritage, and ways of being are a mutual set of values. Māori values are represented in *Tikanga Māori* (Māori values) (Mead, 2016) and include such values as respect, caring relationships, family, community, and relatedness (Graham, Meyer, McKenzie, McClure, & Weir, 2010). A similar list of values is recorded for Pāsifika within the New Zealand Ministry of Education's Pāsifika Education Plan (2015). These values include respect, reciprocity, service, inclusion, spirituality, leadership, love, belonging, and family.

For Pāsifika peoples, relationships, love, and *fānau* (family) are central to their ways of living and being. Other values that are encompassed within notions of *fānau* include trust, respect, reciprocity, collectivism, communalism, and service, and these shape both individual and group identities of Pāsifika peoples within a collectivist cultural frame (Civil & Hunter, 2015). In our own research (Bills & Hunter, 2015; Hunter et al., 2016), we have directed attention to the way in which family is a key element that drives Pāsifika motivation and achievement. This research and the research of others (e.g., Averill, 2012; Hawk, Cowley, Hill, & Sutherland, 2002; Knight-de-Blois, 2015) show the significance Pāsifika learners and their families place on strong teacher-student relationships. Consistently, when interviewed, Pāsifika students report that positive relationships within the educational space are the single most influential factor on their engagement and achievement.

Clearly, on entry to schools in Aotearoa, New Zealand, Pāsifika students bring with them a rich background of values and experiences. But the lived reality inherent in the values and experiences prioritized within educational contexts constructed on those of the dominant Pākeha group can be significantly different. Bartolomé (1996) brings to our attention the need for educational methods to be set within students' cultural world to support their engagement and achievement. He cautions that the learning may not only be alien to their reality but may also be antagonistic to their culture and lived experiences—that is, their cultural capital. In this chapter, we draw on the term *cultural capital* used by Bourdieu and Passeron (1973), which they defined as the general cultural background, knowledge, disposition, and skills that are passed on from one generation to another.

The cultural capital valued in many mathematics classrooms in Aotearoa, New Zealand, reflects the cultural capital of the dominant New Zealand Pākeha group, and so the values prioritized in these settings rather than reflecting a collectivist society tend toward an individualistic and competitive form. As a result, the building of strong and trusting relationships, which Pāsifika students place strong value on within their home context, can be problematic for many within New Zealand educational settings. Within DMIC, significant focus is placed on teachers' talking about and building on the Pāsifika and Māori values in ways that support them to work collaboratively but also to actively engage them in a range of mathematical practices using productive discourse.

METHODS

The research reported in this chapter is part of a large project that involves over 300 teachers in 27 New Zealand urban primary schools. The students come from very low socio-economic home environments and are of predominantly Pāsifika groupings. The teachers come from diverse backgrounds and their length of teaching experience varies although they tend to be less experienced and overseas trained. The teachers and students reported in this chapter participated in the project for more than a year.

Data for this chapter drew on video recorded classroom observations, and student and teacher interviews including the use of a set of open-ended questions, which allowed for multiple responses. Analysis of the data consisted of comparing and contrasting responses from the different teachers and students. Emerging themes and patterns which were consistent across the different participants were determined and analysed using NVivo qualitative data analysis software.

FAMILY SHAPING SOCIAL INTERACTIONS IN MATHEMATICAL ACTIVITY

To establish ways students interact and participate in mathematical activity within DMIC, the Pāsifika concept of family is embedded in the social norms which shape the interactions within mathematical activity. For example, a teacher explains:

> Teacher: One of the biggest things, or the norms that I established, is that we are a family. The concept of family means to say that we are in this together. If someone doesn't understand, we are there to help her or him. Just like the context at home, if you struggle with something at home, you ask a family member to help and build on that. That's how we work in the classroom. There is no one leader in the group, in some cases, some students take a leadership role and when they see someone's struggling, they'll help; it's about the reciprocal learning, the sharing of ideas, ensuring they understand before they carry on with the task.

His explanation emphasizes the collective view applied within the interactions. Each student is held accountable to understand, but also the group is held accountable for ensuring that all members understand the mathematical reasoning.

In the different DMIC classrooms, validating notions of family within the social norms are modelled in multiple different ways. The next example illustrates a teacher affirming that across Pāsifika peoples he knows, they are not a homogenous group and that within individual languages the term *family* might be a different word, but the concept of family is the same and binds them altogether as one unit:

Teacher: What type of group are we?
Students: We are a family.
Teacher: Family, *whānau*. What's the word for family in Samoan?
Students: *Aiga.*
Teacher: *Aiga.* What's the Tongan word for family?
Students: *Famalii.*
Teacher: What do families do?
Students: Take care of each other and help each other.
Teacher: Take care of each other ….

The teachers are proactive in drawing on cultural values embedded in the notions of family to shape how the students interact as they work together. A key feature of DMIC classrooms is that the students work together in small groups to construct one shared solution which they can all explain and justify. In the next example, the teacher has noticed that a group is not working as one unit, and so he initiates discussion around the interdependence he wants to see emerging as they work as a family unit:

> Teacher: What I'm seeing in this group here is you are depending on Sui, and yet you are all supposed to be contributing to the conversation. It is about

> being in a family and understanding what is going on. So anyone in the group can say, "Oh, we did this because …." This idea is one everyone needs to understand. The idea of working in the group is to bring all of you together so that you all have a common understanding.

The idea of working within a Pāsifika family is also used to engage students in asking questions and challenging other members' reasoning in ways that ensure that they feel culturally safe. This is always closely aligned to the value placed on respect, which is earned as members of the family unit work together collectively and each member takes responsibility for their own understandings as well as their contributions. A clear focus is placed on positive outcomes for the wider family rather than for the individual. In the next example, the teacher explains how she explicitly shapes the way the students engage in mathematical argumentation, drawing on their understandings of how to do this in ways that build within their home values:

> Teacher: Respect is a big part of being a family as well. Everyone shows respect because we are a family and they know they all have a say and have a chance to listen to what other people say, and if things are not clear to them they can generate *friendly arguments* and say, "Hey, where did you get that from, because my understanding is this?" and everyone comes away with a broader understanding.

The terms *friendly arguments* or *arguing politely* are heard a lot in DMIC classrooms as the teachers prompt the students to ask questions and sense-make, but in ways which denote respect within their Pāsifika meaning. These terms also serve to support the students to consider inquiry and argumentation positively rather than in what is often negatively based within the media (Hunter, 2008).

The different actions the teachers take to induct students into mathematical activity and mathematical practices, drawing on their home values to support them to take risks with mathematical reasoning and to make mistakes within a culturally safe environment, are recognized by the students. This is illustrated by two students who, when questioned about how they felt when engaging in inquiry and argumentation, outlined the importance of respect within the learning context:

> Sione: Respect is real important. When you have respect, you can have friendly arguments and you argue about the maths, so it's not "Nah, you're wrong" or "You're dumb," eh? It's like "I don't agree with your maths and this is what I think or you have to convince me."

> Grace: You can have friendly debates about maths … and then you have to justify your answer. And then if you made a mistake you learn from your mistake.

What these Pāsifika students have illustrated is that they are able to integrate their lived home values with those constructed within the social norms which shape their interactions in mathematical activity. Furthermore, they are able, through teachers' explicitly embedding these values as part of the classroom culture, to develop a strong and productive mathematical voice.

RESPECT, COLLECTIVISM, AND COMMUNALISM SHAPING MATHEMATICAL ACTIVITY AND WAYS OF BEING

Respect is a deeply embedded value that shapes how Pāsifika students interact and relate to each other within their family and community. However, while drawing on the students' understandings of respect to develop productive mathematical talk, teachers need to remain cognizant of how it is interpreted by the students. In previous research (Bills & Hunter, 2015; Hunter & Hunter, in press; Hunter et al., 2016), we have described how respect is often interpreted by Pāsifika students as precluding them questioning elders (the teacher or more knowledgeable others) or challenging others in ways that may cause Pāsifika-mā (shame or loss of face). In consideration of the different interpretations learners in DMIC classrooms may hold, explicit actions are taken to induct the students into understanding how they can engage in inquiry and mathematical argumentation as part of their active roles as mathematicians. The focus of the teacher's role is to support them to gain a positive mathematical disposition but at the same time also retain a strong cultural identity through learning that they have a different number of required voices to suit the different cultural and social spaces they traverse. One voice they need to recognize is the one in the mathematics classroom which allows them to question and challenge without causing loss of face, or *Pāsifika-mā.*

As part of establishing norms which scaffold students to engage in a range of mathematical practices, the teachers in DMIC classrooms talk about the use of these voices and when they might be appropriate within the different contexts of their daily lives. But at all times, the voices required in the DMIC setting are underpinned by the Pāsifika value placed on respect.

> Teacher: We talk about the value of respect and about accepting others, and that's really strong. We encourage them to participate and take a risk, but it's safe because there is respect and inclusion and love, so nobody is going to put you down if you get it wrong, because we learn from our mistakes. We are all learners, and we have the right to learn and the responsibility to listen to others, because we are family.

Clearly, the teacher is illustrating that the interrelated Pāsifika values, when drawn on in a socially response-able role, allow the learners to work through other values that may have precluded them in previous mathematics classrooms in Aotearoa, New Zealand, in the past.

Within DMIC classrooms, the importance of teachers being able to understand and work within Pāsifika values cannot be emphasized enough. Not only do they need to understand them, but they also need to be critically aware of the dissonance some institutionalized practices may cause Pāsifika students. As we said earlier, this is because many school values presented within common pedagogical practices in Aotearoa, New Zealand, mathematics classrooms most often represent the values of the dominant culture. These institutionalized practices need critical consideration and appraisal if equitable outcomes are a goal for Pāsifika learners. For example, in Aotearoa, New Zealand, the most recent nationwide professional development reforms in mathematics rolled out over 10 years (Ministry of Education, 2004) had teachers grouping students by ability. The use of stratified and streamed groups implicitly encouraged competitiveness and placed importance on individual success. We suggest that this way of stratifying all students by perceived Western-view abilities might go some way to explain why Pāsifika students were most often positioned in the lower mathematics groups (Bills & Hunter, 2015; Hunter & Anthony, 2011).

In DMIC classrooms, the students are grouped socially and so they bring to the mathematical activity different pieces of knowledge and skills to contribute. Within a Pāsifika view, the success of group members is measured by the success of the collective as a whole. Individual contributions are valued as ways of collaboratively constructing the whole. However, ownership of the final solution is shared across the group, and it is the responsibility of each member of this group to be able to explain and justify their reasoning.

The shift from an individual and competitive grouping approach to one in which communalism and collectivism is privileged and how this better supports them as Pāsifika peoples is illustrated by two students:

> Student One: At the start of the year, I would have said being a successful mathematician meant being in the top group and getting the answers right. Now I think it is being a *good person*. Not being the person who is always right but helping others as well. That makes you good at maths.
>
> Student Two: Last year I thought being a good mathematician was getting the answer right. I was in the dumb group, so no good at maths. Now being good at maths is asking questions, sharing ideas, explaining your idea, challenging people, taking risks …. So now I am a good *mathematician.*

The shift in thinking from being in the top mathematics group to considering yourself a *mathematician* because you question and reason, and a *good person* because you work collectively, is significant to these students' mathematical disposition and cultural identity. Another student described the dissonance the stratified grouping system had caused her:

> Student: We all participate and take risks, and it doesn't matter if it is wrong 'cause we are all learning. It is not about being the smartest, we all have to understand. Like even in high group you had top and dumbest, so you could be dumbest in high group [*laughs*], and it's not like that now. I didn't enjoy maths last year in the high group.

The application of Pāsifika values around family, collectivism and communalism also highlights the dissonance the students experience when opportunities to learn mathematics are not equal. For example, one student when asked to compare being in a streamed group the previous year with being in a heterogeneous group in the current year stated:

> Student: This year we are equal and we have an equal right to share. Nobody cares who you are … we only care about ideas. Last year in the dumb group, we didn't do area, volume, ratios, and maths like that …. Only the top group did, but now we all are the same. It is more fun now [that] it is harder.

This student has readily identified that previously, opportunities to learn were reduced if a student was not in the top group. At the same time, their gaze has shifted toward the importance of focusing on mathematical reasoning. Clearly, for these Pāsifika students being able to work together not only supports them constructing a positive mathematical disposition but also supports them retaining a positive cultural identity.

RECIPROCITY SHAPING TEACHERS' AND STUDENTS' MATHEMATICAL VOICES

Notions of the Pāsifika value of reciprocity are threaded through the transformation of mathematics that better meets the needs of Pāsifika learners. Freire (2000) emphasizes the significance of dialogical education built on respect, with people working together in reciprocal relationships. It is the importance of egalitarian dialogue, based on the validity of the argument and not on power imbalances in the teacher–student relationship; a relationship that needs careful consideration and construction. This was illustrated in Boaler's (2002) study, which showed that in mathematics classrooms where teachers hold the balance of control over which methods and procedures are used, students have less opportunity to interact with the mathematics. In such classrooms, the students also perceived that they were secondary to the teacher and became passive recipients of knowledge. In contrast, in classrooms where power was more evenly distributed, students developed positive dispositions, which led to deeper mathematical understandings. In DMIC classrooms, the focus on reciprocity and developing student voice is evident. For example, one teacher described his shifting role:

> Teacher: What I've noticed over the time I have been part of the program is the importance of having the students' voices as opposed to being teacher directed. When it's all teacher directed, you can share till the cows come home, but if it doesn't make sense to them, it's not going to. If you give them a problem and say, "Go away and have a look at that, come back to me. What do you think is happening?" They start sharing their ideas and they feel valued because they are being listened to.

The teacher had shown that an expectation of mutual responsibility built around reciprocity had established a change in the roles and the students no longer waited for the teachers to tell them the mathematics; rather they had assumed wider roles which included them being teachers within the collective. Clearly evident is also their expectation that the teachers could learn from the students within the reciprocal relationship. This is illustrated by another teacher in the project:

> Teacher: There has been a shift in the percentage of students' voices. The locus of control is not so much us, but what the students are discussing and sharing, and that's really powerful, seeing that shift from us having all the power to a shared power … and it's a shared responsibility, and we become the facilitator and the students become teachers to each other, and that's really good to see. Because there are more students' voices and they have more control, you can also see what the students know and how they think; they can explain their reasoning more. We can also learn more about them, like they come out with strategies to solve some of the problems that I haven't thought of and it's like … "Yeah, that works."

The students also identify shared accountability to themselves and to each other as being of importance for their learning and the learning of others. The following response illustrates increased ownership of the mathematical talk and engagement in mathematical practices within collective accountability:

> Student One: Yeah. It's not just one person answering the question; we get others' opinions so everyone understands. It's our responsibility as a team to work together. And it's good because you can talk to people. Before you just had your book and you got a growl if you talked or asked someone a question but now you can.
>
> Student Two: In this maths we have more power. He [teacher] gives us the problem, but the problem is about us … our reality, and we have to figure it out; we are responsible for our own learning and others' learning too. We have control.

These students' comments provide evidence that they have developed a voice and sense of ownership of their mathematical activity through the privileging of their cultural values. This aligns with the seminal work of Paulo Freire (2000), who argued the need for learners' values and beliefs to be affirmed.

FOREGROUNDING THE CULTURAL AND SOCIAL WORLDS OF PĀSIFIKA LEARNERS AS LEARNING TOOLS

Freire (2000) contended that praxis or informed action embedded in values is as important as situating the educational activity in the lived experience of the learners. His focus was placed on the need for education to transform oppressive structures by engaging people who have been marginalized and dehumanized and draw on what they already know: their cultural capital. Other researchers (e.g., Frankenstein, 2010), building on this epistemology through ethnomathematics, outline the need for educators to learn about how cultural practices—daily practice, language, forms of communication, power, and ideology—constitute people's views of mathematics and their ways of thinking mathematically.

In DMIC classrooms, the teachers use problematic tasks that are shaped around and within their Pāsifika learners' cultural and social worlds; that is, within the lived reality of the students, as a powerful feature that connects the students' home and school lives. For example, a teacher is explaining how he draws on common Pāsifika contexts as a vehicle to engage his students in the mathematics:

> Teacher: If you are cooking chop suey for all of your family then how many cans of corned beef will you need … you can also link the strands in … measurement, volume, capacity, how many plates will you need. And the students can relate to it. Cultural contexts might be looking at *tapa* patterns and the types of maths in the patterns, and they get to share back their experience and they can compare what they call it; because in Samoan it's *Siapo*, and in Tongan it's *Ngatu*, and in Fijian it's called *Masi*, because if you just said "tapa," lots of the kids wouldn't know what you were talking about. But they have these discussions and they all have a concept of the patterns, but they hadn't seen the maths in that. You can then pose a problem, and if they have a grasp, you have a foundation for the discussion, and we can then expand the maths ….

The teacher is stating an important aspect of socially response-able mathematics education. In previous mathematics classrooms (and current classrooms which are not within DMIC) in New Zealand, Aotearoa, the problem contexts the Pāsifika students are likely to be asked to do are commonly outside their everyday life and experiences. As a result, the context of the problems as well as the mathematics present a point of struggle for many Pāsifika students. In the DMIC setting, the known problem contexts allow a focus on the mathematics, and teachers are able to increase levels of challenge.

Increasing levels of challenge in the mathematics is important because many teachers hold deficit views, or low expectations, toward their Pāsifika or Māori students (Rubie-Davies, 2016). Turner, Rubie-Davies, and Webber (2015) illustrated that teacher expectations were highest for Asian and Pākehā mathematics students and lowest for Pāsifika and Māori in their New Zealand study. In our own work (Hunter & Hunter, in press), teachers starting to develop DMIC classrooms commonly express deficit views of their Pāsifika students' potential. Moreover, the students themselves state that teachers expect less of them than of their Pākehā and Asian counterparts. Teachers within the DMIC program are required to provide high-level challenging tasks in order to expose teachers to what Pāsifika students can do when given opportunities to work at a challenging mathematical level. One way to support their development of complex high-level problematic tasks within the known Pāsifika contexts of the students in DMIC is for the teachers to meet with parents and ask them to tell them where they see mathematics within their home contexts. Listed below are examples of contexts the parents outlined which would not be presented in mathematics classrooms as problem contexts unless the teachers wrote the problems themselves. They are only able to do so if they draw on community-held knowledge.

- Making clothes e.g., *puletasi*: It takes 2.5 m of fabric to make one dress/*puletasi*. How many dresses/*puletasi* can we make with __m of fabric?
- Family dinner: If one cup of rice feeds 4 people, how many cups of rice will we need to feed __ people?
- Drum lessons: patterns, counting beats, repetition.
- Preparing Sunday lunch/*umukai*: amount of food cooked in the *umukai*, number of plates distributed to neighbours, measurements, etc.
- Making traditional recipes (*takihi, ota*). Baking *nane.* 10 coconuts to 1 x 10 litre pot. 500 g *pia* = 10 big bowls x $10 = $100, or 20 small bowls x $5 = $100.
- How many coconuts (*fua niu pakupaku*) to make coconut cream (*gako niu*)?
- Making traditional necklaces (e.g., 5 *hihi* and 1 *puka* (black beads).
- Planting taro shoots from the baskets (50–70 shoots in each basket). Taro takes 7–9 months to mature. 1/2 hour to plant 1,000 shoots. 1 hour to plant 2,000–2,500. Land size: 1/2 hour = 1 acre; 1 hour = 2 acres. 1/2 hour ploughing takes approximately 40 man hours; 1 hour ploughing takes approximately 80 man hours.

An example of one problem presented to students follows.

> Malakai's mother is making *ta'ovala* for all the big boys in the Tongan group at school. Each boy needs $2\,^7/_9$ metres of mat and she has 100 metres. How many ta'ovala does she make?

The importance of using known contexts not only allows equitable access for Pāsifika students to mathematics within the school setting but also makes mathematics relevant in their lived social and cultural world:

> Teacher: It's not until they make those connections they realize the real life situations they are involved in, like a hair-cutting ceremony or making *ula lole* (lolly necklaces) that maths links in, so when we highlight that, they are like, "WOW, there is maths here." Until we started to bring these types of problems, they didn't make those links and they saw maths as something they did at school that was not relevant. This has been one of the most powerful parts of the maths project. The biggest concept for our kids to know is maths is everywhere … it is not just for maths time. That's the hook in … we practise that with our family … we practise this in our church and in our community,

> so when they make those links and can tie it into … well, maths is everywhere. Maths is part of my culture … the value of maths changes and the idea that maths is hard or alien or random changes as well.

The students also see how the problems connect their home and school mathematical life, but more significantly they recognize the ways in which the problem contexts also normalize them as citizens within their own culture:

> Student One: When the maths is about us, and our culture, it makes me feel normal, and my culture is normal.
>
> Student Two: Yeah like it is normal to be Samoan or Tongan.

In this way, the students are provided with opportunities to see mathematics in their "lived life," while at the same time their "lived life" is affirmed through ensuring that their cultural capital is reflected in both mathematics problems and activity.

AFFIRMING DIVERSE COMMUNICATION PATTERNS AND LANGUAGES

To engage and connect Pāsifika students to mathematics, teachers need to not only consider the values and cultural contexts of the students, but also take into account communication patterns and languages so that the mathematics within their known world becomes visible and affirmed within the lived realities of their lives. Many Pāsifika students within our DMIC classrooms speak English as an additional language and may speak their home island language but also other dialects. Within the classrooms, the language they prefer to speak or which will provide them with richer mathematical understandings is privileged, and this positions them as experts within their culture. In the DMIC classrooms, their expertise has them talk about and explain cultural concepts so that all of the learning community understand. For example, in one lesson a teacher asked a Samoan student to translate the word Sunday feast into Samoan and talk about its meaning to make the context relevant for all the students:

> Student: *Tona'i* pronounced as *Kona-a-ii*, which is just like a Sunday lunch.
>
> Teacher: Thank you, you can talk about this problem in Samoan or your home language so you can have a better understanding.

Through these actions, the teacher explicitly affirmed their right to use the language they considered appropriate and which provided them with a richer way to engage in mathematical talk. Giving learning space or options for students to use their own language in class to discuss mathematics showed the teacher's affirmation of students' cultural heritage, knowledge, and experiences.

Another teacher describes the importance of the use of different Pāsifika languages to engage and deepen understandings:

> Teacher: It's really powerful if they can use their own language because sometimes it might just be that they don't understand the question … even the ones that speak English. There might not be a word in English that represents what they are talking about, or they might be more confident speaking Samoan or Tongan, and then others can translate. Without that, like in the past those kids didn't have a voice and you would just think they couldn't do it. It really helps transfer the power as well, as I don't always understand and they have to translate for me, and their understanding really improves when they do this.

The students also realize the value of their home language being used in the classrooms, not only as a way to support others but also as a way to have a voice:

> Student: Sometimes it helps to explain things in Tongan because some of the Tongans in our class are new and their English isn't that good, but they can understand the maths in Tongan, which is cool because before you didn't really speak Tongan in class.

For Pāsifika learners, like many other minoritized learners, their languages and their cultures are what Kirch (1979) described as inseparable but that extend beyond the oral and written form to include other forms of non-verbal behaviour (Mauheni, 2017). Mauheni (2017), drawing on the work of Metge and Kinloch (1984), outlines the way in which Pāsifika peoples use body language more and verbalization less than their Pākeha (European) counterparts. She argues the need for teachers to be sensitive to the messages within gestures and body language as a valuable form of communication. In DMIC classrooms, teacher attention is focused not only on students communicating their mathematical reasoning, but also on how they participate as they engage in mathematical activity. The goal is for teachers to develop sensitivity toward the messages students are communicating through both spoken and unspoken responses. For example, in one DMIC classroom, a teacher joins a small group and she challenges one of the statements about what they have written on their sheet:

> Teacher: What do you think? Have you any questions?

Noting that all the students are looking uncomfortable and their eyes are on the floor she provides them with a structure to move forward beyond their discomfiture at being asked to challenge an Elder:

> Teacher: Okay, what about if you talk in pairs and you work on agreeing and disagreeing, but remember you must have a maths reason. It's okay to disagree, you know, as long as you do it in a polite way. You are not disagreeing with a person, with me—you are disagreeing with the maths, so right, have a go and start with either "I agree with that idea because" or "I disagree with that idea because …."

In response, the students raised their heads and began to tentatively construct questions and justification. In many other DMIC classroom situations, both the teachers and students are observed to respond to not only spoken responses but also those indicated through body language. For example, a student is explaining a solution strategy and he stops and notices the body language of a peer and suggests:

> Student: You look like you need to ask me a question.

In other DMIC classrooms, the teachers are seen to respond to body language which they recognize as resulting from something which causes a Pāsifika student discomfiture. For example, a Pāsifika student is likely not to continue asking questions of an explainer if he/she thinks that the person being questioned may not know the answer; rather than causing *Pāsifika-mā* (losing face), they will stop. The teacher's response on recognizing the discomfiture shown in the body language is illustrated in the next example. She steps in and redirects the questions to the wider group to allow space for the explainer to reconstruct their reasoning and be able to respond:

> Student: In that bit there, I am just not sure why your group did that with those numbers.

Silence follows and the student, observing the body language of the student they questioned, says, "Oh, it does not matter."

Reading the body language and rather than allowing a withdrawal of an important mathematical idea, the teacher steps in and says to the wider listening group, "Turn and talk to the person next to you about why the group did that with those numbers. In a moment, after you have talked, we will have someone in the explaining group tell us." The teacher's careful reading of the non-verbal communication and the action she takes encourages the students to both ask questions and respond. Through such actions, the students are supported in engaging in both inquiry and mathematical practices, but in ways which match their communication patterns.

CONCLUSION

In this chapter, we have provided persuasive evidence that when teachers draw on the social and cultural backgrounds, languages, values, and ways of being of their Pāsifika students, and use them within a strength-based approach to engage and connect them to mathematics, the students are provided with increased opportunities to engage in higher levels of cognitive reasoning. The approach we describe sits within a culturally responsive (Gaye, 2010) and socially response-able approach to mathematics education (Atweh & Ala'I, 2012). We have illustrated the need for significant reinvention of the practices enacted in Aotearoa, New Zealand, mathematics classrooms if Pāsifika learners are to access high-quality pedagogy with equitable and democratic outcomes. The results presented here of DMIC are transformative in that not only do the teachers bridge the gaps maintained by what is taken as normal or status quo, they also are able to bridge the same gaps for their students and empower them as mathematical learners in equitable ways.

We have shown that when teachers do seriously consider the cultural capital of the students, they carefully consider its influence on their teaching, and so they begin to balance more equitable outcomes for their learners in mathematics. In this chapter, we have shown how the lived reality of the Pāsifika students can guide the ethos of the mathematics classrooms within a set of cultural values which supports them to interact with each other in both culturally respectful and mathematical ways. Through specific teachers' actions, Pāsifika students are able to access challenging and problematic tasks while also being supported to engage with each other mathematically in culturally appropriate ways. Closely aligned are the way in which the views of the students shift toward seeing mathematics as part of the world they inhabit and something they need within that world. The teachers, too, are inducted into a world where they can develop a vision of culturally Pāsifika responsive actions which they can use in their mathematics teaching.

NOTE

[1] Pākeha is a term used to describe non-Māori or New Zealanders who are of European descent.

REFERENCES

Averill, R. (2012). Caring teaching practices in multiethnic mathematics classrooms: Attending to health and well-being. *Maths Education Research Journal, 24*, 105–128.

Atweh, B., & Ala'I, K. (2012). Socially response-able mathematics education: Lessons from three teachers. In J. Dindyal, L. P. Cheng, & S. F. Ng (Eds.), *Mathematics education: Expanding horizons. Proceedings of the 35th Annual Conference of the Mathematics Education Research Group of Australasia* (Vol. 1, pp. 99–105). Singapore: MERGA.

Ball, D., & Bass, H. (2003). Making mathematics reasonable in school. In J. Kilpatrick, J. Martin, & D. Schifter (Eds.), *A research companion to the principles and standards for school mathematics* (pp. 27–45). Reston, VA: National Council of Teachers of Mathematics.

Bartolomé, L. I. (1996). Beyond the methods fetish: Toward a humanizing pedagogy. In P. Leistyna, A. Woodrum, & S. A. Sherblom (Eds.), *Breaking free: The transformative power of critical pedagogy* (pp. 229–252). Cambridge, MA: Harvard Education Press.

Bills, T., & Hunter, R. (2015). The role of cultural capital in creating equity for Pāsifika learners in mathematics. In M. Marshman, V. Geiger, & A. Bennison (Eds.), *Mathematics education in the margins. Proceedings of the 38th Annual Conference of the Mathematics Education Research Group of Australasia* (pp. 109–116). Sunshine Coast: MERGA.

Boaler, J. (2002). *Experiencing school mathematics: Traditional and reform approaches to teaching and their impact on student learning.* Mahwah, NJ: Lawrence Erlbaum Association.

Bourdieu, P., & Passeron, J. C. (1973). Cultural reproduction and social reproduction. In R. K. Brown (Ed.), *Knowledge, education and cultural change*. London: Tavistock.

Civil, M., & Hunter, R. (2015). Participation of non-dominant students in argumentation in the mathematics classroom. *Intercultural Journal, 26*(4), 296–312.

Evans, B. (2011). *Teaching practices that support high achievement in the education of Pacific boys* (Unpublished master's thesis). Unitec Institute of Technology, Auckland, New Zealand.

Frankenstein, M. (2010). Critical mathematics education: An application of Paulo Freire's epistemology with new afterword. *Journal of Education, 165*(4), 315–339.

Freire, P. (2000). *Pedagogy of the oppressed.* New York, NY: Continuum.

Gaye, G. (2010). *Culturally responsive teaching: Theory, research and practice*. New York, NY: Teachers College Press.

Gorinski, R., & Fraser, C. (2006). *Literature review on the effective engagement of Pasifika parents and communities in education.* Wellington, New Zealand: Ministry of Education.

Graham, J., Meyer, L. H., McKenzie, L., McClure, J., & Weir, K. F. (2010). Maori and Pacific secondary student and parent perspectives on achievement, motivation and NCEA. *Assessment Matters, 2*, 132.

Hanushek, E., & Woessmann, L. (2012). Do better schools lead to more growth? Cognitive skills, economic outcomes, and causation. *Journal of Economic Growth, 17*, 267–321. doi:10.1007/s10887-012-9081-x

Hawk, K., Cowley, E. T., Hill, J., & Sutherland, S. (2002). The importance of the teacher/student relationship for Maori and Pasifika students. *Set: Research Information for Teachers, 3*, 44–49.

Hunter, R. (2008). Facilitating communities of mathematical inquiry. In M. Goos, R. Brown, & K. Makar (Eds.), *Navigating currents and charting directions: Proceedings of the 31st Annual Conference of the Mathematics Education Research Group of Australasia* (Vol. 1, pp. 31–39). Brisbane: MERGA.

Hunter, R. (2013). Developing equitable opportunities for Pasifika students to engage in mathematical practices. In A. M. Lindmeier & A. Heinze (Eds.), *Proceedings of the 37th International Group for the Psychology of Mathematics Education* (Vol. 3, pp. 397–406). Kiel: PME.

Hunter, R. K., & Anthony, G. (2011). Forging mathematical relationships in inquiry-based classrooms with Pasifika students. *Journal of Urban Mathematics Education, 4*(1), 98–119.

Hunter, R., & Hunter, J. (in press). Maintaining a cultural identity while constructing a mathematical disposition as a Pāsifika learner. In E. A. McKinley & L. Tuhiwai Smith (Eds.), *Handbook of Indigenous education*. Singapore: Springer.

Hunter, J., Hunter, R., Bills, T., Cheung, I., Hannant, B., Kritesh, K., & Lachaiya, R. (2016). Developing equity for Pāsifika learners within a New Zealand context: Attending to culture and values. *New Zealand Journal of Educational Studies, 51*(2), 197–209.

Kazemi, E., Franke, M., & Lampert, M. (2009). Developing pedagogies in teacher education to support novice teachers' ability to enact ambitious instruction. In R. Hunter, B. Bicknell, & T. Burgess (Eds.), *Crossing divides: Proceedings of the 32nd Annual Conference of the Mathematics Education Research group of Australasia* (pp. 11–29). Wellington, New Zealand: MERGA.

Kirch, M. S. (1979). Non-verbal communication across cultures. *The Modern Language Journal, 63*(6), 416–423.

Knight-de-Blois, L. (2015). *'O mātou'o le fatu'o le fa'amoemoe-fesili mai!: We are the heart of the matter-ask us!* (Unpublished master's thesis). Massey University, Palmerston North, New Zealand.

Lipene, T. (2010). *Education of secondary Samoan students in New Zealand: The road to success* (Unpublished doctoral dissertation). Victoria University, Wellington, New Zealand.

Manu'atu, L., & Kepa, M. (2002, October). *Towards reconstituting the notion of study clinics. A Kakai Tonga Tu'a community based educational project.* Invited presentation to the First National Pasifika Bilingual Education Conference, Alexandra Park Conference Centre, Auckland, New Zealand.

Mara, D., Foliaki, L., & Coxon, E. (1994). Pasifika Islands education. In E. Coxon, K. Jenkins, J. Marshall, & L. Massey (Eds.), *The politics of learning and teaching in Aotearoa New Zealand* (pp. 180–214). Palmerston North, New Zealand: Dunmore Printing Co.

Mara, D. (1999, April). *Why research? Why educational research for/by/with Pacific communities in Aotearoa-New Zealand.* Paper presented to the Pacific Island Educators' Conference, Auckland, New Zealand.

Mauheni, G. (2017). *Classroom teachers' perceptions on the role of non-verbal communication when teaching mathematics to Pasifika children* (Unpublished master's thesis). Auckland University of Technology, Auckland, New Zealand.

Mead, H. (2016). *Tikanga Maori.* Auckland, New Zealand: Huia Publishers.

Metge, J., & Kinloch, P. (1984). *Talking past each other: Problems of cross-cultural communication.* Wellington, New Zealand: Price Milburn.

Ministry of Education. (2004). *Book 3: Getting started.* Wellington, New Zealand: Learning Media.

Ministry of Education. (2015). *Pasifika education monitoring plan 2013.* Wellington, New Zealand: Ministry of Education.

New Zealand Qualifications Authority. (2013). *University entrance.* Retrieved August 20, 2013, from http://www.nzqa.govt.nz/qualifications-standards/awards/university-entrance/

Nakhid, C. (2002). *"Intercultural" perceptions and institutional responses: Explaining Pasifika students' achievement in New Zealand secondary school* (Unpublished doctoral dissertation). University of Auckland, New Zealand.
Passi, J. (2011). *To give back to the hand that fed me* (Unpublished master's thesis). University of Auckland, New Zealand.
RAND Mathematics Study Panel. (2003). *Mathematical proficiency for all students: Towards a strategic research and development program in mathematics education.* Santa Monica, CA: RAND.
Rubie-Davies, C. (2016). High and low expectation teachers: The importance of the teacher factor. In S. T. P. Babel (Ed.), *Interpersonal and intrapersonal expectancies* (pp. 145–157). Abington: Routledge.
Schuster, E. (2008). Who are the best teachers of Pasifika children? *Kairaranga, 9*(2), 10–13.
Tate, W. F. (2013). Race, retrenchment, and the reform of school mathematics. In E. Gutstein & B. Peterson (Eds.), *Rethinking mathematics: Teaching social justice by the numbers* (2nd ed., pp. 42–51). Milwaukee, WI: Rethinking Schools.
Theobald, C. (2013). *Changing our behaviours as teachers in order to meet the needs of our culturally diverse students* (Unpublished master's thesis). Massey University, Palmerston North, New Zealand.
Turner, H., Rubie-Davies, C. M., & Webber, M. (2015). Teacher Expectations, Ethnicity and the Achievement Gap. *New Zealand Journal of Educational Studies, 50*(1), 55–69. doi:10.1007/s40841-015-0004-1
Wood, T., Williams, G., & McNeal, B. (2006). Children's mathematical thinking in different classroom cultures. *Journal for Research in Mathematics Education, 37*(3), 222–255.

ROBYN JORGENSEN (ZEVENBERGEN)

8. CULTURALLY RESPONSIVE PEDAGOGY

Addressing "Shame" for Aboriginal Learners

INTRODUCTION

This chapter draws on data from a large national study into the successful practices of schools in remote and very remote locations across Australia. These schools have significant proportions of Indigenous learners. To be part of the study, the schools must have at least 80% of their students identifying as Indigenous. The study seeks to identify the practices that contribute to the success in mathematics/numeracy in these schools. While there are many practices that have emerged from the study, this chapter focuses on the adoption of culturally responsive pedagogies that are linked to the issue of "shame." Shame is a cultural construct that is remarkably different from Western conceptions and is deeply embedded in the cultural practices and dispositions of Indigenous Australians. Teachers in the study have developed a number of strategies to work with this notion and to bring about success for the learners. One of the key strategies creates opportunities for the students to be "proud" of their language and culture.

CULTURALLY RESPONSIVE PEDAGOGY IN REMOTE AUSTRALIA

The approaches discussed in this chapter are important as they address the potential disconnection between some cultural practices in remote Australia and the usual pedagogies used in modern schooling. For many of the students in remote areas of Australia, their culture, home practices, and dispositions toward learning (and school) are vastly different from mainstream, urban settings. Many of the teachers coming to these regions are new graduates who have little to no experience living and working in remote contexts, and many are working with Indigenous Australians for the first time in their lives. As such, there needs to be ways of working productively and respectfully with First Australians. This chapter describes some of the practices used by teachers and schools that have been successful in bringing about positive outcomes for both students and teachers.

To give some perspective on the Australian context, I provide a very brief overview of some of the issues and history associated with the colonization of Australia, and the concomitant marginalization of the First Peoples. At the time of British settlement of Australia, there were approximately 350 different language groups in Australia.

DOI: 10.1163/9789004415768_008

These language groups were distinct cultural groups who lived nomadic lives and relied on oral languages to communicate and to pass knowledge and histories across generations. The Indigenous peoples of Australia belong to the oldest living cultures in the world. Now, there are only about 100 languages remaining, and only 17 of these are actively spoken. Across Australia, there is considerable diversity in terms of the links with original culture and language.

Settlement has meant that, for many Indigenous groups, their culture has been eroded, and assimilationist policies of the past have meant that many languages and cultures have effectively been extinguished. In other regions where European contact has been more recent and less frequent, language and culture are still very strong. For example, in the Northern Territory, there are 32 Indigenous languages. In many cases, these languages are still used as the main language of communication, and many cultural practices still resonate strongly with those of the past. In some remote areas, it is not uncommon for First Australians to speak a number of Indigenous languages and English is not a shared language, other than to interact with non-Indigenous people. For many remote Indigenous students, coming to school means that they are in an environment where English is a foreign language to them. Alongside issues of language, issues around significant cultural differences also compound the ruptures between the culture of the school and the cultures and language of the home. While English may be the language of government agencies such as health, education, and welfare, it is not the language of the home.

In other parts of the country, particularly in coastal areas or rich grazing areas, there has been a longer history of contact and hence many of the First languages (and aspects of the cultures) have been lost, and creoles have developed. This is the case, for example, in the Kimberley region in the northwest of Australia. A shared language has been developed across the Kimberley—*Kimberley Kriol.* For students in this region, coming to school still demands that they learn in English. Similarly, in the northern parts of Queensland and in the Torres Strait, creoles and Aboriginal English have been adopted as the local languages. Many of the urban Indigenous Australians speak a register of English—Aboriginal English—that is shaped by regional dialects. Indigenous Australians living in the southern areas may speak "Koori English," while those in the North/East regions may speak "Murri English." While there is a greater proportion of Indigenous Australians living in urban and regional areas where English is a shared and common language, this chapter focuses on the experiences of remote Indigenous students.

Education for First Australians requires not only that teachers appreciate and incorporate differences in language as part of their teaching repertoire, but also that they be cognizant of cultural aspects of the local people and build pedagogies that allow entrance into schooling discourses and practices. A range of terms has been used to refer to such approaches, some of which are *culturally responsive pedagogy, two-way education,* and *cultural competence.* These terms, while different in many ways, all share a recognition of the need to acknowledge, respect, and build bridges between the cultures and languages that First Australian students bring to schools and to mathematics classrooms. For the remainder of this chapter, I will use the term "culturally responsive pedagogy" to signify the approaches discussed in this chapter that are being adopted across the schools in the study. While there are studies undertaken in other contexts, such as Alaska (Barnhardt, 2005; Lipka, Wong, & Andrew-Ihrke, 2013), where cultural activities are the centre of the pedagogy, these studies celebrate the overt activities of Indigenous people. It is noteworthy that in this study there was little of this activity being undertaken in the schools. There may be many valid reasons for this, but what is of interest are the more subtle ways in which teachers and schools were changing pedagogies to respect and include the backgrounds of the students.

While there is a widely held belief that teaching practices should support and respond to the social and cultural identities of the students within a given classroom or school, there is less consensus on how to practically realize a culturally responsive pedagogy (Saint-Hilaire, 2014). Taylor and Sobel (2011, p. 21) suggest that a culturally responsive pedagogy is one that embeds the "knowledge, skills, and rich cultural experience that students from diverse backgrounds bring to school" into the practices adopted by the teacher. Others advocate for an approach that considers "academic success, cultural competence, and socio-political consciousness" (Young, 2010, p. 248). The overall approach is in contrast to deficit models of learning where students are often seen to be missing key knowledges and skills. A culturally responsive pedagogy acknowledges the skills, dispositions, and funds of knowledge that learners bring to mathematics and incorporates these into models of learning, curriculum, pedagogy, and assessment. Teachers adopting a culturally responsive pedagogy were seen to incorporate the home and cultural lives of their students into the teaching of mathematics (Harding-DeKam, 2014). The overall approach has been seen to be a positive move away from deficit models of students and learning, as it creates new spaces for promoting equitable practices that enable students to engage with learning while having their cultures validated in and through classroom practice (Schmeichel, 2011).

In the highly regulated teaching accreditation process in Australia, particular teacher qualities are valued and rewarded. The National Professional Standards for Teachers (Australian Institute for Teaching School Leadership, 2011) have provided a national set of standards for Australian teachers. These are seen to benchmark quality teaching (Santoro, Reid, Mayer, & Singh, 2012). While these standards may be treated as "gold standards" in teaching, they may be limited in terms of teaching in non-mainstream settings. Some researchers caution that there is a need to challenge what is seen to be valued in contemporary education (Dinham, 2013), and to incorporate the views of families and communities as to what they see as quality learning environments for their children, school, and community (Lewthwaite et al., 2015). The views of families and communities can be quite different from the mandated qualities identified by the teacher standards frameworks.

A starting point for a culturally responsive pedagogy is to understand the learners and what they bring to the mathematics classroom. For many teachers, a culturally responsive pedagogy in mathematics might seek to incorporate Indigenous knowledges into the curriculum, draw on the language/s that the students bring to their mathematical learning, or draw on aspects of the students' culture as being integral to the teaching practices. However, a real issue for remote education is that many of the teachers coming into remote sites are often new graduates whose background in teaching is already very limited (Roberts, 2005). They are often primary (elementary) teachers whose pedagogical content knowledge and mathematical content knowledge are not strong. It is also the case that most of the beginning teachers are from urban settings, so coming into remote community life is often confronting and alien to them. There is a period where they are inducted into a very different way of life. This is further compounded by the fact that there are issues of teacher retention in remote areas (Storey, 1993), where the period of employment in some regions is measured in months rather than years (Hall, 2012). Collectively, these factors impact a teacher's capacity to understand the culture and learn the local languages of the students in his/her class and then build this into their repertoire of mathematics teaching practices.

There is some sense that a classroom-level culturally responsive pedagogy cannot, alone, redress the educational achievement gaps for Indigenous learners, as issues associated with more macro-level issues (such as attendance) also moderate learning outcomes (Kanu, 2007). But issues such as attendance (or non-attendance) are complex, and often intersect with the cultural norms and values of the families within a community. Perhaps the adoption of a pedagogy that embraces, celebrates, and legitimates cultural practices, including language, into the mathematics norms and social norms of a classroom may, in part, address issues of non-attendance.

While there are many approaches in mathematics education that adopt a culturally responsive pedagogy, or similar approaches known by another name, what is common among these approaches is the recognition that the funds of knowledge that Indigenous learners bring to school may be quite different from those valued and legitimated through the practices within mathematics. As such, a culturally responsive pedagogy is one that recognizes and values the funds of knowledge and dispositions of learners, and incorporates these into a legitimate part of the repertoires of classroom practice.

THE CONCEPT OF SHAME

As a non-Aboriginal woman, my Western concept of "shame" is quite different from, and is overshadowed by, the depth and all-pervasive notion of shame within Aboriginal cultures. I cannot pretend to be an expert in this concept, as it is outside my culture and, as such, my attempt to provide some definition to readers is limited by the lens that I bring to this concept. However, I do have some insights into the power of the concept and the impact that it has in the teaching context. I also draw on the limited literature on shame to try to represent the breadth of this concept, particularly for readers who have not worked in these contexts. What is clear from this literature is that there are marked differences in the use of the term between Western conceptualizations and Aboriginal conceptualizations (Sharifian, 2005) but with a consistent meaning around some form of humiliation. Harkins (1990) provided the following example of shame as being something much too important for Aboriginal people to single out as a distinctive or non-standard vocabulary item. Eades (1996) explains that shame "describes a situation in which a person has been singled out for any purpose, scolding or praise or simply attention, in which the person loses the security and anonymity provided by the group" (p. 293).

In working with teachers across Australia, I have found they are quick to point out how "shame" has a strong presence in classrooms. For example, asking students questions can create shame—if they don't know the answer, or even if they do know the answer, or just from being the focus of attention. There is an obvious embarrassment if they do not know the answer, but shame is also larger than simple renderings of embarrassment or shyness. There is another level of shame that is about social norms—a student may not be able to answer a question even if they know the answer if there is an older relation (sibling or cousin for example) in proximity who may not be able to answer the question. By answering the question, the student may cause shame to the older sibling; thus, shame helps to regulate social behaviours. As a non-Indigenous person, I do not feel it is appropriate to provide a definition, so have drawn on the experiences of an Aboriginal woman who works in teacher education. She told me:

> There is sometimes a reference to it [in the research] but differs based on who is writing, or speaking about it. Non-Indigenous folk tend to define it simply as a lack of confidence or shyness. An emotional response to a situation, or as per theory (Erikson), an assertion of independence. Something that can be "fixed" through encouragement and support. It's about an internal locus of control, by the individual, for an individual outcome.
>
> But from an Aboriginal perspective, it's evidence of a knowledge-moral system that is inter-twined and that encompasses both an internal and external locus of control. It's about both monitoring and managing behaviour based on knowledge of values/morals of the group. Behaviour, thoughts/decisions that were "silly" (minor) or "wrong" (major) in an instance but can be modified and, ultimately, prevented/avoided in the future. It can therefore be used retrospectively. It aligns more with concepts within social-cognition than emotional development. It's therefore about acquiring more knowledge and understanding (i.e., "learning"—not "fixing"). The maintenance of group roles and relationships and also values/morals is paramount. In a worst case scenario, this "shame" can be harsh, and also crippling. The person over-regulates his/her behaviour to avoid this. In a school setting, this can be misconstrued as being difficult, disinterested, anti-social or lacking ability. (K. Martin, personal communication, February 18, 2017)

Martin's comprehensive account highlights the complexity of shame as lived out in the life of classrooms in general, and mathematics classrooms in particular. As there are many language groups across Australia, there is not one Indigenous term for this construct, but the term "shame" is used across Indigenous communities in a relatively common manner. While I will use the term without quotation marks from this point on, the reader should be aware that I use the word to refer to this concept.

Teachers coming into remote communities become aware of shame very early in their induction into the school and community. Most schools have formal induction processes to help teachers transition into the communities. Some systems require new teachers to undertake a state-mandated induction into remote teaching, and then schools offer a further induction that is much more locally based and pertains to that immediate community. The concept of shame for students (and community) is an important aspect of that induction. There is an informal learning that also occurs once teachers enter the classrooms. It does not take long before a teacher will encounter shame at the level of classroom practice. As such, teachers learn quickly the omnipresence of this concept and how it is lived out in the classroom.

In terms of a culturally responsive pedagogy, I draw on the work of teachers in the Remote Numeracy project (Jorgensen, 2015) who recognize the impact of shame when working with Indigenous learners. The project only worked with teachers, leaders, and ancillary staff in the schools, as the project was seeking to describe the practices used in schools that were identified as bringing about success in mathematics/ numeracy learning. As such, the project is ethnographic in its orientation. Due to ethics considerations around the involvement of Indigenous people—students and community members—the project focused only on people associated directly with the schools, as it was deemed that these people would have an intricate and informed knowledge of the practices within the school. In the following sections of the chapter, I discuss how the teachers/educators articulate their understandings of shame and its impact on teaching, and how they have modified their teaching of mathematics to enable access to school mathematics in ways that are respectful of, and inclusive of, the backgrounds of the students.

THE REMOTE NUMERACY PROJECT

The Remote Numeracy Project has been funded by the Australian Research Council through its Discovery Research Program, and approved by the University of Canberra Ethics Committee. The project seeks to develop 32 case studies across Australia of remote and very remote schools that have been successful in mathematics. Success has been defined very broadly, but two key approaches have been employed in the project. In the first instance, performance in the National Assessment Plan of Literacy and Numeracy (NAPLAN)[1] is used to identify successful schools. This data, while problematic (as is the case with any national testing scheme), provides the only comparable data for schools across Australia. Schools are invited to participate in the study if they have been scoring better than similar schools for a number of years.

This suggests that something is working well in those schools. However, as is also recognized, NAPLAN has limitations. One of the main limitations is that the starting level of many remote students is very low, and hence performance on NAPLAN may not reflect the value-adding undertaken at schools. To address this, schools can be recommended to the study. Schools also provide the team with school-based data to provide further evidence of success in mathematics.

Once in the schools, the project is largely ethnographic in its approach. The duration in a site varies depending on the size of the school, the focus of the story, and the availability of staff. As in most remote sites, there are no relief teachers, so the organization within a school is constrained by a range of structural and contextual variables. The shortest period of time has been one day (for a one-teacher school) through to a full week in most of the larger schools. Students are not included in the study, as the focus of the study is on practice. There are many ethical and cultural issues associated with interviewing minors and Aboriginal children when working in remote contexts, so this was a consideration in the research design. The team interviews key staff, including members of the leadership team, teachers, and other staff in the school. Lesson observations are also undertaken, and a productive pedagogies profiling (Gore, Griffiths, & Ladwig, 2004) is undertaken as part of the lesson observations. Any relevant documents (such as policies and curriculum planning) are also collected. Photographs are taken of the classrooms, schools, and surrounding region for the production of the case study reports. All interview data is recorded, transcribed, and then coded in NVivo (QSR, 2004). NVivo uses a grounded theory approach to the analysis of data, so that it becomes possible to see emerging trends across the full data set. Given the size of this project, a software package of this type was needed to manage the data set. The goal of the project is to describe the practices of the schools that are enjoying success in the teaching of mathematics. It is intentionally not interventionist.

At the completion of each site visit, a case study report is generated that documents the practices at the school. These are approved by the nominated person at each school (usually the principal) and once approved, uploaded to the project website (Jorgensen, 2015) so as to celebrate the work of the schools and to enable sharing among schools.

At the time of writing this chapter, the project is now working on the large comprehensive data set to explore the trends across the full data set. There are many themes emerging across the data set. The notion of shame, as seen as a culturally responsive pedagogy, is one of these themes.

EDUCATORS' UNDERSTANDINGS OF SHAME

Across the project, there have been many references to the concept of shame for students as well as Indigenous adults working in the school. Teachers and administrators were able to articulate the impact that shame had on behaviours, particularly those behaviours that were valued within the school, and the need to develop ways to support Indigenous students and staff to be able to work successfully within the school.

At one of the schools, the principal worked with the Aboriginal Education Workers so as to try to give them confidence to engage in behaviours that were outside their cultural norms—such as presenting at assemblies. The school saw this behaviour as desirable, as it provided the students with role models—for example, Aunties presenting certificates to successful students, and Aunties achieving qualifications to teach, and then to organize significant events such as NAIDOC[2] week. Aunties are women in the communities who are held in respect by the community (as opposed to marking a biological connection to children). Having the Aunties play a significant role in the assemblies was seen to give ownership to the community members and validation by esteemed members of the community to the students' work. In the comment below, the principal outlines the roles that the women have taken in the school and the leadership and role modelling that this provides to the children. She equally recognizes that there is an element of shame for the women as they take on roles where they are profiled within the school, and that there is a need to support the women in not being "shamed" when they undertake these roles.

> Principal: Well, obviously all of them now have to have a Cert 3, so they're all trying to, well we've got one going for her Cert 3. There's one that's a little less confident, then the three other confident ones are going for their Cert 4's, so they're actually quite a lot higher. They've been doing that this year. We've got two qualified language teachers, so they're qualified to teach language on their own to students. They've got limited … Then we've got one trainee language teacher as well, so … then obviously the other responsibilities that they all have to do and put in is NAIDOC Day. We don't do NAIDOC day at all, the Aboriginal staff do it and organize it. For instance last year they organized for the whole school to go down to a local billabong and have NAIDOC activities down there for the day and have a big BBQ and got the community to come down. So they organize all of that kind of thing. They have things like assemblies and things like that, they have to give out rewards, certificates, they speak in front of the assembly. So we expect them to model behaviour that we expect from our children because you'll find in a lot of Aboriginal schools that the kids get shame, they call it shame, and they won't stand up in front of people. It's a huge issue, but with our students we are trying to instil in them there's no shame and it's about being brave and getting up there and giving it a go. If our Aboriginal staff are doing it, then there's no reason why they can't do it. And they respond really well to that.

Here the principal is highlighting a particular dimension of shame being lived out in public displays—the assembly—and how the school is working with local people to lead assemblies, particularly in NAIDOC week, which is a celebration of the local culture/s and histories. But, as will be discussed in subsequent sections, there are other ways in which shame is evident, whose implications teachers are working around in the mathematics classroom.

Differentiating

Within any classroom, there is significant diversity in terms of achievement, so it is important that teachers create environments where there is minimal opportunity for the students to feel shame. Shame, in this context, can come about through a range of situations, most obviously when a student does not know the work and there is a public display of knowing (or not knowing). But a more subtle observation of shame can occur when students seek to avoid a situation where another student may feel shame, or a student is the cause of shame to another person, particularly if that person is from their family group. One way that this can occur, as related to this research, is when an older member of the family group may not know the correct response. The younger sibling or cousin is likely to feel pressured into not displaying their knowledge in order to avoid shaming their relative.

A strategy used by many of the teachers in the study was to have similar activities that, from the outside, appear to be the same activity but are differentiated for individual students. In one school, the students were all given the same template for number work that included counting on, skip counting, counting backwards etc. The activity was based on a "magic number," and each student would be given their personal magic number by the teacher. The number varied according to the level of the students, so for all intents and purposes all students appeared to be doing the same activity, albeit working from a different starting point. Some students may be given a number between one and 20, while others might be working with three-digit numbers, and others worked on fractions or decimals, but the original template was the same for all students.

> Teacher: So you're scaffolding them in different ways but differentiating as well so all the kids, particularly Aboriginal kids, all think they're doing the same lesson, they're all doing addition. They don't realize that you're doing something different so there's no shame, this mob doing baby stuff and this mob doing big stuff. That sort of stuff, everybody's doing addition and then pulling the lesson back together and having a reflection on strategies, different strategies people use. There's not one method, you did it this way, and you did it this way. Oh, everyone's done it a different way. That sort of stuff but it's that pulling it back together as well.

The teacher is aware of how shame can impact on students if they are overtly differentiated. In this classroom, differentiation is achieved in a way that, for external appearances, looks like all students are doing the same work as they have the same stimulus page, but the differentiation occurs with the "magic number" given to each student.

As part of many schools' practice in this study, there is a strong emphasis on using student data to inform teaching. As in the previous example with "magic numbers," the teachers were aware of the learning needs of each student and so could target number activities to meet (and extend) their learning needs. The use of data was mandated in many schools, and teachers were required to have data displays in their classrooms. This was seen to be a way of opening up classrooms to students and families so that there was no "secret business" occurring in the schools. With this approach, there were public displays of achievements. Many of the teachers were aware that this meant that there was the possibility for students to be shamed by results. In one classroom, the teachers complied with the directive to display data, but subverted the process by making the displays very small so as to be insignificant among the other displays on the wall.

> Teacher: Yeah, I normally have them displayed up there. I, because of the shame aspect, I shrink them down a little bit. Because I've got such a variety in my class, I've got some who can't count to 10 and I've got some that are in the decimals and millions. So I shrink them to A5, put them up there, and encourage them not to pull them off.

While there is a sense that the public display of student achievement can be problematic in terms of shaming students, this practice is mandated in some states, so teachers are compelled to publicly display students' scores. In part, this has arisen from parental input, where parents have asked to be able to access their students' performance. For some Indigenous families, there is a sense that school offers secret knowledges that they know their children must acquire if they are to walk successfully in the Western world. "Secret business" is a term used by Indigenous people to refer to knowledge systems which they do not know about and which appear to be kept secret from them. Having access to public records is one way in which families are able to see how their children are progressing. As noted in this section, teachers are also aware of the potential impact on students, and so have developed practices through which they are compliant but which have the intent of minimizing the potential for negative shame for their students.

Attendance

While shame was seen as something to be largely avoided, there were cases where the concept was used in a very Western manner, and teachers referred to invoking the concept of shame in a way to elicit desired behaviours from students and families. Attendance, or more specifically non-attendance, is a very big issue in many remote communities. There is considerable pressure on schools to have high attendance rates, as funding is tied to attendance, and there is equally pressure on families to send their children to school. But as noted earlier, there are many reasons for non-attendance in remote communities. Recognizing the complexity of issues impacting attendance, but also the priority of attendance as a precursor to learning, schools rely on many strategies to build, foster, encourage, and demand attendance. In many communities, there are multiple strategies employed to both reward attendance and punish non-attendance. In some communities there is a no-school-no-pool policy, where students who don't attend school are unable to access the local pool (or other resources in the community, including the local store). Some communities will not serve students (or their families) if they come into the store during school hours.

Some communities are proactive in their public accountability, and the community store may have a large sign that indicates how many students are attending the school that day so that community members (including those who do not have children at the school) are aware of attendance rates. In one community, we stopped at the local store where the school attendance was displayed. A community member, curious as to why we were in community, then shared his pride in the growth in attendance and how the community was working with the school to ensure students went to school. Some schools have posters, graphs, and celebratory notices related to attendance. In other schools, there is healthy competition between sectors or grades, within the school, to perform well in attendance.

One principal explained that the school was keen for the community to know who was attending (and who was not) so that this might prompt families to encourage students to attend school. By linking with the store, which is often the hub of the community, the school would display attendance lists, with the community being well aware of the importance of attendance. The principal used the concept of shame to promote active attendance.

> Principal: So I would think that constant, every day having that list in the shop brings a bit of shame I would imagine as well, why aren't you sending your kids to school?

Other principals had developed other strategies to build awareness of attendance. At one school, there was a large Perspex animal in the assembly quadrangle. Students were given a coloured slip to represent their grade. Each student who attended school was given a slip to put into the statue. Once filled, all students were given a reward—often a swim at the local pool, which was undertaken during the allocated sports session. This very visual display alerted students (and families) as to how attendance was being recorded and potentially growing.

Homework and Links with School

In many of the communities, there are few, if any, of the numeracy resources and displays found in mainstream settings—signage, activities, birthdays, etc. There is a sense that the need for numeracy, as found in the school curriculum, is not the same in community as it is in urban settings. For many of the families, the levels of numeracy are quite low (by school standards). While there can often be considerable diversity among parents, it is not uncommon for the students to have stronger numeracy skills than their parents, so sending work home (either to celebrate achievements or as homework) can be a source of shame for the parents, who may not be able to support or understand their children's mathematics. Some schools noted this as an issue of which they need to be cognizant.

> Teacher: I've got a … Some of the women in the town, really big supporters of the, ah, school. And what we're doing. And again, a mixed bag. Some of them are well educated, some of them, you know, to be honest—the kids are going home and, um, are able to do … their numeracy skills are higher than their parents. And, ah, those parents see that as bit of a shame job but see that as great for their kids. But they feel shamed in that they can't help the kids in their learning. They can't do it at home with them and so that's where we have to be careful too 'cause, you know, we try to push, of course, support and do work at home for kids and we have to be careful—it's a fine line as to, you know, you don't want to turn some of our support off.

In some schools, a way around this aspect of shame was to have regular parent evenings or sessions during the day, where teachers or the Aboriginal Education Workers would hold workshops with families to show them what was happening in the mathematics classrooms, and model activities that their children were undertaking. The workshops were undertaken in a friendly and engaging manner, often with tea/coffee and snacks served so that families felt they were not being "taught" but were being included in a more sharing and collaborative environment, where they could see what their children were experiencing. Through these workshops, families not only became familiar with school activities, but were also learning mathematical concepts and were then able to work with their children, and thus avoid feelings of shame.

Behaviour

Behaviour can often be a source of challenge in the classroom. Like attendance, there are many complex reasons for behaviour, but many of the schools have been explicit about expectations of what is seen as appropriate school behaviour, and these behaviours are often explicitly taught. Disciplining students can be a source of great shame for the students, so teachers need to be aware of how to deal with behaviour issues in ways that are culturally sensitive, but that also develop the desired school behaviours. Most of the schools in the study had adopted a mantra of high expectations of students, staff, and community. Building appropriate school behaviours, therefore, needed to be cognizant of the students:

> Teacher: And once you've got their attention then you can talk to them. I don't shout at them very much at all but, um, when I have, it's pulled them into line and they start to listen and then you, you know … Again, I talk to them along the lines of, you know, your future, who you are as a person in the family, how

> you're treating them, how you're … Are you bringing shame on your family by your behaviour here?

In the preceding sections, I have sought to illustrate how teachers have interpreted and recognized issues of shame, and how these issues impact on students' learning. It appears that teachers and leaders are cognizant of the impact of shame in the teaching/learning process, and have sought ways to address the issue.

CLASSROOM PRACTICE AND SHAME

Since a priority for the mathematics classroom is for students to engage with and learn mathematics, teachers acknowledge the negative impact that shame can have on learning. While shame can have a positive influence, such as guiding social behaviour, there is a strong recognition that shame can also hinder learning. The teachers in many of the classrooms have adopted a number of strategies in their daily routines that support the students while keeping shame to a minimum or absent. By way of background, a cultural norm for many Indigenous students is to please or help the teacher, so when a question is posed the class has some responsibility to help the teacher. As a result, calling out is commonplace in many classrooms where the teacher may not have developed other strategies when working in a whole class context. It has been an observation in other work that a teacher will pose a question and students will offer many responses, thus making it almost impossible for the teacher to assess students' understanding. In this situation, it is impossible for a teacher to adopt any semblance of an assessment for learning approach in order to gain an appreciation of whether or not students understand a concept.

Assessing Learning

The calling-out strategy used by the students can resonate with issues of shame where students may not know an answer so will call out any answer. Or they may know the correct answer, but an older relation in the classroom does not know or has made an incorrect response. Then the student may feel an obligation not to shame their older relation by showing that the younger student knows more than the older student. This can curtail considerable activity (and learning/assessment) in this practice.

Across this study, many teachers have adopted practices that are individually based but allow the teacher to assess for learning without shaming students. A dominant strategy has been for the students to use a resource (such as a small whiteboard) on which they record their responses, followed by a class display to the teacher. For example, the teacher may pose an addition question, the students write their answers on their boards, and then display their answer to the teacher. The teacher is then able to check the students' responses without the students knowing who was correct (and who wasn't), and without shaming others or feeling shame themselves.

> Teacher: … feedback is really important. The whiteboard is just a very easy, very quick way that I can very quickly get feedback and I can see weaknesses, strengths or whatever it might be. The kids, if they are training well at it and enjoy doing it and enjoy the routine of it, it's a little more fun, more fun, it's probably not the best English, but than sitting down writing something in a book.

Teachers also used a resource that looked like a ladder in which students predominantly recorded counting activities, where they would count on or back using the grids created by the rungs in the ladder. In other classes, students had other resources (such as putting pegs on a board for counting or addition activities) that they used to show their responses to the teacher who could then quickly survey the classroom and assess students' understanding without creating any fear or shame when students displayed their answers. The process adopted by the teachers was to pose a question, then allow the students time to record their responses. The teacher would monitor their writing, and when it appeared that all the students had written their responses, he/she would ask the students to display their work. The work was visible to the teacher, but not to others, and most often the whiteboards were placed under the students' chins so that it was a very personal display of the response. The teacher was then able to scan the responses and offer a subtle gesture to the students if the response was correct and offer advice if incorrect, but also to assess whole class understandings. If it were the case that too many students had incorrect responses, the teacher was then able to move into further teaching around that concept, or if the opposite were the case, then the lesson could move forward.

Teachers also noted that students were often concerned about having permanent records of incorrect responses or poor writing, so the white boards also enabled the students to carry out mathematics work without leaving a permanent reminder of errors or work with which they felt shame. Building the students' skills and understandings of the expectations around the use of the boards was also a key strategy used by the teachers:

> Teacher: At the start of the year we tell them, or whenever we change our home corner, at the moment we have a hospital, we talk about the things they can play with. So they know that when it's maths time they can only write numbers on the board, that's all they can write, they can't draw faces and they can't do anything else but numbers. Sometimes I would go over there and be like, they're not numbers, rub them off, because they try and get away with it [*laughs*].

Across the schools in this study, the use of tools such as the small whiteboards was found to be a very effective strategy in mathematics for teachers to gain insights into students' mathematical understandings without the pressure, and potential shame, of public displays of knowledge. This resulted in better assessment for learning, and hence better feedback for students.

Language

As noted earlier in the chapter, language is also a large issue in remote Australia. For most of the Indigenous students in remote and very remote schools, the only time they speak English is when they are at school. However, families are aware that English is the language of power, and want their children to be speakers of English. Within this context, it is also important that Indigenous languages are not lost, so there is a process to protect, renew, or revive local languages. At schools where the home language is still strong, or the local creole is spoken, many of the schools have developed proactive processes to maintain or rebuild the local languages while scaffolding students into Standard Australian English.

At most of the remote schools, the teachers and administrators recognized the importance of valuing the home language/s, particularly in the early years, as well as the importance of having high expectations of the students to learn Standard Australian English (SAE) so that they would be able to communicate confidently and competently outside the school and community:

> Teacher: Valuing their home language, especially for the early childhood, that's what they bring into the classroom so we need to value that and push that as their knowledge. That's what they already bring so if we build on that and take the known into the unknown. The [tribal name] kids, they hardly ever hear Standard Australian English and we are the models for that so that needs to be explicitly taught, because from what I understand a lot of people in the days gone by didn't know they were speaking creole, they didn't know, so they would go down to Perth and speak creole and be like, why can't these people understand what I'm saying? They had no idea. So it needs to be explicitly taught. Creole is a separate language, we value this language. The curriculum is Standard Australian English, we model that, we need to teach that, that's the way of the world.

Teachers and administrators at some of the schools also recognized that students come to school strong in their home language, and that as part of valuing the home language this should be part of the practice within the school. At one school, the home language was integral to teaching in the early years (first 3 years of schooling), but as the students progressed through the middle years of primary school there was a transitioning process. By the time they were in upper primary school, they should be speaking (and fluent in) SAE. This process valued the home language while ensuring the students were empowered in and through SAE, and explicitly aware of the differences between the two codes.

> Teacher: Mainly lower primary, but the last principal, [name], thought, I think it was … I can't remember what grade … they shouldn't be speaking creole at all it should just be Standard Australian English. In the upper years they should be well aware of home language vs. school language.

One of the ways in which this valuing of the home language in the early years was achieved was through the employment of local families as teacher support staff. As noted earlier in the chapter, many teachers do not remain in remote communities for long periods of time, and so do not learn the home languages. The local community members are the "experts" in language and are employed to support teachers and students with code-switching and scaffolding the students.

One of the ways in which this is undertaken is through the production of resources. One of the resources is the production of big books where the AIO (Aboriginal Indigenous Officer) will take photographs of the students or the local environment (which is very familiar to the students) and then create texts for the books, highlighting the synergies in the home language and the mathematical language.

> Teacher: There used to be, I'm not sure if they have it any more, the L1 club where all the AIOs from the [area name] schools get together and they sit at (name of centre) adult ed centre in town and type up these books and come up with ideas. It's really generated from the AIOs and they bring the books back. And once they get confident and Miss [name] is normally here and she's a gem [name]. She'll take the kids out, take the photos and we have a slow day like we have now, [name] will go and publish another book. It's brilliant.

An example of one of the books created by the AIOs was on positional language, where children were placed in various situations (under the slide, on top of table, next to the table, etc.) and the description of the locations was recorded in both the home language and SAE. Children were also named in the books, so that there was an ownership of the story.

Another process is in the teaching and modelling of code-switching between the teacher and AIO. Here the teacher will say something, and then the AIO will translate and model the translation from SAE/Mathematics to the home language.

> Deputy Principal: If there's AIOs, try to get them to model code-switching. So if teachers are speaking in Australian School Language, then the AIOs will do it in home language or creole.

In one school, the deputy principal was an Aboriginal woman from the larger region, but not the community in which she was teaching. She reflected on the changes within schools as they have become more culturally responsive to the language needs of the students (and communities):

> Deputy Principal: I know, where I'm from, the traditional language or home language is what our parents would teach us. We'd be having Aboriginal English, our home language, and creole, but when we're at school, it's Standard Australian English. These days we have schools now doing the LOTE (Languages Other Than English) where we are doing the teaching of the home language, the traditional language whereas when we grew up that was our parents' responsibility.

Being culturally responsive was also about ensuring the students could be proud of, rather than shamed by, their language/s. Schools stressed that the languages *were* different rather than being valued differently. One of the schools was developing projects where the students wrote their stories in three languages—SAE, Aboriginal English, and then in the home language:

> Teacher: … so we're going to write a two-way book on it and, like, you know, just on the kids' experiences up here and send it down to them. You know, because our kids like mustering and shooting and swimming in swimming holes, and it's really cool, like, what they do. And so they write, in black we're going to write it in Standard Australian English, and in red underneath we do it in Aboriginal English. And then they're going to do it, I think they're doing it in blue, they're going to do their Aboriginal language down there, which is Noongar, and then we're going to, like, swap the books over …. Yeah, so that's how, yeah, it's just a, they need to be proud of, you know, their identities and their culture.

Language inclusion was a dominant approach in many of the schools in this study. Teachers were aware of the issues around language and instruction being in a foreign language for most remote Indigenous learners. The highly nuanced language of school mathematics posed a particular set of challenges, so creating resources and teaching approaches that build bridges between the home and school languages helped facilitate learning mathematics specifically while also celebrating the differences in the languages, rather than creating situations where students could feel shame about their home language as something inferior and/or that they were not able to engage successfully with school language.

CULTURALLY RESPONSIVE PEDAGOGY: SUBTLE DIFFERENCES MAKE THE DIFFERENCE

What I have sought to undertake in this chapter is an exploration of the subtle differences that teachers and leaders have been discussing in this larger study. When I commenced the study, I had anticipated that there would be many examples where there would have been learning activities that incorporated activities undertaken in community, not dissimilar to the ones used in the Alaskan work of Lipka and his colleagues (Lipka et al., 2013), or the work of language being incorporated into the classroom in significant ways (Lunney Borden, 2013). While initially struggling with the lack of these forms of cultural responsiveness, I came to realize that the schools, teachers, educators, and leaders were taking more subtle steps in how they undertook a culturally responsive pedagogy. This may, in part, be due to the uniqueness of the Australian context, where the duration of a teacher's time in a community is often small and there is limited time to develop deep understandings of the cultural practices or the language of the students and their families. At the same time, there are few members of the very remote communities who have achieved formal education qualifications, and hence they are limited in their capacity to influence educational change. However, what has been influential in this study has been the practices adopted by those working in the field, influenced by the practical wisdom of the community members, to embrace ways of working that address many aspects of culture. This chapter has provided some examples of how this is being achieved.

What was clear from many of the practices observed across the schools in the study was an awareness of the impact of shame on learning in mathematics. Teachers and others in the schools have shown how shame can impact learning, and the importance of respecting and validating the culture and language of the students. To address situations where shame hinders learning, teachers have devised strategies that help reduce (or eliminate) practices that can contribute to feelings of shame and a subsequent withdrawal from mathematics. What has been equally apparent in the study is the promotion of practices—such as the celebration of the language/s students bring to the mathematics classrooms—and the creation of an ethos within the schools that allows space for students to be proud of their heritage and language/s.

So, while it initially appeared that there was an absence of overt culturally responsive pedagogies, this study has shown that teachers can adopt practices that address the subtle, less obvious aspects of culture and create practices that are mindful and respectful of the funds of knowledge that students bring to the mathematics classroom. Being subtle can be empowering.

NOTES

1 NAPLAN is the national testing scheme that operates across all states and territories in Australia. School performance is made publicly available through the MySchool website.

2 NAIDOC Week is celebrated across Australia in July each year. The week is dedicated to the celebration of the history, culture, and achievements of Aboriginal and Torres Strait Islander people. NAIDOC stands for National Aborigines and Islanders Day of Observance Committee. Local celebrations are encouraged, with schools being a major player in the local celebrations.

REFERENCES

Australian Institute for Teaching School Leadership (AITSL). (2011). *Australian professional standards for teachers*. Canberra: Education Services Australia (MCEECDYA).

Barnhardt, R. (2005). Indigenous knowledge systems and Alaska Native ways of knowing. *Anthropology and Education Quarterly, 36*(1), 8–23.

Dinham, S. (2013). The quality teaching movement in Australia encounters difficult terrain: A personal perspective. *Australian Journal of Education, 57*(2), 91–106. doi:10.1177/0004944113485840

Eades, D. (1996). Communicataive strategies in Aboriginal English. In J. Maybin & N. Mercer (Eds.), *Using English from conversation to canon* (pp. 28–31). London: Routledge.

Gore, J. M., Griffiths, T., & Ladwig, J. G. (2004). Towards better teaching: Productive pedagogy as a framework for teacher education. *Teaching and Teacher Education, 20*, 375–387.

Hall, L. (2012). The "come and go" syndrome of teachers in remote Indigenous schools: Listening to the perspective of Indigenous teachers about what helps teachers to stay and what makes them go. *The Australian Journal of Indigenous Education, 41*(2), 187–195.

Harding-DeKam, J. L. (2014). Defining culturally responsive teaching: The case of mathematics. *Cogent Education, 1*(1). (Online open access). doi:10.1080/2331186X.2014.972676

Harkins, J. (1990). Shame and shyness in the Aboriginal classroom: A case for "practical semantics." *Australian Journal of Linguistics, 10*(2), 293–306. doi:10.1080/07268609008599445

Jorgensen, R. (2015). *Celebrating success: Numeracy in remote Indigenous contexts*. Retrieved from http://www.canberra.edu.au/research/faculty-research-centres/stem-education-research-centre/research-projects/remote-numeracy

Kanu, Y. (2007). Increasing school success among Aboriginal students: Culturally responsive curriculum or macrostructural variables affecting schooling? *Diaspora, Indigenous, and Minority Education, 1*(1), 21–41. doi:10.1080/15595690709336599

Lewthwaite, B. E., Osborne, B., Lloyd, N., Boon, H., Llewellyn, L., Webber, T., & Wills, J. (2015). Seeking a pedagogy of difference: What Aboriginal students and their parents in North Queensland say about teaching and their learning. *Australian Journal of Teacher Education, 40*(5). doi:10.14221/ajte.2015v40n5.8

Lipka, J., Wong, M., & Andrew-Ihrke, D. (2013). Alaska Native Indigenous knowledge: Opportunities for learning mathematics. *Mathematics Education Research Journal, 25*(1), 129–150. doi:10.1007/s13394-012-0061-4

Lunney Borden, L. (2013). What's the word for …? Is there a word for …? How understanding Mi'kmaw language can help support Mi'kmaw learners in mathematics. *Mathematics Education Research Journal, 25*(1), 5–22. doi:10.1007/s13394-012-0042-7

QSR. (2004). *NVivo 2.0* [Computer software]. Doncaster, Australia.

Roberts, P. (2005). *Staffing an empty schoolhouse: Attracting and retaining teachers in rural, remote and isolated communities*. Surry Hills: New South Wales Teachers Federation.

Saint-Hilaire, L. A. (2014). So, how do I teach them? Understanding multicultural education and culturally relevant pedagogy. *Reflective Practice, 15*(5), 592–602. doi:10.1080/14623943.2014.900026

Santoro, N., Reid, J.-A., Mayer, D., & Singh, M. (2012). Producing "quality" teachers: The role of teacher professional standards. *Asia-Pacific Journal of Teacher Education, 40*(1), 1–3. doi:10.1080/1359866X.2012.644508

Schmeichel, M. (2011). Good teaching? An examination of culturally relevant pedagogy as an equity practice. *Journal of Curriculum Studies, 44*(2), 211–231. doi:10.1080/00220272.2011.591434

Sharifian, F. (2005). Cultural conceptualisations in English words: A study of Aboriginal children in Perth. *Language and Education, 19*(1), 74–88. doi:10.1080/09500780508668805

Storey, V. J. (1993). Factors, issues and problems in the recruitment and retention of teachers for rural schools. *Journal for Research in Rural Education, 9*(3), 160–169.

Taylor, S. V., & Sobel, D. M. (2011). *Culturally responsive pedagogy: Teaching like our students' lives matter*. London: Emerald Group.

Young, E. (2010). Challenges to conceptualizing and actualizing culturally relevant pedagogy: How viable is the theory in classroom practice? *Journal of Teacher Education, 61*(3), 248–260.

A. J. (SANDY) DAWSON, BEATRIZ AGUON CAMACHO, TAUVELA FALE, DORA BORJA MIURA AND AMANDA FRITZLAN

9. MATHEMATICS AND CULTURE IN MICRONESIA

INTRODUCTION

The 22 participants in Project MACIMISE[1] are from the state of Hawai'i, and nine US-affiliated islands of the Pacific—the Republic of Palau; Guam; the Commonwealth of the Northern Mariana Islands (CNMI); the Federated States of Micronesia (FSM), which includes Yap, Chuuk, Pohnpei, and Kosrae; the Republic of the Marshall Islands (RMI); and American Samoa. Twelve of the participants work in their local departments/ministries of education, eight work for local colleges, one works for the state's justice department, and one is a state legislator. The latter two were with the local education authority when selected for the project. All participants are currently enrolled in doctoral (10 participants) or master's (12 participants) programs in the Department of Curriculum Studies at the University of Hawai'i-Mānoa. They call themselves the Macimisers. MACIMISE is a National Science Foundation[2] funded project that focuses on the languages and cultural practices of the islands noted above. The islands encompass at least 10 distinct language groupings that are spread across 1.5 million square miles of the Pacific Ocean yet have a total landmass of less than 1,000 square miles. MACIMISE is a five-year Project that began in September 2009 and concluded August 2014.

The first goal of the project is the development of elementary school mathematics curricula sensitive to local mathematical thought and experience. A necessary prerequisite for the achievement of this first goal is to recapture and honour the mathematics developed and practised in the various communities. This is the project's second goal. The recapture of local mathematical thought and its transformation into school curricula requires local experts in the teaching and learning of mathematics who are familiar with the mathematical practices in their own cultures and who, in the years ahead, will provide leadership in the development of curricula sensitive to local mathematical thought. In order to maximize (MACIMISE) its impact, the goals of the project are to

- develop and assess local mathematics curriculum units for Grades 1, 4, and 7;
- rediscover/uncover the Indigenous mathematics of 10 Pacific Island language communities; and
- build local capacity by offering advanced degree opportunities to local mathematics educators.

 | DOI: 10.1163/9789004415768_009

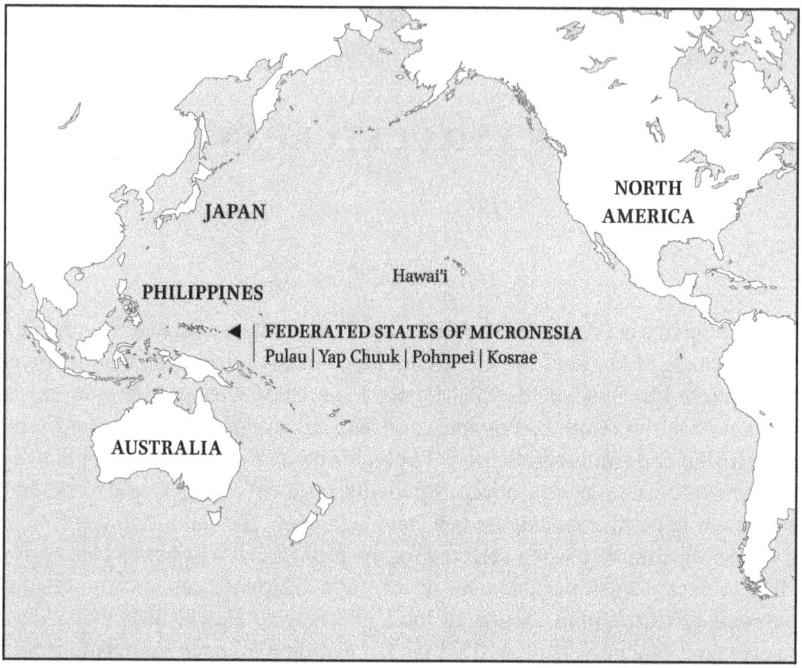

Figure 9.1. Map of the Pacific Region where Project MACIMISE is situated

The dynamics of how these goals are to be achieved consists of three phases: (1) to educate the local mathematics educators to be trained collectors of and (2) documenters of the mathematical practices found in their language and culture, and (3) to develop these local mathematical practices into curriculum units that will be implemented and assessed in schools on the noted island communities.

At the time of writing this chapter, two years of the five-year project are almost complete. During those two years, participants have been taking the courses required for their doctoral and master's programs, and working with local Elders (aka persons of knowledge) to uncover/recover the mathematics extant in various Indigenous practices.

MAKING SENSE

In an attempt to make sense of the undertaking that the Macimisers have embarked on, a book by Wheatley and Frieze (2011) entitled *Walk Out Walk On* is used to provide a conceptual framework for the work being done by the MACIMISE participants.

Among many ideas discussed by Wheatley and Frieze (2011) that seem germane to the Macimisers' endeavour is the concept of scaling across. *Scaling across* means "releasing knowledge, practices, and resources, and allowing them to circulate freely so that others may adapt them to their local community" (p. 32). This concept seems particularly appropriate to describe what is taking place in Project MACIMISE, and is offered in distinction to the idea of scaling up that dominates many Western educational reforms. According to Wheatley and Frieze, *scaling up* "creates a monoculture that relies on replication, standardization, promotion, and compliance" (p. 35). Scaling up is the opposite of what the Macimisers wish to accomplish. They are interested in preserving and promoting cultural practices and knowledge that is in danger of being lost. They are not interested in designing one approach, or one interpretation, or one body of knowledge to encompass the vibrant cultures found in each of their 10 jurisdictions: "But respect for the invisible forces of place—to which we could add its social and cultural heritage—is hardly conventional wisdom when it comes to taking things to scale" (p. 35).

The Macimisers would support Wheatley and Frieze's contention that *scaling up* does not fit well with the ways of working, respect for the wisdom of their Elders, and the power found in their cultural practices that has enabled them to thrive on their remote Pacific Islands. The "conventional" wisdom of Micronesian cultures is not the conventional wisdom of Western education systems.

This is not to imply that the practices found on one island are not similar to those found on other islands. There are some at least superficial similarities in language, in methods of building canoes and houses, forms of dance, and other Indigenous activities. But there is not the standardization and replication that Western educational reform movements seek. "Best practices" are not the Holy Grail for these Pacific Island nation/states that they are for many Western educators. As Wheatley and Frieze (2011) note, "Exchanging best practices often doesn't work. What does work is when teams from one organization travel to another and, through that experience, see themselves more clearly, strengthen their relationships, and renew their creativity" (p. 35).

In the case of Project MACIMISE, the "organizations" that travel are the Macimisers themselves. Table 9.1 provides a description of where the Macimisers are from, their local employers, the roles they currently play, and the degrees they are pursuing. The Indigenous languages of Guam and the CNMI are Chamorro and Carolinian. Hence, there are 10 Indigenous languages represented by the 10 island nation/states, with at least two participants from each language group.

Eleven of the Macimisers were involved for eight years prior to the inception of MACIMISE with previously NSF-funded projects in which PREL was the lead organization. It was the practice in those projects to bring the participants together at least once per year, rotating the meetings among the islands each year. This practice has continued with MACIMISE; the first meeting was held in July 2010 on Saipan in the CNMI, and the second occurs July 2011 on Pohnpei. A significant aspect of each meeting is the time given over to cultural sharing, thereby giving participants an opportunity to experience and learn about other cultural practices, to build strong interpersonal relationships bound together by a common interest in mathematics education, and to devote their considerable expertise and creativity to

Table 9.1. Macimisers by organization, role, and degree sought

Nation/State	*Employer*	*Role*	*Degree*
Palau	Palau Community College Ministry of Education	College Math Instructor School Vice Principal	Doctorate Master's
Yap	Department of Education State Legislature	Math resource teacher Senator	Master's Master's
Guam	Department of Education	Secondary math teacher	Doctorate
CNMI	Public School System	Secondary math teacher	Doctorate
Chuuk	College of Micronesia/Chuuk College of Micronesia/Chuuk Chuuk State School System Chuuk State Attorney General's Office	College Math Instructor College Math Instructor State Math Specialist Assistant to AG	Doctorate Master's Master's Master's
Pohnpei	College of Micronesia/Pohnpei College of Micronesia/Pohnpei Department of Education	College Math Instructor College Math Instructor State Math Specialist	Doctorate Master's Master's
Kosrae	College of Micronesia/Kosrae Department of Education Department of Education	College Math Instructor Elementary Principal Elementary Teacher	Doctorate Doctorate Master's
Marshall Islands	Ministry of Education Ministry of Education Ministry of Education	Elementary Teacher Elementary Teacher Elementary Teacher	Master's Master's Master's
American Samoa	University of Hawai'i American Samoa Community College	Math Teacher Educator College Math Instructor	Doctorate Doctorate
Hawai'i	University of Hawai'i	Math Teacher Educator	Doctorate

the development of teaching and learning situations based on the local children's experiences and practices.

Macimisers take the ideas they encounter during regional meetings and learn from them. They carry these learnings back to their home islands and develop them in their own unique ways. As Wheatley and Frieze (2011) note, "Scaling across invites communities to learn from one another and solve their own problems in their own particular way" (p. 36). They would also use the term *trans-local* to describe what has happened when Macimisers carry an idea from one place back to let it loose in their home environment, often fostering its growth into something quite different.

WHERE TO BEGIN

Working with peoples from 10 different language groups, 10 cultural settings spread over 1.5 million square miles of the Pacific Ocean, presents many challenges. Where to begin was certainly a question that faced the project staff. Wheatley and Frieze (2011, p. 91) talk about *starting anywhere, following it everywhere* which is more or less what happened when the project was launched.

Because of the Macimisers' varied backgrounds, it was decided their first course should deal with how to undertake anthropological research, preferably taught by an anthropologist who knew the Pacific region and at least some of the island cultures that were part of the project. A professor at the University of Guam filled that requirement and was recruited to offer the first course. This was the *start*, and the professor followed that *everywhere* by sharing with the Macimisers materials and experiences he had from his work on Yap and Chuuk. Based on his own experiences, he was well aware of the fact that "each place is an interdependent web of relationships, which is why you can start anywhere" (Wheatley & Frieze, 2011, p. 93). So he started with the weaving patterns found in materials produced by the peoples on the island of Fais, which is part of the state of Yap.

The course was taught via an Internet-based, synchronous interactive computer system called Elluminate.[3] Though the Macimisers used whatever computer they had available for this first course, during the first summer institute (July 2010), each was given a computer designed to run open-source software. The decision to use these machines was based on equity issues with a view that the computers should not be expensive to keep up to date, since many of the Macimisers live in subsistence cultures. Wheatley and Frieze (2011) quote Desmond Tutu talking about Ubuntu, which

> means [peoples who are] generous, hospitable, friendly, caring, and compassionate. They share what they have. It also means my humanity is caught up, is inextricably bound up, in theirs. We belong in a bundle of life. We say, "A person is a person through other people." It is not "I think therefore I am." It says rather "I am human because I belong." I participate, I share. (p. 82)

The operating system of the computers provided to the Macimisers was called Ubuntu—they are called Ubuntu computers—and during that first summer institute, the Macimisers bonded together and quickly displayed that "generous, hospitable, friendly, caring and compassionate" spirit so characteristic of Pacific Island peoples. As Wheatley and Frieze (2011) conclude, "This, too, is Ubuntu, an invitation to each other and every one of us to recognize that we are inextricably bound up together in a bundle of life" (p. 95).

Part of each day of all summer institutes is given over to cultural sharing, a time when the group of Macimisers from a particular language group presents an aspect of their culture to their colleagues. This sharing may take the form of a chant, a song, or a dance that is taught, and usually enthusiastically engaged in by fellow Macimisers; or the sharing may describe certain cultural ceremonies performed at, for example, weddings, funerals, or first birthday celebrations. Often gifts are presented: lava-lavas from the American Samoans, latte stones from the Chamorro group, leis from Yap, black pepper from Pohnpei, and so on. Prior to the launch of the project, many of the Macimisers never had the opportunity to interact or visit with other Pacific Islanders.

Figure 9.2. Macimisers, July 2010

Knowledge of each other's practices and beliefs was minimal, so it was with the Ubuntu spirit that the cultural sharing aspect of regional meetings was initiated. Figure 9.2 shows the assembled MACIMISE group just after their American Samoan colleagues had been presented to them and had taught them how to tie a lava-lava made from very colourful Samoan material. The picture was taken during the July 2010 summer institute held at the Saipan Southern High School in the CNMI.

After 10 days of intense work together, the Macimisers had bonded and undertook to provide support for each other academically, physically, and spiritually. All Macimisers were given unlimited Internet access. Perhaps not surprising was the fact that Facebook became a daily venue for the sharing of successes, challenges, and frustrations experienced by group members. This, too, is the Ubuntu spirit expressing itself in the workings of the project. So the project started *somewhere*, and in the intervening two years, the *Macimisers* were leading it *everywhere*.

FROM INTERVENTION TO FRIENDSHIP

The history of Micronesia is replete with examples of programs and projects being introduced by outside people who claim to be able to contribute to providing a solution for whatever challenge brought them to the islands. No longer do Pacific Islanders necessarily welcome such intrusions. When visiting the islands for the first time many years ago, one educational leader said to me, "So, you have a new program, and three years from now you will pack up and head back to the States, and what will be left to show from your program?" He was clearly sceptical, no doubt based on previous experiences, of the long-term benefits of the program being proposed. Such interventions are short-term strategies for current challenges, but it is clear that Pacific

Islanders realize that any longer-term impact requires the active engagement over time of significant people. Wheatley and Frieze (2011) posit the following relationship between the short-term interventionist programs, and the philosophical underpinnings of the outside educator's vision: "This empty-vessel paradigm of learning is one of the foundations of [the] interventionist mind-set. It posits that the trainer is full, the trainee is empty, and it is only a matter … of pouring knowledge from one into the other" (p. 172). They go on to argue that "the empty-vessel paradigm of learning is fundamentally founded on the inequality between the professional and the amateur, the expert and the ignoramus, the so-called developed and underdeveloped" (p. 77).

The genesis of the MACIMISE Project came from the Pacific Islanders themselves, and arose because of the intimate connections that project staff had developed with the islanders over the previous 10 years of working across the Pacific region. Back in 2000, one of the current Macimisers became part of a previous NSF-funded project (DELTA)[4] almost by accident. He happened to be standing in the College of Micronesia: FSM Yap campus offices one day when I was there seeking a college math person to be part of the year-old DELTA Project. Eight years later he had taken part in both the DELTA Project and Project MENTOR.[5] Over a meal on Yap when I was making one final trip across the region as Project MENTOR was wrapping up, his response to a question about "what should we do next if money could be found" elicited this thoughtful and heartfelt reply:

> For eight years we've studied Western mathematics, mainland mathematics, and teaching approaches that are suited to mainland children. Why don't we ever look at Yapese cultural practices and languages, examine them for the embedded mathematical knowledge, and then create lessons and units of work for our children that are based on things they've experienced? Not many Micronesian children have ever experienced snow, but they sure know about fishing in the vast Pacific Ocean.

His response was the germ of the idea that eventually blossomed into Project MACIMISE.

In hindsight, it is clear that Project DELTA was interventionist. It met and fulfilled a short-term need that was to assist Pacific Islanders in developing their capabilities to provide in-service education to their local teachers. Project MENTOR worked with these newly empowered in-service providers as they moved into the schools to support teachers new to the teaching of mathematics. The suggestion noted above offered by the Yapese mentor arose because of the relationship that had developed between the mentor and me. "Perfect friendship is a relationship between equals who offer good will to one another" (Wheatley & Frieze, 2011, p. 182). Because such a relationship existed between the *mentor* and myself, he felt comfortable in pointing out the weakness with Project DELTA and MENTOR, and in offering an alternative goal for the fledgling Project MACIMISE.

The Yapese mentor's response to my query was the conversation opener for the remainder of that 2008 trip across the region. On each of the islands visited (Palau, CNMI, Guam, the FSM, and the RMI), the reaction was the same: "Yes, why don't we look at our own cultural practices, our own languages?" (Some of the islands use as many as four or five different dialects.) They said that not only would their children relate more easily to examples and illustrations from things they know, but also such an approach would help to preserve some aspects of their cultures and languages that are being lost. One gentleman on Chuuk lamented the fact that at one time there were more than 50 ways of counting, and now there were only three and even those were in danger of being lost. Conversations with community Elders (persons of knowledge who may or may not be old in terms of age) reinforced the desire that traditional practices not be lost, that younger generations be introduced to and educated in the ways of living on isolated Pacific Islands that enabled survival, protected the land and waters surrounding their islands, and kept the people locally smart as well as world smart. MACIMISE is not an interventionist project. It is a project conceived of by Pacific Islanders to serve the perceived needs of Pacific Islanders. It supports Pacific Islanders in engaging in examination of their own practices, to recover any Indigenous knowledge on the verge of being lost, and to reframe the mathematical attributes of that knowledge into locally developed classroom experiences and investigations.

Project MACIMISE is an exemplar of "what becomes possible as we work together on what we care most about, freed from overbearing control, curious about one another's talents and knowledge, discovering the wisdom and wealth revealed when we turn to one another" (Wheatley & Frieze, 2011, p. 219). The Macimisers are fully engaged in bringing to life the possibility of locally developed curriculum units for mathematics that are based on local cultural practices.

FROM HERO TO HOST: CIRCLING AND THE ART OF HOSTING

Though gradual, the transition from Project DELTA (an interventionist endeavour) to Project MACIMISE (a collaborative, Pacific Islander-led enterprise) was envisioned from the outset of Project DELTA. An approach initiated at the first meeting of the Project DELTA participants in January 2000 was that of the *wisdom circle* and the use of a *talking stick*. Christine Baldwin (1998) has written extensively about the circle and its origins with Indigenous cultures. She says,

> A circle is not just a meeting with the chairs rearranged … A circle is a way of doing things differently than we have become accustomed to. It is a return to our original form of community. In circle, we rediscover an ancient process of consultation and communion that, for tens of thousands of years, has held the human community together and shaped its course. (p. 26)

It was with a good deal of trepidation that circling was introduced at the first meeting of the DELTA participants. At the conclusion of DELTA's three-week workshop, acknowledging first that I was leery of using the circle approach, participants were asked their reactions to the use of the circle. One man from American Samoa spoke up, a big smile on his face, saying,

> In my village at home, we use circle as our way of working when we have community meetings. We sit in circle, and each person in turn around the circle is given the opportunity to express his or her views of whatever we are discussing. Circle was not new to me. It is actually quite old and has been used for a long time in my village.

And so began the circling process for Projects DELTA, MENTOR, and MACIMISE that is still being used. During the MACIMISE summer institutes, each day will conclude with circle where Macimisers will "talk story" about the events of the day, or respond to some questions brought forward from the participants.

Circling facilitates the transference of responsibility from a leader to the participants. Leaders are supposed to have the answers. Wheatley and Frieze (2011) state, "When we believe this, we willingly give away our power. We wait for leaders to direct us, assuming they know what they are doing" (p. 207). But in circle, as the talking piece travels around, every voice is heard, "even those that for reasons of age or gender or politics have been silenced" (p. 103). The talking stick is a powerful tool. Initially, a sacred eagle feather (presented to the lead author by an Indigenous band in northern Canada) was used, but it deteriorated over time and was replaced by a hand-carved Haida Gwaii talking stick. The use of the stick changes the conversation, changes the participants' focus, and though resisted initially by some who are not used to its use, has a powerful impact on the group.

> But when people get engaged in the conversation, the process sells itself. It's easy to do. What I like about the talking piece is it slows the conversation down. You put the rules on the wall—speak with intention, listen with intention—it changes the dynamic. All of a sudden, you put the Blackberry down and pay attention. You don't talk over people. You learn to deeply respect other points of view. (p. 193)

Wheatley and Frieze (2011) compare the art of hosting to the computer code—Linux—that is the basis of open-source software. "Like Linux," they state, "the Art of Hosting operating system encourages experimentation and sharing worldwide" (p. 199). Working within circle engenders an environment of "hosting" rather than one led by "heroes." There is a code surrounding the use of the circle, just like the code that Wheatley and Frieze claim surrounds the Art of Hosting: "Its code is a set of principles and practices of how to host conversations that matter: setting intention, creating hospitable space, asking powerful questions, surfacing collective intelligence, trusting emergence, finding mates, harvesting learning, and moving into wise action" (p. 199).

The project presents the Macimisers with multiple opportunities: opportunities to interact with other Pacific Islanders, opportunities to interact with international experts from the field of ethnomathematics (e.g., Ubi D'Ambrosio, Bill Barton, Jerry Lipka, Arthur Powell, and others—all members of the project's advisory board), opportunities to share with colleagues on their home islands, and opportunities to learn from a broad range of people in a variety of environments including but not limited to community Elders. Furthermore, as one Macimiser said, "There is a giving of respect to those who went before—Elders, researchers, and scholars—and gaining respect for and confidence in the work we are doing." The vision of the project is to maximize the potential of Pacific Islanders to become educational leaders on their home islands, and to further the growth and understanding of Indigenous-based mathematics curriculum as found in the cultural practices of their island nations.

As this first part of the chapter is prepared, Project MACIMISE is just completing the second of a five-year journey. The *scaling across* the 10 island language groups is underway, the Macimisers are identifying *start*ing points and are following them *everywhere*, they have forged *friendships* that grow in depth and richness with each interchange that occurs, whether face-to-face or via the Internet, and when they come to sit together in a hospitable space for the first circle at the July 2011 Institute, they will ask each other powerful questions, allow their collective intelligence to emerge, harvest the learning that occurs, and move forward to wise action.

AN UPDATE ON PROJECT MACIMISE

The following update endeavours to describe some of the accomplishments as well as ongoing practices and projects that have manifested from the expansive and creative Project MACIMISE. Information has been drawn mainly from reports written for the project's primary funding body, the National Science Foundation in the United States, and from the comprehensive descriptions of developed units of mathematics education that are sensitive to local cultural practices and languages. These units are available online through the Pacific Resources for Education, PREL.

Communication with a few of the participants, the Macimisers, who continue to live Project MACIMISE through their professional education practices and community and collegial relationships gives a small glimpse into the complex web of knowledge and community that connects Indigenous mathematical thinking of nine Pacific Islands and Hawai'i to culturally responsive curricula through Project MACIMISE.

A major hardship and challenge during this project was the resignation of Dr. Sandy Dawson in November of 2014 from his role as principal investigator. He passed away in January of 2015. The rapidity of his illness did not allow for transition processes with Macimisers or other facilitators of the project (Rubinstein, 2015, p. 14). Co-Principal Investigator Dr. Donald Rubenstein (PREL; University of Guam) stepped in as the principal investigator. UHM, faculty Dr. Thomas Craven, became the new co-principal investigator (PREL, 2015). Together, participants, instructors, directors, and funders co-ordinated an effort to support remaining doctoral candidates, organize conference participation, and further develop, assess and publish Macimiser-made mathematics modules (Rubinstein, 2015, pp. 3–6).

The Macimisers' Ongoing Practices

At the time of this publication, 11 out of the 12 Macimisers pursuing master's degrees have been awarded degrees by the UHM. Seven out of 10 have received doctoral degrees from UHM, with the rest nearing completion. During their graduate programs, Macimisers gained further professional experience by attending or presenting their research at one national or international mathematics education conference each year (Rubinstein, 2015, p. 2). One example of this is the Pacific Education Conference, July 2015, in Majuro, Republic of the Marshall Islands, at which 20 Macimisers "gave presentations on their math units and other aspects of their professional work to wellattended sessions" (p. 5).

By offering opportunities to pursue graduate degrees in mathematics education, Project MACIMISE held the goal of increasing local capacity in the Pacific Islands, of supporting the development of educators in their own communities. This was achieved through the ingenious and creative use of "a hybrid on-site and synchronous online graduate program over an area noted for limited internet services" (Rubinstein, 2015, p. 7). The effect of the Macimisers' professional development is evidenced in the leadership roles that they have assumed. These include, among others, election or appointment as members of the state legislature; state director of education; associate director of the Department of Education; principal; vice-principal; school board chair; and college or university teacher training instructor (p. 14). "Macimisers are in positions to influence policy, teacher preparation and serve as master teachers in their schools" (p. 7).

Resources for Educators

"Each Macimiser developed a week-long culturally related unit to teach Western linked mathematical skills such as counting, fractions, pattern recognition, addition, subtraction, and geometry" (Rubinstein, 2015, p. 7). In total, 23 teaching modules were developed targeting Grades 1, 4, and 7. The primary goal for developing resources for mathematics education was to engage with local knowledge. Indigenous cultural practices formed the starting points to guide mathematical learning. Macimisers consulted with traditional knowledge holders and with elders on their home islands. Four areas of knowledge were addressed: 2D designs; 3D constructions; language; and social ceremonies. Fourteen of the modules developed were piloted by mathematics teachers in local classrooms, and six were chosen for further testing (p. 2). All of the resources developed through Project MACIMISE are free and available for educators online at http://macimise.prel.org/research-area/.

2D designs found in local Indigenous day-to-day and cultural practices have been linked to mathematical thinking. Traditional Samoan motifs, *mamanu*, have been connected to geometric concepts of line, angles, and image transformations in a unit that gives attention to performing *siapo* (painted cloth) and *pe'a* (tattoos for men) design (Fale & Tauiliili, 2017b). Another unit starts with weaving Marshallese *jaki* (mats), an activity traditionally performed by females. The process of teaching this practice allows for consideration of symmetry, patterns, and the use of traditional units of measurement: *jetan, makoj, nene*, and *ba* (Langidrik, 2017). As well, mathematical principles of counting, measuring, and patterning are found in a unit inspired by the practice of Marshallese necklace making (Term, 2017).

One of the modules developed that considers 3D constructions focuses on the ruins of the ancient stone city Nan Modal on the island of Pohnpei. The cultural significance of Nan Modal as well as mathematical concepts of measurement, ratio, scale, and maps shape this resource for educators (Etse, 2017). Another unit was developed from the Traditional Feast House, Nahs Pohnpei, to include identifying geometrical shapes, lines, angles; and measuring using Pohnpeian units of measurement *ngahp* and *tupwengen peh* (Liwy, 2017). Macimiser Hachibmai identified that the mathematical skills necessary for house building on the Outer Island of Yap included proportions, patterns, and algebraic thinking. Frequent storms and natural disasters make this knowledge essential to local residents (PREL, 2017a). Research by Miuty E. Nokar has resulted in the creation of a storybook, *The Little Crooked House* (Lee, 2015), a video, and nine supporting lessons that link building a traditional Chuukese house to hands-on experience of measuring and forming geometric shapes. The unit is a reflection of the traditional Chuukese gestural measurement system (PREL, 2017b).

Modules developed specifically connect Indigenous vocabulary for measurement and counting systems to mathematics education. Counting and culture are linked in a module developed around the *salulima*, a traditional Samoan hand-held broom. Using *tuaniu*, midrib fibres, from the coconut leaf, this lesson introduces place value through using bundles of 10 for construction (Fale & Tauiliili, 2017a). All of the modules stemming from local Indigenous knowledge are further connected to local communities by including stories and traditional ceremonies or events.

Significance of Research in the Field of Ethnomathematics

Ethnomathematics has been defined as the practice of mathematical thinking particular to a cultural group (D'Ambrosia, 1985). It can be interpreted further as the study of "the mathematics extant in the customary practices of Indigenous and other minority or identifiable cultural groups" (Rubenstein, 2015, p. 12). This field has a history of mainly non-Indigenous, Western researchers, particularly in the Pacific Islands (p. 12). "MACIMISE demonstrated that Indigenous teachers can be researchers of their own cultures and develop culturally based curriculum units" (p. 7).

> This project has opened my mind and my heart to gaining a deeper appreciation for my culture and the intelligence behind my ancient ancestors. Often we

> think of our ancestors being more primitive and less intelligent than we are, however I now know this to be false as evidenced in the technology they used to build our latte homes, to navigate the seas, and to find food for themselves. (D. Muira, personal communication, June 1, 2017)

UHM courses delivered qualitative and ethnographic methodologies as well as current theories of ethnomathematics to the Macimisers (Rubenstein, 2015, p. 12).

The Macimisers consisted of a culturally diverse group of participants from 10 widely spread out US Affiliated Pacific Islands. Significantly, the populations and political and economic histories of Indigenous groups varied widely throughout the islands where research took place. However, all of the participants' research was connected through "the concept of the *social positioning* of Indigenous knowledge" (Rubenstein, 2015, p. 12, emphasis in original). Bringing the mathematical thinking found in the cultural practices of local Indigenous groups to the theory and practice of mathematical education in formal educational contexts necessarily shifted the social connections of the knowledge. "These units use many different aspects of the culture to teach math and illustrate the rich potential of using culture in schooling to more effectively teach Western math while maintaining and promoting cultural knowledge" (p. 7).

Further Findings and Directions

Teacher and student responses to the piloted culturally based mathematics education modules were positive. It was found that students learned the mathematical skills embedded in the units for Grades 1, 4, and 7 and that culturally familiar activities increased interest. Teachers "recognized the value of using cultural references to teach math as well as other subject" (Rubenstein, 2015, p. 3). PREL dissemination of the modules developed during Project MACIMISE provides incentive for mathematics educators to make connections to cultural knowledge in their classrooms. Improving the performance of Indigenous students in standard related mathematical skills has been identified as a potential outcome (p. 13).

During Project MACIMISE, "a strong sense of cohort and shared purpose among the Macimisers, instructors, and project staff" was developed (Rubenstein, 2015, p. 2). The practice of community connection and cultural grounding that participants experienced during Project MACIMISE continue to inform and enrich their professional practices as well as their interactions with colleagues and students.

> My experiences with Macimise and the Macimisers have made me appreciate how important it is when people make connections, not just in mathematics, but in everything. We make sense of information, we make connections, then we have a deeper understanding and knowledge base. Macimise made me appreciate and truly value those connections with mathematics, culture, and community. (B. Camacho, personal communication, April 30, 2017)

NOTES

[1] MACIMISE, pronounced as if spelled *maximize*, stands for *M*athematics *A*nd *C*ulture *I*n *M*icronesia: *I*ntegrating *S*ociety *E*xperiences.

[2] The project is supported by a National Science Foundation grant (0918309). This material in this chapter is based on work supported by that grant. The content does not necessarily reflect the views of the NSF or any other agency of the US government. The Project is a collaborative effort between Pacific Resources for Education and Learning (PREL) and the University of Hawai'i-Mānoa (UHM) with PREL as the lead organization.

[3] Elluminate (rebranded as Blackboard Collaborate) is Web-based, interactive instructional software. The software includes several visual tools, including a whiteboard, application sharing, file transfer, and Web tour. The software also includes a record feature that allows the moderator to record the class for others to watch later as well as a graphing tool, breakout rooms for group work, and a timer. The whiteboard supports the uploading of presentations for viewing during the class or meeting.

[4] Developing Effective Leadership Team Activities, NSF grant (9819630) 1999–2002.

[5] Mathematics Education for Novice Teachers: Opportunities for Reflection, NSF grant (0138916), 2002–2008.

REFERENCES

Baldwin, C. (1998). *Calling the circle: The first and future culture*. New York, NY: Bantam Books.

D'Ambrosio, U. (1985). Ethnomathematics and its place in the history and pedagogy of mathematics. *For the Learning of Mathematics, 5*(1), 44–48.

Etse, S. (2016). *Nan modal: Culture and mathematics*. Retrieved from http://macimise.prel.org

Fale, T., & Tauiliili, P. (2017a). *Counting and the salulima*. Retrieved from http://macimise.prel.org

Fale, T., & Tauiliili, P. (2017b). *Samoan design and geometry*. Retrieved from http://macimise.prel.org

Hachibmai, C. (2017). *Local house building*. Retrieved from http://macimise.prel.org

Langridrik, A. (2017). *The weaving mat*. Retrieved from http://macimise.prel.org

Lee, L. (2015). *The little crooked house*. Paliker: Island Research and Education Initiative.

Liwy, T. (2017). *Nahs: Pohnpei feast house*. Retrieved from http://macimise.prel.org

PREL. (2015). *Mathematics and culture in Micronesia: Integrating societal experiences*. Retrieved from http://prel.org/programs/macimise/

PREL. (2017a). *Local house building*. Retrieved from http://macimise.prel.org

PREL. (2017b). *Nine lessons from: The little crooked house*. Retrieved from http://macimise.prel.org

Rubinstein, D. (2015). *Mathematics and culture in Micronesia: Integrating societal experiences* (Preview of award 0918309 – final project report). Retrieved from https://reporting.research.gov/rppr%C2%ADweb/rppr?execution=e1s60

Term, S. (2017). *Patterns, counting, measures*. Retrieved from http://macimise.prel.org

Tutu, D. (1999). *No future without forgiveness*. London: Rider.

Wheatley, M., & Frieze, D. (2011). *Walk out walk on: A Learning journey into communities daring to live the future now*. San Francisco, CA: Berrett-Koehler Publisher.

JULIE KAOMEA

10. HAWAIIAN MATH FOR A SUSTAINABLE FUTURE

Envisioning a Conceptual Framework for Rigorous and Culturally Relevant 21st-Century Elementary Mathematics Education

INTRODUCTION

> I dare you to find an 11-year-old in America who wants to be an engineer today. (Friedman, 2005a, p. 30)

I open this chapter with a quotation from the Pulitzer Prize-winning author of *The World Is Flat: A Brief History of the Twenty-First Century* (Friedman, 2005b). Commenting that few young people in the United States today seem interested in the fields of science and engineering, Thomas Friedman challenged readers of his national best-seller to locate American youth interested in pursuing an engineering career.

As Friedman (2005b) pointed out, the number of jobs requiring science and engineering skills in the United States labour force is growing by almost 5% per year, while the number of Americans prepared for these careers is dwindling rapidly. More than half of America's scientists and engineers are 40 years old or older, and the average age is steadily rising (National Science Board, 2010).

According to the National Science Board (2010), only 5% of all bachelor's degrees earned in the United States today are in engineering. This is in contrast to 20% in Asia and 33% in China. Science and engineering degrees now represent more than half of all bachelor's degrees earned in Japan (63%), China (53%), and Singapore (51%). By contrast, the percentage of bachelor's degrees in science and engineering in the United States remains roughly at 32% (National Science Board, 2010).

After an extensive review of trends in the United States and abroad, a committee of the National Academies has expressed concern that "the [United States'] scientific and technological building blocks … are eroding at a time when many other nations are gathering strength" (National Academy of Sciences, National Academy of Engineering, & Institute of Medicine, 2007). This is particularly alarming in light of our rapidly changing, and increasingly technological, global society (Seeley, 2009).

As Friedman's book *Hot, Flat, and Crowded* (2008) pointed out, we are facing an era focused on energy and climate, making it even more urgent to prepare students so that they can use mathematics and science to tackle these and other critical problems of the future. Tom Luce, CEO of the National Math and Science Initiative, asserted:

> This new generation will have the opportunity to solve many global issues: healthcare, energy … and the global food crisis to name just a few. Given this, the missing ingredient is a better knowledge of math and science and its power to provide solutions to these problems. Technology can and will change the world. For this younger generation to be the force for good they want to be, they need to understand that the new literacy of the 21st century includes math and science. (National Math and Science Initiative, 2010)

Luce's assertion is especially true for Native Hawaiians and other Pacific Islanders who are disproportionately under-represented in the fields of math and science (Native Hawaiian Science & Engineering Mentorship Program, 2009) and whose island communities are particularly vulnerable to these impending global crises. For instance, scientists predict that, within this century, Hawai'i and other island communities will experience a three-foot rise in sea level that will bring saltwater seepage into freshwater tables and will lead to the loss of beaches and coastal regions, which serve as the home to already threatened native plants and animals. It is predicted that the increasing acidity of the oceans will have a devastating effect on Hawai'i's coral reefs and nearshore fish, and changes in rainfall patterns will bring not only more droughts but also more hurricanes and heavy rains (Fletcher, 2009). While these climatic changes will impact all of Hawai'i, their effects will be felt most dramatically by Native Hawaiian communities who depend on native plants and animals, and reef and nearshore sea life, for their livelihood and sustenance, and for whom the beaches and low-lying coastal regions hold important cultural, economic, and spiritual significance. To ensure a sustainable future for our Hawaiian community and our Hawaiian culture, we will need Hawaiian engineers and scientists who are prepared to deal with these and other impending crises.

Fortunately, thanks to the foresight of university programs geared toward recruiting and mentoring Native Hawaiian students in science, technology, engineering, and mathematics (STEM) careers, the number of Hawaiian university students studying math, science, and engineering is on the rise. Since the inception of the cohort-based Native Hawaiian Science & Engineering Mentorship Program (NHSEMP) in 2001, the enrolment of Native Hawaiian and Pacific Islander students in the College of Engineering at the University of Hawai'i–Mānoa has steadily increased. Engineering is now the third most popular undergraduate major among the university's Native Hawaiian students, with 112 Hawaiian undergraduates enrolled in the College of Engineering in 2008 (Freitas & Balutski, 2009). Moreover, NHSEMP students are completing their engineering degrees in record numbers, with a program retention rate of 70%, compared with a 56% undergraduate engineering retention rate nationwide (Fortenberry, Sullivan, Jordan, & Knight, 2007) and a 27% retention rate for Indigenous Americans (Kaakua, 2010).

In collaboration with the Kamakakūokalani Center for Hawaiian Studies, NHSEMP integrates studies in renewable energy, smart development, and sustainable environmental practices with related service-learning projects in *loko i'a* (fishpond) restoration, *heiau* (temple) restoration, and *kalo* (taro) farming. Through participation in this culturally relevant mentorship program, NHSEMP students come to realize how important their work as Hawaiian engineers is to the future of our island community (Kaakua, 2010).

With mentorship programs to recruit and retain Hawaiian students in STEM fields now in place at the university level, the onus is on those of us who work with Native Hawaiian students at the elementary school grades to ensure that our community's future leaders emerge from grade school with both the preparation to pursue further studies in math and science and the inclination to utilize math and science to become a "force of good" (National Math and Science Initiative, 2010) in the world and our Hawaiian community.

Elementary educators play a critical, foundational role in the development of students' skill and interest in mathematics. This is particularly true for Native Hawaiian and other students of colour, for whom math is conventionally taught in ways that are unconnected to their lives and experiences. William Tate (2006), Gloria Ladson-Billings (1995), and Carter Woodson (1993), among others, have argued that far too often conventional math pedagogy persuades students from historically marginalized communities to consider school mathematics as a subject divorced from their personal and cultural experiences, and consequently leads these students to perform poorly in math classes and internalize negative self-images about their knowledge and ability in mathematics. Culturally relevant mathematics pedagogy, they contended, which connects school mathematics to the experiences of culturally diverse students, is thus essential to creating equitable conditions in mathematics education.

Culturally relevant mathematics pedagogy for Native Hawaiian students has been the thrust of my work in elementary teacher education for many years. However, over time I have come to the realization that developing a Hawaiian mathematics curriculum that is simultaneously culturally relevant, temporally relevant (i.e., applicable to students' present-day experiences), and academically rigorous is a formidable undertaking.

Drawing on my 20 years of experience in elementary mathematics education, along with recent research in ethnomathematics (Anderson, 2006; Ortiz-Franco, 2006), social justice mathematics (Gutstein & Peterson, 2006), and critical mathematical literacy (Frankenstein, 2006), I offer below a framework that elementary teachers might consider when conceptualizing and implementing culturally relevant 21st-century mathematics pedagogy for Native Hawaiian students.

EPISTEMOLOGICAL UNDERPINNING

Different cultures have distinct ideas about how to best prepare their children for the future. In *Native Land and Foreign Desires*, Lilikalā Kame'eleihiwa (1992) discussed the traditional Hawaiian worldview regarding the relative significance of the past, the present, and the future. Kame'eleihiwa (1992) explained:

> In Hawaiian, the past is referred to as Ka wā mamua, or the "time in front or before." Whereas the future … is Ka wā mahope, or "the time which comes after or behind." It is as if the Hawaiian stands firmly in the present, with his back to the future, and his eyes fixed upon the past, seeking historical answers for present-day dilemmas. (p. 22)

Kame'eleihiwa elaborated that this orientation is an "eminently practical" one, for while the future is always unknown, our Hawaiian past is "rich in glory and knowledge" (p. 22).

Following Kame'eleihiwa, I contend that to prepare young Hawaiian leaders for an unknown future, it is essential that elementary educators go beyond conventional methods for teaching mathematics. We should go beyond teaching basic skills, beyond requiring students to know how to perform procedures, and beyond offering recipes for solving problems that look alike. If we are to prepare our students to seek sustainable, culturally appropriate solutions to global and societal problems that no one yet knows how to solve—or that no one has ever seen before—they will need a firm grounding in the accumulated wisdom of our ancestors, coupled with excellent creative thinking and mathematical problem-solving skills.

In the sections that follow, I outline a general framework for conceptualizing Hawaiian mathematics curricula that can inspire Hawaiian students to appreciate the mathematical accomplishments of our rich cultural past, while simultaneously preparing them to meet the mathematical challenges of the present and future. Drawing from Kame'eleihiwa (1992), I suggest the need for Hawaiian math lessons in three temporal categories of instruction: *ka wā mamua* (the past), *kēia wā* (the present), and *ka wā mahope* (the future). Arguing that lessons in each of these three categories are necessary for a rigorous and comprehensive Hawaiian math curriculum, in the three subsequent sections I pair theory in ethnomathematics, social justice mathematics, and critical mathematical literacy with preliminary suggestions for corresponding lessons in past, present, and future Hawaiian mathematics.

KA WĀ MAMUA: LEARNING FROM HAWAIIANS' RICH MATHEMATICAL PAST

In an impassioned speech in the critically acclaimed film *Stand and Deliver* (Musca & Menendez, 1988), teacher Jaime Escalante (played by actor Edward James Olmos) inspires and challenges his Mexican American mathematics students by exclaiming, "You burros have math in your blood!" Escalante explains that neither the Greeks nor the Romans were capable of using the concept of zero. It was the ancestors of Mexicans, the Mayans, who first contemplated the zero or the absence of value.

In an article with a similar title, mathematics professor Ortiz-Franco (2006) simultaneously outlined the long and distinguished Mayan and Aztec heritage in math and science (which is largely unknown or untaught in U.S. classrooms), and decried the comparatively poor performance of Mexican American students in U.S.

math classrooms and the relatively miniscule number of Mexican Americans who are motivated to pursue mathematics-based careers in the United States today.

Arguing that the impressive history of mathematical achievements of non-European civilizations has been almost entirely ignored by U.S. schools, Ortiz-Franco (2006), Anderson (2006), and other ethnomathematicians have asserted that it is time to "shatter the myth that mathematics was or is a 'white man's thing'" (Anderson, 2006, p. 44). Anderson (2006) contended that it is important for students to know that "Europe is not now, nor was it ever, the 'civilizing center' of the world surrounded by wilderness and chaos" (p. 44). Since the beginning of time, people around the world developed mathematical systems for quantifying and understanding the phenomena in their lives. Many early civilizations, including Native Hawaiians, achieved a high level of mathematical sophistication that allowed them to classify, order, count, measure, and otherwise mathematize their environment. However, the mathematical achievements and contributions of non-European cultures are largely overlooked in U.S. classrooms today.

Following Anderson and Ortiz-Franco, I believe that Hawaiian students could likewise benefit from explicit lessons in the rich mathematical heritage of our Hawaiian ancestors. Prior to European contact, Hawaiians had a thriving, self-sufficient, and self-sustaining civilization. Recent estimates suggest that our ancestors were able to support upward of 800,000 people on these small islands (Stannard, 1989). This was accomplished, in part, through sophisticated forms of community-based resource management developed from close observations of the moon and season cycles.

By noting recurring cycles that happened each season, year after year, Hawaiians were able to develop planting and fishing patterns in alignment with lunar patterns to produce optimum yields (Hui Malama o Mo'omomi, 2009). Our ancestors observed that each day, month, and season had its individual characteristics that would yield favourable or unfavourable results in fishing and farming, and thus they did all their fishing and farming in accordance with the various moon phases. They knew which days (or nights) would be good for fishing, because particular fish would be abundant, and on which days certain varieties of fish should be left alone because they were spawning. They knew which days were good for planting particular varieties of plants, and which days should be set aside to allow the gardens to rest.

Through keen observational processes and problem-solving strategies, Hawaiians were able to meet their present food needs without compromising the ability of future generations to meet their needs. They developed intricate systems for sustainable resource monitoring and management that would rival any system of resource management in existence today (Friedlander et al., 2000).

Our Hawaiian ancestors had a long and rich tradition of impressive mathematical achievements, and it is important for Hawaiian students to know this. The sample lesson that follows below was adapted from Hui Malama o Mo'omomi's (2009) excellent curriculum on *Cycles and Hawaiian Traditions*. It is one of many possible examples of a ka wā mamua lesson in traditional Hawaiian mathematics that can instil within Hawaiian students a sense of pride in Hawai'i's mathematical heritage and enable them to see that we, too, have math in our blood.

Example: Moi Fishing in Ka Wā Kahiko

Imagine that you are a *lawai'a* (fisherman) living in *ka wā kahiko* (ancient times). Your *kuleana* (responsibility) is to catch *moi* (threadfish) for the *ali'i* (chief). Moi is a reef fish that is best caught during low tide. You need to plan your fishing carefully because moi are rare fish that are fairly difficult to find. You must find the best time for catching moi while being careful not to disturb the moi's spawning cycle so that they can reproduce and replenish.

It is now the first day of Nana, and you will need to bring moi to the ali'i once a month for the next three months (through Ikiiki). Using the relevant season, moon, and tide data from your local Hawaiian moon calendar and a local tide chart, follow the steps below to select the three best days (or nights) and times to fish for moi within the months of Nana, Welo, and Ikiiki:

> Step 1. Using your local Hawaiian moon calendar, first find the months in which moi fishing is *kapu* (prohibited) because the fish are spawning. Cross out those months on your calendar so that you are sure to leave the fish alone during that time.
>
> Step 2. Next, using the same moon calendar, find the good months for moi fishing. Circle the favourable moi fishing months on the calendar.
>
> Step 3. Using the moon calendar once again, circle the days (within the months of Nana, Welo, and Ikiiki) on which moi or reef fishing will be particularly good.
>
> Step 4. Now use the local tide chart to find the best tide periods for moi fishing. (Remember that low tide is the best time to catch moi.) On each of the days that you circled on the moon calendar in Step 3 above, record the hour of low tide.
>
> Step 5. Finally, mark the best three days (or nights) and times on your calendar, and explain why the weather and tide conditions for those days and times are ideal for moi fishing. (Adapted from Hui Malama o Mo'omomi, 2009)

This lesson provides students with problem-solving experience in data analysis while simultaneously giving them a glimpse into the complex moon and season cycles that were central to the planning and practice of traditional Hawaiian conservation fishing. Students are challenged to analyze, interpret, and juxtapose data from both a local Hawaiian moon calendar and a local tide chart, and they are encouraged to make and justify predictions based on the data. In the process, they develop an understanding of the cyclical nature of tides and learn to appreciate the seasonal, monthly, and daily rhythms and patterns of nature, which our early ancestors have keenly observed and noted in the prediction of ocean and fishing conditions since ka wā mamua.

KĒIA WĀ: CONNECTING MATH TO STUDENTS' EVERYDAY LIVES

At the same time that I suggest that lessons in the mathematical accomplishments of our ancestors are necessary to enable Hawaiian students to connect math with our cultural and community histories, I do not mean to imply that lessons on historical achievements alone are sufficient to enfranchise Native Hawaiian students in mathematics. Like William Tate (2006), I believe that an academically rigorous and culturally relevant math curriculum should also connect mathematics to students' present-day lived experiences. It should give them opportunities to incorporate mathematical reasoning into their everyday lives.

Drawing once again from Kame'eleihiwa (1992), I suggest that in order for Hawaiian students to "stand … firmly in the present" (p. 22), they should also be given opportunities to engage with a second category of math lessons that are grounded in their current circumstances or kēia wā. There are many possible ways in which this can be accomplished. One possibility is through lessons that enable students to simultaneously embrace both traditional Hawaiian culture and contemporary global-technological culture.

As Kamana'opono Crabbe and Kaliko Baker suggested in a recent Ka Wai Ola article (Dashefsky, 2010), young Hawaiians are literally "wired in to the present" (p. 13) through iPods, video games, cell phones, and the Internet. The following kēia wā lesson capitalizes on contemporary Hawaiian students' interest in computer games and social networking technology, while challenging them to use mathematics to create more opportunities to practise the Hawaiian language in their everyday, technological lives. Students begin by designing and conducting a small-scale statistical study to deepen their understanding of the order of usage of letters in the Hawaiian language (i.e., which letters are used most often and which letters are used least often). They are then challenged to apply the findings of their studies to one of two possible contemporary situations.

Example: Creating a Hawaiian Scrabble Application and Designing a Hawaiian Language Keyboard

Option 1: Creating a Hawaiian scrabble app. What if you wanted to create a Hawaiian language Scrabble game app for your iPhone or iPad? How many tiles would you make available for each letter and what would be their respective point values?

Option 2: Designing a Hawaiian language keyboard. What if you could alter the design of the onscreen keyboard on your iPhone when sending Hawaiian language text messages to your Hawaiian-speaking friends? What letters and diacritical markings would you place on your newly designed Hawaiian language keyboard and where would you position them?

Suggested procedure. Before proceeding with either of these projects, design and conduct a small-scale statistical study to explore which letters of the Hawaiian alphabet are used most frequently and which are used least frequently. For instance, you could take a single piece of Hawaiian language literature and tally the number of times each letter appears on the page. Or you might decide to gather a diverse sampling of Hawaiian language resources and tally the letters from a single sentence or paragraph in each of the different resources. Collect the data for your study, and then interpret and apply your study findings to the design of your Hawaiian language keyboard or Scrabble game.

This lesson engages students with several concepts in the area of statistics. They are posed with a question and are challenged to collect, organize, and analyze data in an attempt to answer that question. In the process of designing and conducting their statistical studies, students grapple with issues of sample size and sampling bias and consider the inferences or generalizations that can reasonably be made from their particular data sample. Finally, and perhaps most significantly, students apply the findings of their statistical studies to contemporary real-world situations.

Too often, math is taught in ways divorced from the real world. Students learn and practise mathematics skills without seeing the connection between these skills and the situations that they encounter in their daily lives. When math is isolated to pages upon pages of abstract textbook problems, the not-so-subtle message is that math is basically irrelevant except for achieving success in future math classes. The alternative is to teach mathematics in a way that helps students to see math as a powerful and useful tool for solving problems that confront students in their everyday lives. Activities such as this kēia wā lesson make mathematics more lively, accessible, and personally meaningful for students, who in turn are more engaged and learn math in more depth (Gutstein & Peterson, 2006).

KA WĀ MAHOPE: PROBLEM-POSING FOR A SUSTAINABLE FUTURE

As Kameʻeleihiwa (1992) suggested, the future that lies "behind" us is out of sight and largely unknown. Consequently, preparing young Hawaiian leaders to solve global and societal problems that no one yet knows how to solve—or that no one has ever faced before—will require creative thinking and mathematical problem-solving skills far beyond rote or formulaic problem-solving experiences. Hawaiian leaders of the future will need experience in what the late Brazilian educator Paolo Freire termed "problem-posing" education.

In *Pedagogy of the Oppressed*, Freire (1993) contrasted the oppressive, "banking" approach to education with the more liberatory, "problem-posing" approach. According to Freire (1993), the banking approach to education, in which teachers deposit knowledge into the heads of students, produces "automatons" (p. 55) who receive and accept the world (and their lot in it) as "passive entities" (p. 57). By contrast, Freire noted that through problem-posing education, students are challenged to understand and eventually transform the problems in their communities and the broader world. In the process, they learn to view the world not as a "static reality" (p. 64), which they have no choice but to accept, but as a "reality in process" (p. 64), which they can transform through creative problem solving. Whereas the banking method reinforces students' "fatalistic perception[s]" (p. 66) of difficult situations, the problem-posing method presents these very situations to them as problems which, through creativity and collective action, they, themselves, can control.

Freire's concept of problem-posing education emphasizes that problems with neat, pared-down data and clear-cut solutions give students a false picture of reality (Frankenstein, 2006). Real-life problems are messy. They intersect and interact and may not have a single, clear-cut solution. In real-life problems, it is rare that a person will be given all the information that he or she needs in one tidy package. Instead, the person has to collect the data from a variety of sources. There is rarely one possible method or strategy for solving real-life problems; instead the person must select one strategy from several possible options. Moreover, one does not always know for sure if the solution one comes up with is the "right" one; one decides on a solution that one thinks is "best," and it may be only later, if ever, that one can evaluate one's choice. Whereas conventional math problems "isolate and simplify particular aspects of reality in order to give students practice in techniques" (Frankenstein, 2006, pp. 21–22), Freirean problem-posing is intended to reveal the "inter-connections and complexities of real-life situations" (Frankenstein, 2006, pp. 21–22).

Furthermore, Freire inspired us to look for solutions to problems in a variety of intellectual traditions. Freire reminded us that far too often "the intellectual activity of those without power is … characterized as non-intellectual" (Freire & Macedo, 1987, p. 122). However, all people have rich intellectual traditions, and all people are continually creating knowledge and doing intellectual work (Frankenstein, 2006).

As Frankenstein (2006) attested, up until now, American society's near-exclusive reliance on Western perspectives and intellectual traditions has left most of the "burning social, political, economic, and ethical questions of our time" (p. 28) unanswered. Although the United States is a society of enormous wealth, it still has significant hunger and homelessness. Although American doctors have engaged in medical and scientific research for hundreds of years, they are not much closer to changing the prognosis for most cancers (Frankenstein, 2006). Following Freire (Freire & Macedo, 1987), problem solvers of the future will benefit from opening themselves up to learning from the perspectives and philosophies of a variety of intellectual and cultural traditions and bringing these fresh perspectives to bear upon persistent societal problems.

But how do we begin to prepare young Hawaiian students for this future reality? Surely, we cannot expect them to be solving world problems from their early years in grade school. However, we can begin to lay the foundation by engaging the students in in-depth mathematical explorations that are rigorous and meaningful, and that empower them to understand and transform the problems of the world rather than passively accept them. We can challenge them with complex, real-life problems to which answers are not yet known, problems that intersect and interact, which require the collection of data from a variety of sources, and which may not have a single, clear-cut solution. Finally, we can encourage them to look for solutions in a variety of intellectual traditions, and to consider how the lessons of our ancestors' rich mathematical past can help us to "more wisely build the future" (Freire, 1993, p. 65).

Example: Subsistence Farming in Ka Wā Mahope

The year is 2040. Fuel prices have skyrocketed and, correspondingly, so have the prices of shipped-in produce. To offset the high produce prices, over the past year, your *'ohana* (family) has relocated from Honolulu to Kainaliu, eight kilometers south of Kailua-Kona, where you have plans to start a small, subsistence farm, with dry-land kalo and *'uala* (sweet potato), in your backyard. After consulting a Honolulu-based moon calendar, you note that the calendar suggests planting kalo and 'uala in Kaulua (in the middle of the wet season) and harvesting them in 'Ikuwā (near the end the dry season and before the start of the winter rains). However, after spending the past year in Kainaliu, you have observed that the dry and wet seasons in Kainaliu seem to be the reverse of what is indicated on the Honolulu moon calendar. As far as you can tell, it seems that the summers in Kainaliu are wetter than the winters.

Part 1. Working in your cooperative 'ohana groups, use the National Oceanic and Atmospheric Administration (NOAA) graphs (NOAA, 2018) to compare the average precipitation data in Kainaliu from month to month. Do the data support your conjecture that the dry and wet seasons in Kainaliu are the reverse of what is indicated on the Honolulu-based moon calendar? Is the wettest time of year in Kainaliu indeed in the summer? Juxtaposing the Honolulu moon calendar with the Kainaliu average precipitation data, explain any modifications that you would make to the Honolulu moon calendar farming schedule when planting and harvesting your subsistence farm to account for the unique seasonal rainfall pattern in Kainaliu.

Part 2. After a few years of success with your modified planting and harvesting schedule, you have recently been informed by NOAA that the coming year will be a strong El Niño year, which is predicted to significantly alter plant growth and agricultural production in the Kona district over the next two years. Working again in your cooperative 'ohana groups, use the NOAA graphs (NOAA, 2018) to compare the "normal" or average monthly precipitation data for Kainaliu with the monthly precipitation anomalies for Kainaliu in El Niño years. Using these two sets of data along with the Kainaliu planting and harvesting calendar, which you developed in Part 1, explain any changes that you will make to your Kainaliu planting and harvesting schedule for both the coming El Niño year and the year after the El Niño event.

Whereas this chapter's initial, ka wā mamua, lesson focused on appreciating Hawaiian knowledge of the past, and the second, kēia wā, lesson was firmly grounded in the present, this ka wā mahope lesson challenges students to combine elements of traditional Hawaiian knowledge with present-day scientific knowledge to forecast a course of action for the future.

In Part 1 of the lesson, students begin with traditional harvesting and planting data from a Honolulu moon calendar and make modifications to the calendar based on monthly average precipitation data for the Kainaliu district. This is similar to a task that one might do in ka wā kahiko but with the addition of contemporary scientific data. Hawaiians were keenly aware that, given the number and variety of microclimates on each island, there simply was not one universal planting and harvesting calendar that would work for all the islands. Hawaiian months and planting calendars were exceedingly place-based and differed from one island to the next, and sometimes from one *ahupua'a* (land division) to the next, based on close observations of the unique climatic patterns of each particular district. In Part 1 of this lesson, students access and interpret 30-year average precipitation data from a NOAA website to draw conclusions about the typical climatic patterns in Kainaliu and develop a corresponding planting and harvesting schedule for this particular community.

Part 2 of the lesson adds another layer of complexity to the problem, as it urges students to consider how Kainaliu's typical climatic patterns may be affected in a strong El Niño year and challenges them to adapt their planting and harvesting schedule accordingly. Students once again access the NOAA website—this time to examine and interpret data about rainfall variations for the Kainaliu region during past El Niño years—and they apply the data to make predictions about the effects of an upcoming El Niño on agricultural production in their region.

This is a complex, real-life problem with real-life consequences, which continues to engage the interest of NOAA scientists today. There is considerable concern that global warming produced by increased carbon dioxide emissions may be contributing to greater fluctuations in sea temperatures and more intense and severe El Niño events, causing distinct and dramatic changes in local and regional weather patterns around the globe (NOAA, n.d.). Predicting, and being proactive in planning for, local and regional changes in precipitation and temperature during El Niño (and La Niña) events is essential to agricultural interests in the Kona district and worldwide, as the growing process for many crops is highly sensitive to rainfall and temperature variations.

In addition to being significant, this problem is also deep, complex, and messy. It has the potential to challenge students and scientists alike for hours, days, weeks, months, or years as they continue to identify and explore additional technological tools and external mathematical resources to deepen their understanding of the problem and develop possible solutions.

In the face of gloom-and-doom future forecasting, this Freirean-inspired problem-posing lesson empowers students to be proactive in dealing with the challenges that await them in the future. As Freire (1993) suggested, the oppressive, "banking" model of education presents the world as "a fixed entity, as something given—something to which people, as mere spectators, must adapt" (p. 120) and conform. In contrast, Freirean, problem-posing education is a "practice of freedom," a "means by which men and women deal critically and creatively with reality and discover how to participate in the transformation of their world" (Freire, 1993, p. 16). In a similar spirit, this ka wā mahope lesson denounces the fatalism of gloom-and-doom future forecasts and boldly asserts that Hawaiians are "transformative beings and not beings for accommodation" (Freire & Freire, 1997, p. 36).

CONCLUSION

The most pressing goals and challenges of the future, both globally and in Hawai'i, will include mitigating the effects of climate change, finding local sources of renewable energy, and sustainably managing food resources. The future leaders who will be challenged to solve these global and local crises 30 years from now are today's children.

To devise creative and sustainable solutions to these challenges, Hawai'i's future leaders will need a solid foundation in the accumulated wisdom of our Hawaiian ancestors, a firm grasp of current technological and mathematical advancements, and exceptional skill in complex, mathematical problem solving. Through thoughtfully conceptualized lessons in past, present, and future Hawaiian mathematics, contemporary educators can engage Hawaiian students with a rigorous and culturally relevant mathematics education—one that will prepare these future leaders to "stand ... firmly in the present" (Kame'eleihiwa, 1992, p. 22) while applying historical wisdom to future dilemmas.

ACKNOWLEDGEMENT

This chapter originally appeared in *Hūlili: Multidisciplinary Research on Hawaiian Well-Being*, Volume 7, pp. 289–306, 2011. Reprinted here with permission from the publishers.

REFERENCES

Anderson, S. E. (2006). Historical, cultural, and social implications of mathematics. In E. Gutstein & B. Peterson (Eds.), *Rethinking mathematics: Teaching social justice by the numbers* (pp. 43–44). Milwaukee, WI: Rethinking Schools.

Dashefsky, H. (2010, July). 'Aha kāne: Embracing the past, to ensure a better future. *Ka Wai Ola, 28*, 13.

Fletcher, C. (2009). Sea level by the end of the 21st century: A review. *Shore and Beach, 77*(4), 4–12.

Fortenberry, N., Sullivan, J., Jordan, P., & Knight, D. (2007). Retention: Engineering education research aids instruction. *Science, 317*, 1175–1176.

Frankenstein, M. (2006). Reading the world with math: Goals for a critical mathematical literacy curriculum. In E. Gutstein & B. Peterson (Eds.), *Rethinking mathematics: Teaching social justice by the numbers* (pp. 19–28). Milwaukee, WI: Rethinking Schools.

Freitas, A., & Balutski, B. (2009, October). *Native Hawaiian student profile*. Honolulu, HI: University of Hawai'i–Mānoa, Kōkua a Puni.

Friedlander, A., Poepoe, K. K., Poepoe, K., Helm, K., Bartram, P., Maragos, J., & Abbott, I. (2000). Application of Hawaiian traditions to community-based fishery management. In *Proceedings of the 9th International Coral Reef Symposium*, Bali, Indonesia, 813–818.

Friedman, T. (2005a). Globalization 3.0. *Blueprint Magazine, 2005*(2). Retrieved from http://www.dlc.org/print.cfm?contentid=253354

Friedman, T. (2005b). *The world is flat: A brief history of the twenty-first century*. New York, NY: Farrar, Straus and Giroux.

Friedman, T. (2008). *Hot, flat, and crowded: Why we need a green revolution and how it can renew America*. New York, NY: Farrar, Straus and Giroux.

Freire, P. (1993). *Pedagogy of the oppressed*. New York, NY: Continuum.

Freire, P., & Freire, A. M. A. (1997). *Pedagogy of the heart*. New York, NY: Continuum.

Freire, P., & Macedo, D. (1987). *Literacy: Reading the word and the world*. South Hadley, MA: Bergin & Garvey.

Gutstein, E., & Peterson, B. (Eds.). (2006). *Rethinking mathematics: Teaching social justice by the numbers*. Milwaukee, WI: Rethinking Schools.

Hui Malama o Moʻomomi. (2009). *Cycles and Hawaiian traditions*. Honolulu, HI: Pacific American Foundation.

Kaakua, J. (2010). Native Hawaiian Science & Engineering Mentorship Program: A look at increasing student diversity at the University of Hawaiʻi. *Electrical and Computer Engineering Department Heads Association Source Newsletter, 3*, 1–3.

Kameʻeleihiwa, L. (1992). *Native land and foreign desires: Pehea lā e pono ai?* Honolulu, HI: Bishop Museum Press.

Ladson-Billings, G. (1995). Making mathematics meaningful in multicultural contexts. In W. G. Secada, E. Fennema, & L. B. Adajian (Eds.), *New directions for equity in mathematics education* (pp. 126–145). Cambridge: Cambridge University Press.

Musca, T. (Producer), & Menendez, R. (Director). (1988). *Stand and deliver* [Motion picture]. United States: Warner Brothers.

National Academy of Sciences, National Academy of Engineering, and Institute of Medicine. (2007). *Rising above the gathering storm: Energizing and employing America for a brighter economic future*. Washington, DC: National Academies Press.

National Math and Science Initiative. (2010). *STEM quotes: Quotable quotes about math + science*. Retrieved from http://www.nms.org/News-and-Views/Blog/2013/January-2013/QUOTABLE-QUOTES-ABOUT-MATH-SCIENCE.aspx

National Oceanic and Atmospheric Administration (NOAA). (n.d.). *Education and outreach: Frequently asked questions*. Retrieved from http://www.esrl.noaa.gov/gmd/infodata/faq_cat-1.html

National Oceanic and Atmospheric Administration (NOAA). (2018). *Kona/Keahole (Big Island) temperature and rainfall graphs*. Retrieved from http://www.prh.noaa.gov/hnl/pages/cli_graphs.php?station=PHKO&mo=

National Science Board. (2010). *Science and engineering indicators 2010* (NSB 10-01). Arlington, VA: National Science Foundation.

Native Hawaiian Science & Engineering Mentorship Program. (2009). *NHSEMP mission*. Retrieved from http://nhsemp.eng.hawaii.edu/

Ortiz-Franco, L. (2006). Chicanos have math in their blood: Pre-Columbian mathematics. In E. Gutstein & B. Peterson (Eds.), *Rethinking mathematics: Teaching social justice by the numbers* (pp. 70–75). Milwaukee, WI: Rethinking Schools.

Seeley, C. (2009). *Faster isn't smarter: Messages about math, teaching, and learning in the 21st century*. Sausalito, CA: Math Solutions.

Stannard, D. E. (1989). *Before the horror: The population of Hawaiʻi on the eve of Western contact*. Honolulu, HI: University of Hawaiʻi, Social Science Research Institute.

Tate, W. F. (2006). Race, retrenchment, and the reform of school mathematics. In E. Gutstein & B. Peterson (Eds.), *Rethinking mathematics: Teaching social justice by the numbers* (pp. 31–40). Milwaukee, WI: Rethinking Schools.

Woodson, C. (1993). *The mis-education of the Negro*. Trenton, NJ: Africa World Press.

JERRY LIPKA, DORA ANDREW-IHRKE AND
EVA EVELYN YANEZ

11. SYMMETRY AND BODY PROPORTIONAL MEASURING

Contributions of Yup'ik Elders to Mathematics Education

INTRODUCTION

Math in a Cultural Context (MCC) is a long-term collaborative curriculum, professional development, and research project with Yup'ik Elders, teachers, and Alaskan school districts. MCC arose in response to one of the most vexing and enduring issues in education: the underperformance of schools serving American Indian/Alaska Native (AI/AN) communities, particularly concerning students' mathematics performance (Grigg, Moran, & Kuang, 2010). Further, the federal involvement in AI/AN has its basis in the U.S. constitution (Article 1, Section 8, Clause 3), which establishes a government-to-government relationship between the U.S. federal government and sovereign tribes. Presidential Executive Orders affirm that relationship and urge improved academic performance while meeting the "unique educational and culturally related academic needs of AI/AN students … consistent with tribal traditions, languages, and culture" (Bush, 2004). Federal reports have, for almost a century, pointed toward educational approaches that connect school and community (Institute for Government Research, 1971; White House Conference on Indian Education, 1992; Clinton, 1998; Beaulieu, 2000; Reyhner, 2009).

MCC has successfully responded to this challenge by developing a 2nd-through-6th grade supplemental mathematics curriculum series based on everyday activities and Indigenous knowledge (IK). The series includes 10 supplementary elementary school math modules and 11 storybooks. The storybooks provide a contextual background for the mathematics. Further, MCC has conducted numerous random, quasi-experimental, designed studies between 2002 and the present. The MCC program has been noted by *Education Week* as one of the few culturally based programs to provide evidence of its efficacy (Zehr, 2008; Kisker et al., 2012). Also, we have published an extensive number of qualitative case studies describing how teachers effectively implemented MCC by bridging and connecting the culture and wisdom of the community to school mathematics (see Lipka, Hogan et al., 2005; Lipka, Sharp et al., 2007; Lipka, Yanez et al., 2009). Despite this volume of work and despite the federal government's legal responsibility to improve the academic performance of AI/AN, MCC remains marginalized, in part because our efforts are

 | DOI: 10.1163/9789004415768_011

overshadowed by a larger socio-cultural and political context in which the others' knowledge and power pale in contrast to the larger society surrounding them. This chapter speaks to these issues. In short, how can IK inform curriculum and instruction and become more relevant for its target group and for the larger society?

One example from Alaska underscores why many of us have been fighting for cultural inclusion in education. Dora Andrew-Ihrke, a former bilingual coordinator, Alaska teacher of the year, and long-term (MCC) colleague puts this struggle into perspective: "When I was a school-aged youngster I was told to, wipe my feet at the school door, symbolically wiping out my culture and language." These personal experiences are attributable to institutional and some governmental policies and practices that translate into the systematic exclusion of Indigenous peoples' knowledge, cultures, and languages in the processes and content of schooling (Watson-Gegeo, 1994), despite the federal government's responsibility.

In this chapter, we intend to argue that including Indigenous knowledge and practice can not only strengthen the education of Indigenous people, but also that the mathematical principles learned from Yup'ik Elders' everyday practice can provide powerful lessons for all students and teachers. The latest lessons learned from this long-term ethnographic collaborative curriculum project have opened up a potentially powerful and alternative pathway for teaching and integrating aspects of measuring, geometry, rational number reasoning (RNR), and algebraic thinking. This new approach based on IK resonates with cutting-edge approaches to teaching children numbers, ratios, proportions, and fractions (Confrey, Maloney, Nguyen, Mojica, & Myers, 2009) as well as geometry and early algebraic thinking (Dougherty, 2010). It creates the potential for a math curriculum that can provide not only AN but other students as well with a strong mathematical foundation

We will provide concrete examples of how lessons learned from Yup'ik Elders provide alternative and fresh ways of perceiving, understanding, and teaching measuring, geometry, RNR, and early algebraic reasoning in school. Some Yup'ik ways of solving everyday problems may well have the potential to offer educators ways of understanding the underlying connections between geometry, measuring, patterns, and numbers. Furthermore, as we will show, these ways of engaging in mathematically oriented problems relate well to the National Council of Teachers of Mathematics (2000) process standards of problem solving, reasoning and proof, communication, representations, and, maybe most of all, connections to different mathematics strands (National Mathematics Advisory Panel, 2008). This chapter illustrates that the knowledge contributed by Yup'ik Elders represents a potentially elegant, systematic way to teach multiple math strands through a number of powerful, interrelated mathematical concepts, practices, and processes.

This chapter will first present a brief review of literature that represents alternatives to the struggle to include the culture of "others" in schooling. This will be followed by a brief review of current obstacles to the inclusion of the culture of "others." We describe the MCC program and discuss in some detail the lessons that we have learned from Yup'ik Elders and the potential applicability and resonance of these lessons more widely to both in-school and out-of-school mathematics.

SUPPORTS AND OBSTACLES

Researchers across a variety of academic disciplines have identified everyday out-of-school knowledge and the potential it has to connect the knowledge of the community and students' everyday context to school mathematical knowledge (e.g., Brenner, 1998; Carraher, Carraher, & Schliemann, 1985; Lave, 1977, 1982; Saxe, 1988a, 1988b). These mostly ethnographically oriented studies share Bruner's (1996) notion of culture. Bruner states, "Culture shapes mind … it provides us with the tool kit by which we construct not only our worlds but our very conceptions of ourselves and our powers." He further states that "you cannot understand mental activity unless you take into account the cultural setting and its resources, the very things that give mind its shape and scope. Learning, remembering, talking, and imagining: all of them are made possible by participating in a culture" (Bruner, 1996, pp. x–xi). Wiping out culture, erasing the very foundations that Bruner speaks of, is blatantly not the way to build mind and build upon cultural resources. Psychologists such as Sternberg and his colleagues have empirically shown that connecting everyday cultural knowledge to schooling by building activities that tap into students' practical, academic, and creative abilities results in improved student performance when compared to comparable groups of students who were taught without making these connections (Grigorenko et al., 2004; Sternberg, Lipka, Newman, Wildfeuer, & Grigorenko, 2006; and Sternberg et al., 1998). Further, ethnographically oriented research conducted at the intersection of education, linguistics, and cross-cultural psychology applied to schooling (Au & Jordan, 1981; Cazden & Leggett, 1981; Mohatt & Erickson, 1981; Erickson & Mohatt, 1982; Lipka, Mohatt, & The Ciulistet Group, 1998) have also shown the importance of out-of-school communication, particularly in Indigenous communities, and how in-school classroom communication can reflect community norms. One of the early effective programs to develop from this perspective was the Kamehameha Early Education Project (K.E.E.P.). Although a program known for reading and literacy by adapting classroom communicative patterns to reflect those of the students' home community (Au & Mason, 1993; Au & Carroll, 1997), the approach was also successfully applied to school mathematics (Brenner, 1998). Furthermore, a small but growing number of programs have connected out-of-school learning to in-school mathematics learning such as that described by Harris (1991) in Australia. In the U.S. there are a number of projects that have effectively connected in-school and out of-school mathematics learning: Moses and Cobb (2001) in rethinking education for African-Americans through the Algebra Project; Moll, Amanti, Neff, and González (1992) and Civil and Kahn (2002) with the Funds of Knowledge project; and MCC's work in Alaska (Kisker, Lipka, et al., 2012; Lipka, Hogan, et al., 2005; Lipka, Sharp et al., 2005; Lipka, Yanez, et al., 2009). This body of research provides insights on how to link and strengthen the connections between everyday practice and knowledge to schooling in Indigenous and minority contexts.

The work of Nasir, Hand, and Taylor (2008) views the blurring of boundaries between in-school and out-of-school learning, or between Indigenous and minority knowledge and in-school knowledge, as further indication of the increasing legitimacy of the role of out-of-school knowledge in schooling. Although the field of ethnomathematics has evolved since the days of the Merriam report, its work is not complete, as the resistance to the inclusion of Indigenous and minority cultural/ linguistic knowledge still persists.

Across the U.S., including Alaska, the passage of President Bush's No Child Left Behind (NCLB) Act (United States Department of Education, 2002) was intended to increase states' and school districts' accountability as the key way to improve student achievement. Although some of NCLB goals are laudable, such as "closing the academic achievement gap" and "increasing academic achievement for all," this law has resulted in an over emphasis on high-stakes testing, including punitive measures for "low performing schools" (p. 9). Inadvertently, NCLB threatens the inclusion of Indigenous language and culture (Castagno & Brayboy, 2008) in schooling, and it places undue pressure on English Language Learners (ELLs), those students whose first language is not English or students who are developing proficiency in English (Abedi, 2004). Some rural Indigenous school districts are considered to be "failing" based on single high-stakes tests. Caught in this high-stakes testing cycle, districts increase their efforts on test taking, ushering in frequent assessments, and devoting enormous chunks of time so that schools will be taken off the dreaded "Not Making Adequate Yearly Progress" list. These practices, almost obsessive in nature, can sacrifice and separate learning from achievement and subordinate student cultural identity to one of academic inadequacy (Gutiérrez, 2008). These narrow attempts at improving students' performance on high-stakes tests can increasingly marginalize culturally and linguistically based programs, communities, and students. In Alaska, this occurs even in school districts that have adopted policies and practices to include culturally relevant curriculum, including MCC's program. Now, these districts find themselves caught between what they wish they could do and what they feel forced to do. This contentious and litigious context has been further exacerbated by a lawsuit first filed against the State of Alaska in 2004 by a group of parents and plaintiffs. The presiding judge, Gleason, ruled that the "[State] failed to exercise adequate supervision and oversight of chronically underperforming schools … [the State has failed to provide] a meaningful opportunity to acquire proficiency in the subject areas tested by the State" (*Kristine Moore et al. v. State of Alaska*, 2010, p. 1). Not only are these mostly rural Indigenous students not meeting state standards, but, again, Indigenous culture and language may be what is left behind. One district involved in this suit, the Yupiit School District, is one of MCC's key partner districts, whose Elders provided a substantial amount of cultural, pedagogical, and mathematical knowledge that comprise MCC. Implementation research on MCC modules conducted in Yupiit shows substantially improved students' mathematical knowledge (Lipka, Hogan, et al., 2005). Yet it is problematic for such districts to implement culturally based curriculum.[1]

In addition to the external threats to developing ethnomathematics curriculum and professional development such as NCLB, we are faced with internal threats that further place us on our collective heels. Are "culturally relevant" approaches valid? Where is the evidence? The literature on "culturally based" curriculum and pedagogy is contentious with arguments about its validity (Castagno & Brayboy, 2008) particularly when viewing the scant empirical evidence of its efficacy (Demmert & Towner, 2003). In their review of over 8,000 culturally based studies, Demmert and Towner (2003) found only a handful of "culturally based" efforts that met their criteria of utilizing rigorous research methods. Despite these arguments, it does not take much imagination to envision why these students would struggle if "regular" school teaching and curricula materials are taught in ways that use unfamiliar communicative norms and practices, and do not build on prior or shared knowledge. They often produce learning experiences that Indigenous and linguistic minority students cannot identify with.

In part, we need to continue to make the argument beyond the fact that teaching students in unfamiliar ways, in some instances in unfamiliar languages and with values in conflict with the students' culture, is patently inadequate. We need positive examples that reflect the wisdom of Elders' practice and honour the U.S. government's constitutional relationship with AI/AN students. It simply does not make good pedagogical sense to continue failed practices. Thus, the rest of this chapter proffers grounded possibilities on meeting the challenges of improving AI/AN education while integrating the wisdom of Elders' practice, and outlines how this approach can improve education more broadly.

We find arguments against culturally based education to be specious and misleading. All curricula are culturally based; the question is whose culture is it based on? Why is one set of standards used for "culturally based" education and another for mainstream education? For example, "culturally based" education needs to justify its underlying assumptions from how communication in the classroom is organized, to the inclusion of familiar contextual background, and the inclusion of mathematically related knowledge from everyday contexts. From a mainstream perspective, many of these "issues" are supposedly culturally free assumptions about learning, ways of organizing classroom communication, what mathematical knowledge counts as legitimate, and what approaches to teaching mathematics are considered appropriate. Of course, the curriculum in place is considered the "standard" because it reflects mainstream cultural norms. These arguments are about whose knowledge, power, and identity count. Education needs to be understood as a social cultural production, and the exclusion of deep cultural knowledge from different cultural and ethnic minority groups is an educational loss to those groups. However, more importantly, for mainstream education the loss or lack of serious integration of knowledge beyond mainstream arenas results in an educational loss for all. Our argument is not only that Indigenous and minority knowledge can be of obvious importance to the "targeted" cultural groups but also that lessons from these "peripheral" groups may well benefit teachers and students in general.

We have learned from MCC's work that even meeting the supposed "gold standard" of rigorous research methods and yielding statistically significant findings for educational research do not necessarily pave the way for the inclusion of mathematics curriculum based on Indigenous knowledge. MCC is still fighting for inclusion in the very districts where it has yielded its positive results. We are not suggesting resignation, but adding new approaches to the inclusion of curriculum based on Indigenous knowledge. We can reframe the contribution that these programs can make to education in general. The following finding from MCC's data provides an insight into reconceptualizing our argument for inclusion. It is worth noting that when we analyzed MCC data collected in multiple studies, particularly our "gold standard" study (Kisker et al., 2012) with over 900 students across locations (urban vs. rural), we observed that both urban and rural treatment students improved performance in relatively similar ways. More specifically, from 2007 to 2009, using *Picking Berries: Connections Between Data Collection, Graphing, and Measuring* (Lipka, Parmelee, & Adams, 2004), urban students gained 32% while rural students gained 36%, and Alaska Native second-grade students gained 35% while all other students gained 32%. Similarly, on MCC's pre- and post-tests, we have found that across all categories, such as students considered to be developing their English language skills or students identified with special needs by experienced and inexperienced teachers, MCC students outperform their counterparts who use their school district's regular curriculum. Although the curriculum materials were developed with Yup'ik Elders and teachers in conjunction with math educators, these materials had positive impacts on mainstream students as well as on rural, mostly Indigenous, students. Could it be that lessons we learned from our long-term ethnographically oriented work with Yup'ik Elders can be applied to schooling more widely? Could it be that educational, psychological, and cognitive studies with distinct cultural groups and in multiple settings and situations may well yield ways of perceiving and pursuing mathematical problems and solutions that are legitimate in their own right and can possibly inform mainstream educational practices? We argue that this is precisely what such work can yield.

MATHEMATICS IN A CULTURAL CONTEXT PROGRAM

Mathematics in a Cultural Context (MCC) began in 1995 with funding from the National Science Foundation's Instructional Materials Development program. MCC's approach was a direct and deliberate response to include Yup'ik culture and language in the processes and products of schooling. MCC's curriculum development paradigm was based on Yup'ik Eskimo Elders' subsistence life cycle. Themes of collecting or gathering resources usually included measuring, sorting, storing, using, and navigating. These categories are similar to Bishop's (1988) counting, locating, measuring, designing, playing, and explaining. To develop this curriculum, we worked in situ with Yup'ik Elders, expert and novice Yup'ik teachers, other teachers, mathematicians, mathematics educators, and educational researchers. We explored and analyzed subsistence-related activities, particularly performance, thought processes, and tool construction, and developed school-based mathematics curriculum (see Lipka et al., 1998) based on these activities. We developed mathematics modules on key everyday or seasonal activities such as collecting eggs and berries. We used the context of "egging" as a narrative in which, after collectors have climbed steep cliffs, the eggs are placed in buckets, sorted, distributed, used, and stored. These Alaskan activities provided a familiar context for students to learn about grouping and place value. The mathematical challenge for students was to sort, group, and use place value in the Yup'ik base 20 sub base 5 system, providing a direct contrast to the base 10 system used in school. We believed that contrasting base 20 with base 10 would make understanding grouping and place value more challenging, interesting, and provide students and teachers with a deeper, more flexible understanding of numeration. MCC's professional development workshops challenge teachers to think more deeply and systematically about such topics as place value. This resulted in a teacher telling the authors that she had an epiphany at the workshop. Prior to the workshop, she didn't think it was possible to teach a subject like place value in a culturally based way. Now she had insights on how she could link math and culture in the teaching of grouping and place value.

Mindful of the linkages between culture and cognition (Cole & Scribner, 1974; White, 1959; Bruner, 1964) and the almost ubiquitous use of spatial abilities and geometrically related constructions in Elders' everyday practice, our curriculum development work included school-based activities that incorporated spatial abilities and manipulation with connections to geometry. Although this earlier phase of work identified mathematical concepts embedded in activity, we did not identify underlying mathematical connections across activities. We continually noticed that Yup'ik Elders' mathematics incorporated geometry and spatial relations. For example, the module *Star Navigation: Explorations into Angles and Measurement* (Adams, Kagle, & George, 2007) is linked to the mathematical concepts of measuring angles, in part through body measures. This is not surprising, since scholars have long identified spatial abilities and manipulation as a relative cognitive strength of many Indigenous groups, particularly Yup'ik and Inuit (Berry, Poortinga, Segall, & Dasen, 2002). Cognitive preference develops through a process involving the emphasis of a particular set of skills and abilities within a culture as part of its response to the ecological demands of the social and physical environment it occupies (Berry, 1976).

This phase of our work produced 10 supplemental mathematics modules. Also, our earlier worked with Yup'ik Elders and teachers demonstrated not only how to use everyday knowledge in the construction of curriculum but also how to coordinate this knowledge with pedagogical approaches used by Yup'ik Elders and some expert Yup'ik teachers (Lipka & Yanez, 1998). Repeatedly, we were taught through an expert-apprentice modelling, where tasks were situated in a context of purposefulness. Initial observation was followed by practice, with the expert distributing some of their knowledge to those students who were "ready to learn" (Lipka & Yanez, 1998). More advanced novices instructed less advanced novices until all students had completed the task. Elders and Yup'ik teachers performed purposeful tasks—teachers in schools thus connected learning to community-valued tasks. Learning and purpose were united in some of the best examples observed in experienced and expert Yup'ik classrooms (Lipka, Sharp, et al., 2005). We observed Yup'ik teachers using cognitive apprenticeship similar to Lee's (1995) work. While, unbeknownst to us, Tharp and his colleagues (Doherty, Hilberg, Epaloose, & Tharp, 2002) coined the phrase "joint productive activity," we too were using the phrase as a way to describe teachers and students working jointly on the same classroom project (Lipka, Sharp, et al., 2007). Thus, this phase of work included the specific mathematics of everyday activities, purposeful tasks or projects, activities that included spatial thinking, and a pedagogical approach that included expert-apprentice modelling.

MCC's latest work is the process of understanding everyday activity, systematically connecting underlying mathematical concepts, procedures, and processes. This is the work that we believe holds promise for teaching mathematical concepts across contexts. Because of the authors' long-term relationship, we were able to revisit and reanalyze everyday activities in a way that allowed us to perceive the underlying mathematical connections across activities.

To explain these underlying and powerful connections, Dora and Evelyn state that from a Yup'ik perspective, mathematics begins with our bodies. Bodies are the first instruments used to measure. Lakoff and Núñez (2000) state that the way in which people learn mathematical concepts "is through concepts in our minds that are shaped by our bodies and brains and realized physically in our neural system … mathematics *is* embodied mathematics" (p. 146). They go on to state that the embodied mathematics is not axioms and proofs but image schemas, metaphors, and conceptual blends, among others.

Working with Elders over many years, we've noticed that the Elders consistently repeated the importance of centre points. Dora added that Yup'ik perceive a centre point at the top of the head running through the middle of the body. Thus, the body is perceived with centre points located on the surface and extending through the centre of the body meeting between the front (belly button) and the back. Simply, the body is seen somewhat analogously to a sphere composed of intersecting planes.[2] The concept of centre points and the body is also used in body regions such as the hand. Further, the body appears to be the sensing/moving instrument that provides us with a sense of space, distance, speed, and proportionality. We follow the hypothesis and empirical findings of Dehaene, Spelke, Pinel, Stanescu, and Tsivkin (1999) that approximating (number sense, estimating, etc.) relies on non-verbal visual-spatial cerebral networks, while exact numbers are more language dependent. Thus, the approach we take from the Elders uses the visual-spatial pathway to teach number relations and operations. Further, another fundamental building block to the processes that we are about to describe is the notion of symmetry. We have observed the use of bilateral symmetry across a variety of everyday Yup'ik activities, including and most importantly, body proportional measuring. Interestingly, symmetry (Feng, 2002) appears to be connected to attractiveness in both animals and humans, including the manner in which people perceive facial beauty. Giurfa, Eichmann, and Menzel (1996) note that bees can be trained to discriminate bilaterally symmetrical and asymmetrical patterns and generalize this training to novel stimuli. Because of the body's bilateral symmetry, we posit that symmetry may be a fundamental way in which we perceive the world, construct objects, make art, represent objects, and are potentially predisposed to think mathematically. This supposition receives support from Wynn (2002): "Given the ubiquity of reflectional symmetry in the natural world, and its correlation with successful ontogenetic development in many, many organisms, it is not at all surprising that perceptual systems should have evolved a sensitivity to symmetry" (p. 391).

THE POWER OF SYMMETRY, PROPORTIONAL MEASURING: CONNECTIONS FROM GEOMETRY TO NUMERATION

Yup'ik Worldview, Epistemology, and Practice: Symmetry and Body Proportional Measuring in Its Cultural Context Applied to School Mathematics

Over the past few years, we slowly began to realize that there are powerful connections between Yup'ik worldview, epistemology, practice, and school mathematics (Lipka, Andrew-Ihrke, & Yanez, 2011). Figure 11.1 encapsulates our understanding of these relationships. Although we originally did not see the connections between these components, they now seem obvious. This figure depicts important aspects of Yupiaq worldview, such as reciprocal relationships between human beings (*Yupiaq* means the "real people") and animals, the need to achieve harmonious relationships between these "worlds," and the need to please the "spirits" to maintain harmony; harmony is, in part, maintained by following important Yup'ik cultural rules such as not wasting, caring for, and respecting animals as a way to fend off famines. Animals are not caught as much as they "give themselves up" to the meticulous hunter. Dora and Evelyn speak of the importance of making ceremonial artifacts, such as the headdress, as a way to demonstrate that the maker is precise, meticulous, and worthy of receiving the animals' offering.

To produce precisely and carefully made products such as the headdress (see Figure 11.2) required mathematical precision without the use of Western instrumentation. Culture influences perception and behaviour (Berry et al., 2002), and the way Dora was taught to make her headdress and other Yup'ik projects relates to the importance of centre points, lines of symmetry, and body proportional measuring. Measuring and symmetry are the centre point of this learning trajectory. The measuring and symmetry are the generative processes used in constructing a square. These are cohesive processes; the square is viewed dynamically. From Dora's perspective, the square embeds a circle (Figure 11.2). This depicts the fluidity and transformational nature of regular polygons; a square can be transformed into

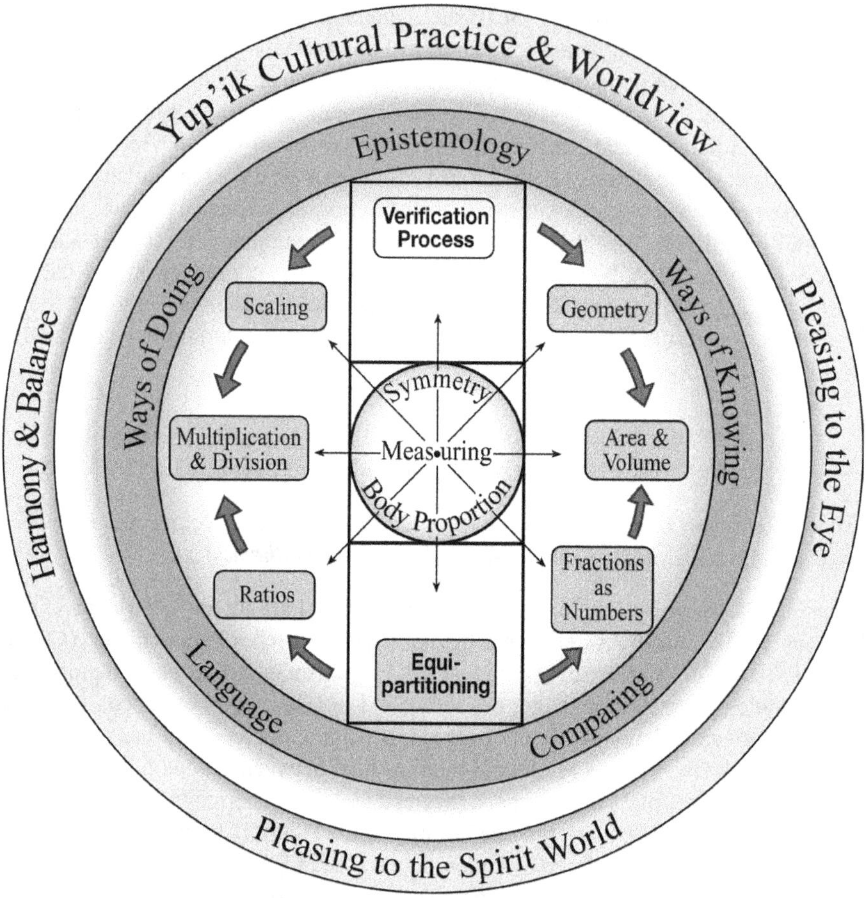

Figure 11.1. Yup'ik worldview, cultural practices, and connections to rational number reasoning and geometry

a circle or morphed into other shapes. Similarly, geometry relates to numbers, operations, and ratios.

Dora talks about the importance of the square in making many different projects. Her way of constructing the square is instrumental for its heuristic value, as the process reveals a few powerful concepts and processes that can be applied mathematically across many projects and mathematics strands. Again, in a world without Western instrumentation, oftentimes the human body becomes the default tool. The body from a Yup'ik perspective embeds four powerful concepts: (1) bilateral symmetry, (2) centre points, (3) measuring, and (4) proportionality. Since constructions need to fit the end user, the headdress, in this example, needs to be made for that person. Body proportional measuring and symmetry ensures that products fit the user.

Figure 11.2. Dora wearing her ceremonial Yup'ik headdress

The rectangular box in the centre of Figure 11.1 depicts the mathematically generative aspects of Yup'ik cultural activity: equi-partitioning, measuring, proportionality, symmetry, and verification. This is the generative constellation of concepts and processes that relate to the mathematics strands depicted in Figure 11.1 Symmetry can be used to create equal partitions and to verify if the portions are equal. As the Elders would say, "Is this side equal to that side?" This elegant set of concepts, principles, and processes represents what we believe is a generative, powerful, and integrative way to build a mathematical foundation for young students. The body proportional measuring, symmetry/splitting, and verification process relate directly to the construction of geometrical shapes. Thus, this approach can support young students' initial understanding of shape and property through constructing their own shapes. At first, they construct shapes with assistance, for example, embedding hints within the materials and later unassisted. In addition to geometrical properties and shapes, creating shapes within shapes shows not only how shapes are related geometrically but also how geometry can connect to numbers and multiplicative thinking. The use of symmetrical splitting, proportionality, and halving/doubling connects geometrical shapes to one another, for example, squares to triangles to rectangles, and rectangles to rhombi. Choosing a shape from the constellation of shapes constructed by symmetrical halving or thirding as a unit of measure provides a way to develop the concept of area as a comparison of measurements (see Figure 11.11). These processes can lead to students' explorations into scaling, ratios, fractions, and multiplication/division. This takes place in a relational measuring context.

Dora's headdress construction highlights some of these processes and mathematical concepts. Starting with uneven irregular material, she begins the process of making her headdress, reflecting how her mother and other Elders constructed the headdress and other projects out of raw materials. Dora typically contextualizes the process of making products or ceremonial artifacts; see Lipka, Andrew-Ihrke, and Yanez (2011) for a fuller description of the Yup'ik context.

The Square: The Centre of Yup'ik Constructions

The headdress construction begins with the fashioning of a square out of raw materials. At workshops or similar venues, Dora demonstrates the step-by-step process, but for the purpose of this chapter we will only explore the first phase in the square-making process. She began by creating a square out of uneven material. Because natural materials do not come with edges or right angles, she folded a crease on uneven material—for the demonstration she used uneven paper—to create a straight line, then cut along the straight line (see Figure 11.3).

An essential element of the Yup'ik process of constructing a square is the use of body proportional measuring (a specific body measure related to the end product, e.g., knuckle length; see Figures 11.4 and 11.5). She used the knuckle measure because she wanted an end product to fit her body, that is, a product in proportion to her body. Dora adopted this practice in the construction of the square. She measured one knuckle length from the cut edge of the material and marked it (see Figure 11.4).

She then measured another knuckle length from the previous point and marked the new length. She repeated the measures at two other points along the line. Each point in the set was equidistant from the straight line.

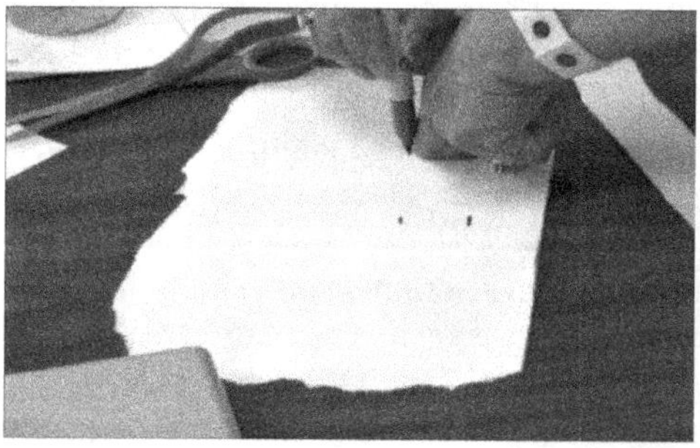

Figure 11.3. Making a square from uneven material, using the knuckle measure

Figure 11.4. Using body proportional measures to create a custom made square

These three points established what would become the opposite side or edge of the square from the straight line. Then, she folded the straight line (cut) onto the three dots and, using a straight-line edge, drew in the line representing the opposite edge of the square and cut along the line. Then, she folded the paper in half perpendicular to the previous fold. This created a temporary centre point. She repeated the knuckle measurement process from the centre point in both directions. This established an outline of the square; she envisioned the square from the centre point out, not from the outside corners. Then, she cut the outline of the square (see Figure 11.5).[3]

Perceiving the square from the centre out is a shift from the classical definition of a square: four right angles (viewing the corners) and four equal sides. The four right angles can be perceived at the centre point, with four 90-degree angles rotating around the centre and with equal lengths from the centre point to the edge. Parenthetically, a Yup'ik demonstrative "indicates an entity by its spatial or temporal relationship to the speaker" (Jacobson, 1984 p. 653). Thus, from the perspective of the speaker, these demonstratives indicate location or direction, thus reinforcing the notion that culture and language shape perception and hence a cultural preference for perceiving centre points.

She folded along the four lines of symmetry, verifying that the halves (the two diagonals as well as on the vertical and horizontal planes) of the square were congruent (see Figure 11.6). This simple and elegant process or folding algorithm is the basic building block for constructing other geometrical shapes (including three-dimensional shapes), creating pattern sets that adorn clothing, measuring, and partitioning areas into fractional parts.

The square is the foundation of Yup'ik geometrical constructions. Elders have reinforced the importance of centre points and lines of symmetry. Symmetry and measuring create geometrical centres where plane shapes transform into one another, such as squares to circles, circles into stars, and stars into flowers. Figures 11.6 through 11.8 show a square transforming into a circle. In Figure 11.7 view the

Figure 11.5. Finding the centre and verifying a square

Figure 11.6. Measuring from the centre and forming right angles

Figure 11.7. Dora creates an instrument to measure from the centre point to the four cardinal directions

line segment from the centre point to a side; this length becomes the radius of the "embedded circle." Dora envisions the circle in the square.

Dora continues the folding process until she has created a sufficient number of folds, usually 16, allowing her to cut out the "embedded" circle. Figures 11.8 and 11.9 show the symmetrical and binary process of splitting an n-gon into a circle. Further, squares can be transformed into rectangles, and rectangles into rhombi. In fact, she can make any plane shape that I have challenged her to make—parallelograms, trapezoids, and kites. What lessons can be taken from these examples? For teachers of elementary school mathematics and for students, these processes—dynamic measuring (proportionality), the use of transformational geometry (symmetry and congruence)—shift the perceptual field from the corners and external sides to the centre and may tap into preexisting cognitive pathways, or, at the least, tap into students' familiarity with symmetry.

Teachers in our professional development workshops are quick to see how certain geometrical plane shapes are related. This approach provides teachers and students with flexible thinking and movement away from crystallized knowledge into more adaptive thinking as they explore shapes, their properties, and how to see similarities across different shapes. As will be elaborated, Yup'ik solutions to everyday problems embed multiple mathematical strands, and in a school context, provide teachers and students with an integrated, connected, and potentially effective approach to learning mathematics.

From Geometry and Measuring to Early Algebraic Thinking to Numbers and Number Relations

In addition to the Yup'ik headdress, another common project is making squares, "stripping the squares" (into smaller rectangular strips), and making pattern pieces

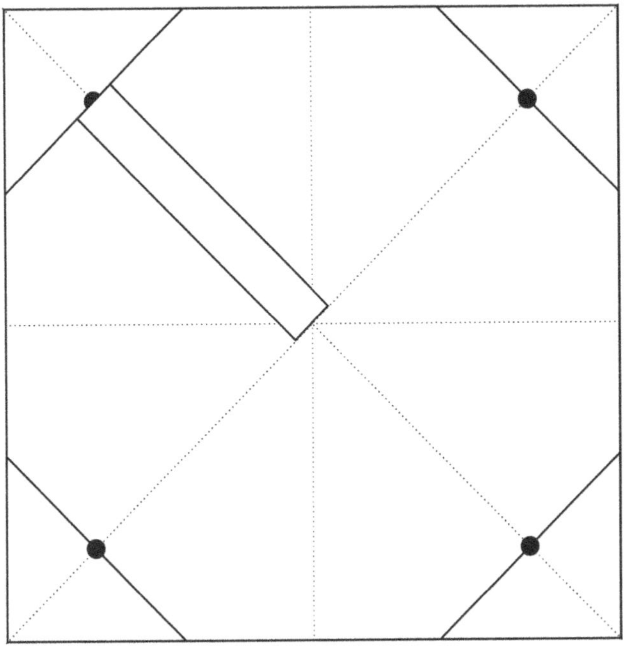

Figure 11.8. Square becoming an octagon

as shown in Figure 11.10. These were used to make the smaller, intricate, and pleasing geometrically shaped pieces that are decoratively placed on clothing. Dora uses symmetry to verify the precision of each piece.

Yup'ik activity includes comparing and measuring. Starting from the use of the knuckle measure to make a particular square and on to arm length measures used to make a kayak, body proportional measures by definition establish ratios and proportions. For purposes of school mathematics, the Yup'ik approach establishes a relational measuring context.

The knuckle measure is sometimes compared to the three middle fingers (which in the case of the square made from the knuckle measure is equal to two knuckle measures) (see Figure 11.11).

Simple body proportions and simple classroom projects provide young students with an opportunity to ground early algebraic reasoning. For example, students can explore the relationship between these body measures by labelling the knuckle measure "K" and the three-finger measure "L." They can easily compare the length of these measures. With little difficulty, they will notice that the lines are not the same length. They will, in time, incorporate formal mathematical notation by naming each measure, K and L for example: $K < L$; $L > K$; and $K \neq L$. Further, students will explore physically and spatially that L/K (division) requires 2K (multiplication) to

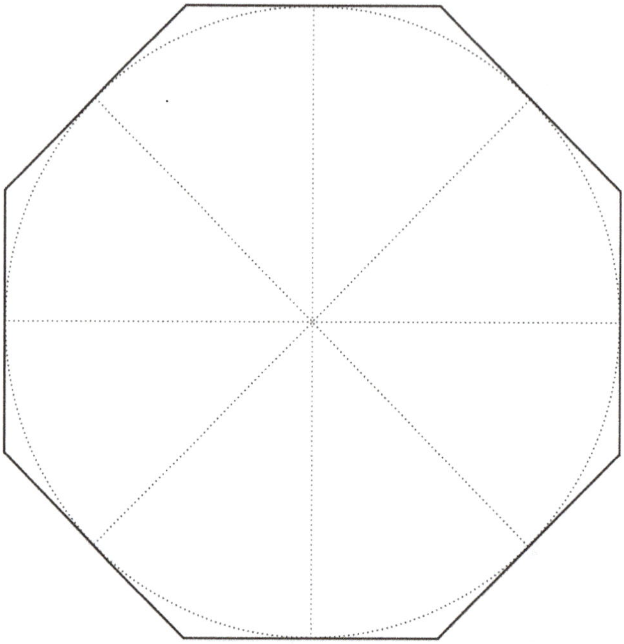

Figure 11.9. Square transforming into a circle

Figure 11.10. Practical everyday work reveals shapes within shapes, the relationship between constructing patterns and fractions

measure L; or K + K = L; or L – K = K. Connecting measuring and early algebraic reasoning follows the work of Davydov (1975) and Dougherty (2010).

Similarly, we can connect geometry (geometrical similarity) and rational number reasoning, in part through multiplicative reasoning, by using the symmetry/splitting approach, similar to Confrey and Scarano's (1995) notion of halving/splitting. We can do this by taking the constructed square and making four recursive symmetrical

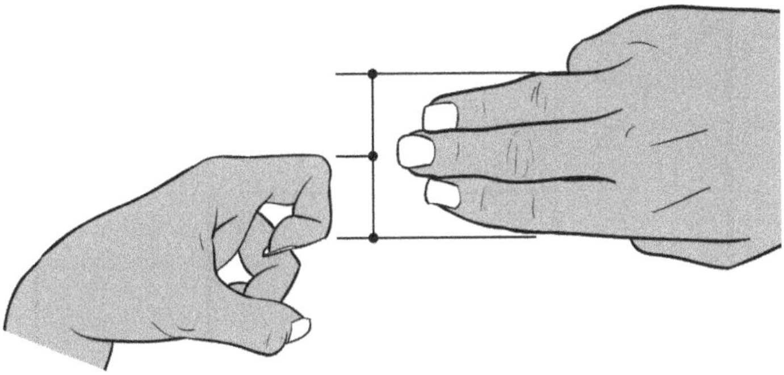

Figure 11.11. Physically comparing lengths—body proportionality

folds; the resulting geometric shape (Figure 11.11) reflects scaling and geometrical similarity (Confrey et al., 2009). For example, Figure 11.12 illustrates geometrically similar triangles and squares. These examples link the multiplicative action of paper folding and a Yup'ik halving algorithm applied in one dimension (focusing on length while using paper strip lengths) and applied in two dimensions to geometrical shapes (squares, triangles, and other shapes); encouraging additional explorations into how halving/doubling behave differently in one and two dimensions.

Further Explorations into Numbers

The halving and doubling process becomes another way in which the Yup'ik relational measuring context connects to numbers and number relations. At various professional development institutes held by MCC, we take a square and create four rectangular strips. Dora, following what she learned from her mother, would partition the strips into 2 to 12 parts. She then arranged the strips into two piles: strips with partitions of 2, 4, and 8 parts and all other partitions. Her mother labelled halving/doubling as easy folds and all others as hard-to-make folds. This demonstrates the fundamental and accessible nature of this halving process (Confrey & Scarano, 1995). Dora's folding algorithm relies directly on symmetry as an action, splitting or partitioning pieces (Pothier & Sawada, 1983; Pothier, Dalya, & Sawada, 1990; Tzur & Simon, 2005). Arranging fractional partitions in this schema results in a series, $1/2^1$, $1/2^2$, $1/2^3$, … and so on. Thus, folding or partitioning a whole into a series of $1/2$ partitions provides students with experiences in multiplicative thinking as opposed to simply repeated addition as a basis for multiplication, and this is part of a larger alternative pathway in which students construct numbers.

Subsequently, Dora demonstrated how to construct fractions that seemingly cannot be created by recursive halving. She estimates the first part, in effect subtracting one from an odd number of folds. For example, to construct a fifth, Dora observed the $1/4$ piece and estimated a smaller amount for the $1/5$ piece. With the rest of the strip, she used the halving/splitting algorithm to create four pieces folded on top of one another

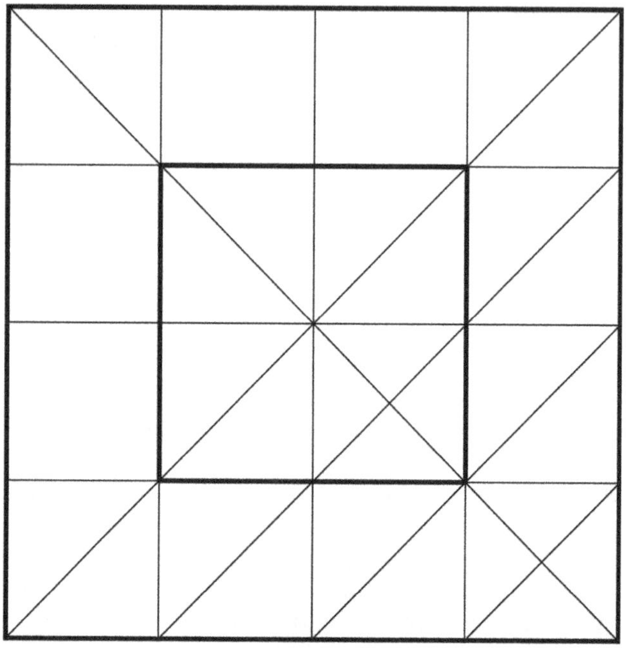

Figure 11.12. Symmetrical halving and scaling

(see Figure 11.13). Stated differently, she subtracted a 1/5 piece by folding it under, leaving a smaller strip that is folded using the halving/splitting algorithm. The next step reverts back to the simpler and more accessible halving folding algorithm. The final step is verifying, using symmetry. Is this piece (the folded-under piece) equal to the other four pieces? Are all pieces of equal size? Once verified, you have five equal lengths (as length is the focus) and these lengths can be thought of fractional parts or whole numbers.

Then Dora said, "Once you know these folds, you can make any number of folds." You can use this method to fold any prime within the limitations of the material being used and the person's facility, using analogous reasoning. This is another intersection between out-of-school and in-school learning, where everyday Yup'ik processes intersect with school mathematics. When presenting this challenge to teachers, we ask, "Can you make a set of strips from 1/2 to 1/12?" Many teachers find making the 1/5 and 1/7 particularly problematic. Thus, these have been highly instructive exercises for teachers, as the folding itself reveals a multiplicative structure.

To further illustrate how Yup'ik everyday knowledge can impact the teaching of fractions, the next example illustrates insights into number relations/number theory. These strategies in a Yup'ik context occur when some Yup'ik Elders create patterns (taking a strip and creating fractional pieces for a design). At professional development education workshops, Dora or one of the MCC faculty members will

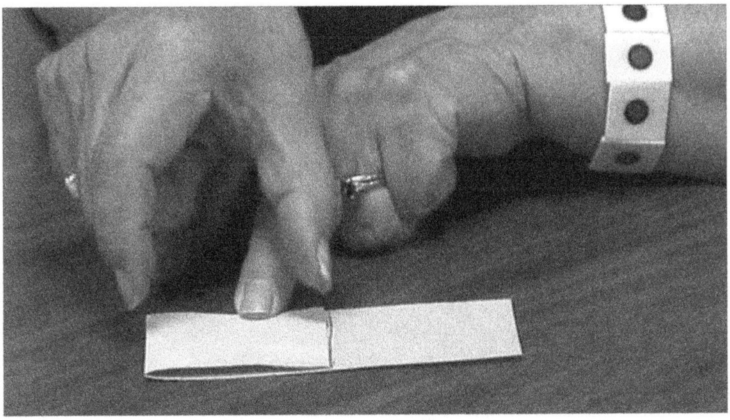

Figure 11.13. Dora demonstrates thirds through halving and symmetry

create two piles of strips. One will have 1/2, 1/4, and 1/8. The other pile will have 1/3, 1/5, 1/6, 1/7, 1/9, 1/10, etc. Teachers will be asked, "How were these strips organized? What is the organizing principle?" Usually this exercise and the previous one will be very engaging for teachers. Teachers will see that the principle is not odd and even, but those strips that can be made by doubling/halving process in the first pile and all others in the other pile. This is somewhat reminiscent of Deborah Ball's student's explanation of why six could be an odd or an even number "because it is constructed with from both odd and even numbers" (Ball, 1993, pp. 385–386). Similarly, Dora explained that six was not included in the 2, 4, 8 pile because, although it is an even number, it is constructed from both even and odd numbers, which makes it a composite number.

We have found that mathematics educators, mathematics textbooks for math educators, and teachers in general have a difficult time determining how to create a 1/5. One textbook (Brumbaugh, Rock, Brumbaugh, & Rock, 2003, pp. 97 & SM 14) suggests making a 1/8 and cutting off three subunits to create a 1/5! The procedure and ways of thinking about numbers generated from Annie Andrew, Dora's mother, provide insights into constructing numbers and understanding number relations, potentially strengthening students' cognitive/mathematical structures.

The methods described in making fraction strips include the same fundamental cognitive and mathematical strategies employed in constructing a square. These repeating and integrated strategies begin with notions of estimation, measuring, the importance of the centre point, and the notion of verifying: Is this side equal to the opposite side? Figure 11.14 summarizes this integrated approach to teaching mathematics. These strategies have applicability to fractions, composite, and prime numbers. This is a potentially powerful way to teach fractions from unit fractions to division of fractions; other mathematical topics all generated from the splitting/partitioning approach.

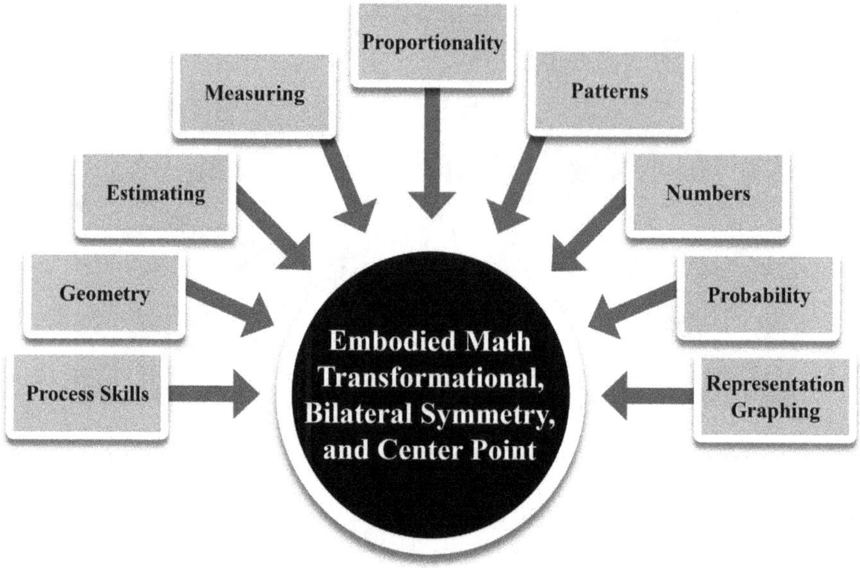

Figure 11.14. Transformational geometry connects geometry, measuring, patterns, and numbers

As the above examples have indicated, MCC's work integrates key mathematics strands, as noted in Figure 11.13. The centrality of measuring, proportionality, and the processes of comparing and verifying through symmetry create connections between math strands.

WORK UNDER DEVELOPMENT

We believe that the way Yup'ik people construct their everyday artifacts and their use of expert-apprentice modelling can translate successfully into the elementary school classroom. Thus, MCC's new work advances our previous work in a number of important ways. First, the elegance of everyday Yup'ik activity, with its emphasis on symmetry and measuring, centre point, and partitioning becomes a critical component in the curriculum development process, for example, students' constructing their own mathematical tools, such as geometrically related shapes or fraction strips, through a set of consistent principles. This establishes the possibility for mathematical explorations into a wide variety of mathematical concepts. The approach appears particularly powerful regarding the exploration of rational number reasoning in a context of materials that students construct. The use of a constructivist approach and paper folding provides students with a record of their actions/operations on physical quantities such on lengths, areas, and number lines. This supports a way for students to become increasingly autonomous as they explore, review, and identify patterns and structures behind folding. This should enable the student to see that symmetrical splitting actions on raw material not only transform material into useable classroom tools but link student actions to rational number reasoning, and assist in developing mathematical representations and concepts. Confrey, Maloney, Nguyen, Mojica, and Myers (2009) note that equal partitioning is foundational to understanding rational number reasoning. Students' use of recursive binary folding provides a powerful way to introduce students to exponents. We have extended and connected binary folding and powers of two in a measuring context as a way to model place value understanding (see Lipka, Andrew-Ihrke et al., 2015, for a more detailed discussion). Although we begin with a binary measurement system as a way to model base two, this approach can be used for other bases, as the underlying multiplicative structure of place value is revealed through such exercises.

Further, paper folding support students' understanding of equivalent fractions $\left(\frac{a}{b}\times\frac{n}{n}=\frac{a}{b}\right)$; also, as Kieren (1995) observed, paper folding can be used to model fraction operators. The cumulative effect of this type of folding paper embodies a multiplicative structure (Empson & Turner, 2006) which is essential for understanding fractions (Thompson & Saldanha, 2003). Similar to Empson and Turner, we envision sets of tasks and assessments in which the use of prime factorization using half-folds and third-folds gives rise to composite numbers. Clearly, these methods have applicability beyond the Yup'ik context.

Teaching students the folding algorithm to create $1/5$s or $1/7$s, or even $1/11$s as performed by the Yup'ik, is a rich learning task. The process of students constructing their own geometrically related shapes and fraction strip sets challenge and further develop their multiplicative thinking. This process results in products (the fraction strips can be used as culturally mediating math tools) that will further support students' understanding of number relations and multiplication. This process also supports partitioning and measurement division. As students create fraction strip tools, they will represent units and investigate the relationship of subdividing the iterated unit (Bright, Behr, Post, & Wachsmuth, 1988).

We have used the fraction constructions described above as a way for teachers to distinguish between additive and multiplicative thinking. For example, asking students to make a $1/8$ piece can be used to distinguish students who think additively from those who think multiplicatively. As Empson and Turner's (2006) findings suggest, "Given a task structure that provides multiple opportunities to test and revise ideas, shifts from additive to emergent multiplicative reasoning are likely" (p. 55). Empson (1999) noted that prepartitioned representational tools did not facilitate students' understanding of fraction equivalence, thus supporting the folding operations used by the Yup'ik.

Further, we have explored these tools particularly for the teaching of division of fractions. At first, we observed how Dora divided fractions during everyday tasks such as working with material and or making pattern strips. Dora demonstrated using what she learned from her mom and other Elders and how it might apply to teaching

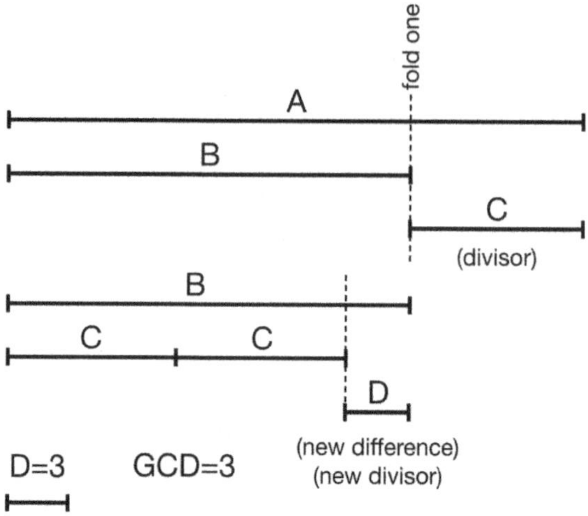

Figure 11.15. Non-numerical comparison of two lengths resulting in GCD

division of fractions in a school context. We realized that she was using alternative algorithms to invert and multiply—an algorithm more in line with Euclid's theorem as well as a common denominator approach to dividing fractions. This is briefly illustrated in Figure 11.15.

The difference between the measure of Length A and B is Length C. Length C becomes the divisor. Length D, the new difference, divides Length C into 3 equal parts. Length D is the Greatest Common Divisor; GCD equals 3.[4]

The same fundamental processes of measuring as comparing, symmetry, partitioning, and verifying are applied to making n parts of a length which can be used in teaching fractions, creating number lines, and geometrical constructions such as the square. These processes support arithmetic operations and early forms of algebraic thinking. Similarly, these processes support rational number reasoning and geometrical similarity. We recognized that when Dora and other Elders accomplished these tasks, they were also using transformational geometry as a method for constructing everyday artifacts. This approach appears to appeal to teachers who did not know how to connect the culture of the community to the culture of school. In more mainstream contexts, teachers have reported how the use of halving (used to produce 1/3s, 1/5s, and so on) has been used to teach an array of elementary school topics from partitioning a strip into 10 equal parts and create proportional money to partitioning a length into 12 equal parts to teach fractions, common denominator, factors, and prime factorization.[5]

Thus, these examples from MCC's work begin to shift the argument about culturally based/responsive education from the margins to mainstream. Here are strategies that can be applied widely and effectively. Here are ways of engaging teachers in professional development workshops that can be engaging and challenging. Further, Elders have also shown an elegant and systematic way to approach the teaching of mathematics that integrates multiple mathematical strands and ways of communicating, teaching/learning, and verifying. Expert-apprentice modelling assists in building knowledge in novices. Therefore, we argue that lessons learned from the margins may help improve mainstream mathematics education. Can we afford not to listen to each other?

SUMMARY AND CONCLUSION

The vision of this work rests on a deep appreciation for the lessons that we have learned from Yup'ik Elders over many years. The examples provided in this chapter show the power of symmetry and measuring, and how mathematical approaches to solving everyday problems have applicability in the larger school context beyond Yup'ik communities.

Significantly, based on Dora's demonstrations, her unschooled mother and other Elders intuitively understood the structure of numbers—recognizing the power of halving, primes, and composites. They accomplished this in the context of making clothing and pattern pieces while developing methods to create equal partitions. From this model, we observe how Yup'ik actions, culturally preferred ways of perceiving and manipulating objects in space, result in a set of procedures resting on the importance of symmetry and balance, centre point, and connecting measuring with proportionality (similar to Davydov, as noted by National Research Council, 2001, p. 257).

One of the interesting conclusions about this culturally specific work in Alaska is that it appears to be simultaneously applicable to a wide range of people and settings. The very core notions of the approach outlined in this chapter—body proportional measuring, symmetry, equi-partitioning, and verification—are processes that have developed across cultures and times. These processes are both culturally specific as developed and learned in one cultural context but may also be universally applicable. All people use body proportional measuring, use symmetry in their artwork, clothing, or artifacts (Washburn & Crowe, 1998), probably based on the predisposition people have to see beauty in symmetry and being surrounded by living things that embed symmetry. Thus, the approach being learned from Yup'ik Elders has a number of attributes that are accessible much more widely. For purposes of schooling, the learning trajectory being established in this chapter shows how one action results in many mathematical strands. The approach outlined is a potentially efficient and effective way to establish students' mathematical foundation.

As Dora and Yup'ik Elders have instructed us about the importance of balance and centre point, this chapter seeks to support the aforementioned practices in becoming a seamless part of schooling, taking advantage of some of the unique perspectives that Yup'ik people have to offer to instruction within school settings. Our current work will ensure that these processes become an integral part of schooling, a part that accommodates students whose culture and language have been previously excluded. Yet, the approaches to teaching mathematics in this chapter provide symmetry—symmetry between small Indigenous groups and the contributions that they may make to the larger mainstream culture. This symmetry is also one of power and respect, potentially creating more space for the contributions of others who are currently perceived as marginalized.

NOTES

1 This is a very complex issue including many other factors, such as teacher and administrator turnover rates, Alaska's ongoing budget crisis, and problems associated with despair, partially resulting in increased use of drugs, alcohol, and incidents of suicide.

2 This was illustrated and discussed in our most recent meeting on April 5, 2017, in Anchorage with Sassa Petersen, Evelyn Yanez, and Dora Andrew-Ihrke, each describing spherical space. They also described two-dimensional space as well as one-dimensional space as they reviewed making clothing and weaving a basket.

3 MCC has archived a video collection on YouTube which includes the square making described above. See https://www.youtube.com/watch?v=755sBTe5Dcw&list=PL9qLy5US9b8JRSOU_Ya6_Jt7uOJBP6o4T and, more generally, https://www.youtube.com/channel/UCpNLxDROxgsMnzDyx2Q2heg, where other constructions are videoed as well sample curriculum tasks and cultural events).

4 For a similar view, see Euclid's algorithm: https://en.wikipedia.org/wiki/Euclidean_algorithm

5 Based on classroom observations of teachers implementing aspects of this curriculum in rural and urban parts of Alaska.

REFERENCES

Abedi, J. (2004). The no child left behind act and English language learners: Assessment and accountability issues. *Educational Researcher, 33*(1), 4–14.

Adams, B. L., Kagle, M., & George, F. (2007). *Star navigation: Explorations into angles and measurement.* Calgary: Detselig.

Alaska Native Education Program, Annual Report Award #S356A060004. (2009). *Annual Report to U.S. Department of Education, Alaska Native Education Program.* Fairbanks, AL: University of Alaska Fairbanks.

Au, K., & Carroll, J. (1997). Improving literacy achievement through a constructivist approach: The KEEP demonstration classroom project. *The Elementary School Journal, 97*(3), 203–221.

Au, K., & Jordan, C. (1981). Teaching reading to Hawaiian children: Finding a culturally appropriate solution. In H. Trueba, G. Guthrie, & K. Au (Eds.), *Culture in the bilingual classroom: Studies in classroom ethnography* (pp. 139–152) Rowley, MA: Newbury House.

Au, K., & Mason, J. (1981). Social organizational factors in learning to read: The balance of rights hypothesis. *Reading Research Quarterly*, 115–152.

Ball, D. L. (1993). With an eye on the mathematical horizon: Dilemmas of teaching elementary school mathematics. *The Elementary School Journal, 93*(4), 373–397.

Beaulieu, D. L. (2000). Comprehensive reform and American Indian education. *Journal of American Indian Education, 39*(2), 29–38.

Berry, J. W. (1976). *Human ecology and cognitive style.* New York, NY: John Wiley & Sons.

Berry, J. W., Poortinga, Y. H., Segall, M. H., & Dasen, P. R. (2002). *Cross-cultural psychology: Research and applications* (2nd ed.). Cambridge, MA: Cambridge University Press.

Bishop, A. J. (1988). Mathematics education in its cultural context. *Educational Studies in Mathematics, 19*(2), 179–191.

Brenner, M. E. (1998). Adding cognition to the formula for culturally relevant instruction in mathematics. *Anthropology and Education Quarterly, 29*, 214–244.

Bright, G. W., Behr, M. J., Post, T. R., & Wachsmuth, I. (1988). Identifying fractions on number lines. *Journal for Research in Mathematics Education, 19*(3), 215–232.

Brumbaugh, D. K., Rock, D., Brumbaugh, L. S., & Rock, M. (2003). *Teaching K-6 mathematics*. Mahwah, NJ: Lawrence Erlbaum Associates.

Bruner, J. S. (1964). The course of cognitive growth. *American Psychologist, 19*, 1–15.

Bruner, J. S. (1996). *The culture of education*. Cambridge, MA: Harvard University Press.

Bush, G. W. (2004, April 30). *Executive Order 13336 – American Indian and Alaska Native Education*. Retrieved from http://www.presidency.ucsb.edu/ws/?pid=61430.

Carraher, T. N., Carraher, D. W., & Schliemann, A. D. (1985). Mathematics in the streets and in school. *British Journal of Developmental Psychology, 3*, 21–29.

Castagno, A. E., & Brayboy, B. M. J. (2008). Culturally responsive schooling for indigenous youth: A review of the literature. *Review of Educational Research, 78*(4), 941–993.

Cazden, C. B., & Leggett, E. L. (1981). Culturally responsive education: Recommendations for achieving Lau remedies. In H. T. Trueba, G. P. Guthrie, & K. H. Au (Eds.), *Culture and the bilingual classroom: Studies in classroom ethnography* (pp. 69–86). Rowley, MA: Newbury House.

Civil, M., & Kahn, L. H. (2002). Everyday mathematics, mathematicians' mathematics, and school mathematics: Can we bring them together? In M. Brenner & J. Moschkovich (Eds.), *Everyday and academic mathematics in the classroom. Journal of Research in Mathematics Education Monograph #11* (pp. 40–62). Reston, VA: National Council of Teachers of Mathematics.

Clinton, W. J. (1998, August 6). *Executive Order 13096 – American Indian and Alaska Native Education*. Retrieved from http://www.presidency.ucsb.edu/ws/?pid=54747.

Cole, M., & Scribner, S. (1974). *Culture and thought: A psychological introduction*. New York, NY: Wiley & Sons.

Confrey, J., Maloney, A., Nguyen, K., Mojica, G., & Myers, M. (2009). Equipartitioning/splitting as a foundation of rational number reasoning using learning trajectories. *Proceedings of the 33rd Conference of the International Group for the Psychology of Mathematics Education* (pp. 345–353). Thessaloniki, Greece.

Confrey, J., & Scarano, G. H. (1995, October). *Splitting reexamined: Results from a three-year longitudinal study of children in grades three to five*. Paper presented at the North American Chapter of the International Group for the Psychology of Mathematics Education, Columbus, OH.

Davydov, V. V. (1975). The psychological characteristics of the "prenumerical" period of mathematics instruction. In L. P. Steffe (Ed.), *Soviet studies in the psychology of learning and teaching mathematics: Childrens' capacity for learning mathematics* (Vol. 7, pp. 109–205). Chicago, IL: University of Chicago Press.

Dehaene, S., Spelke, E., Pinel, P., Stanescu, R., & Tsivkin, S. (1999). Sources of mathematical thinking: Behavioral and brain-imaging evidence. *Science, 284*(5416), 970–974.

Demmert, W. G., & Towner, J. C. (2003). *A review of the research literature on the influences of culturally based education on the academic performance of Native American students*. Portland, OR: Northwest Regional Educational Laboratory.

Deyhle, D., & Swisher, K. (1997). Research in American Indian and Alaska Native education: From assimilation to self-determination. *Review of Research in Education, 22*, 113–194.

Doherty, W., Hilberg, S., Epaloose, G., & Tharp, R. (2002). Standards performance continuum: Development and validation of a measure of effective pedagogy. *The Journal of Educational Research, 96*(2), 78–89.

Dougherty, B. J. (2010). A Davydov approach to early mathematics. In Z. Usiskin, K. Andersen, & N. Zotto (Eds.), *Future curricular trends in algebra and geometry* (pp. 63–69). Charlotte, NC: Information Age Publishing.

Empson, S. B. (1999). Equal sharing and shared meaning: The development of fraction concepts. *Cognition and Instruction, 17*(3), 283–342.

Empson, S. B., & Turner, E. (2006). The emergence of multiplicative thinking in children's solutions to paper folding tasks. *Journal of Mathematical Behavior, 25*(1), 46–56.

Erickson, F., & Mohatt, C. (1982). Cultural organization and participation structures in two classrooms of Indian students. In G. Spindler (Ed.), *Doing the ethnography of schooling* (pp. 131–174). New York, NY: Holt, Rinehart, & Winston.

Feng, C. (2002). Looking good: The psychology and biology of beauty. *The Journal of Young Investigators, 6*, n.p.

Giurfa, M., Eichmann, B., & Menzel, R. (1996). Symmetry perception in an insect. *Nature, 382*, 458–461.

Grigg, W., Moran, R., & Kuang, M. (2010). *National Indian education study. Part I: Performance of American Indian and Alaska Native students at grades 4 and 8 on NAEP 2009 reading and mathematics assessments* (NCES 2010-462). Retrieved from https://eric.ed.gov/?id=ED510598

Grigorenko, E. L., Meier, E., Lipka, J., Mohatt, G. V., Yanez, E., & Sternberg, R. (2004). The relationship between academic and practical intelligence: A case study of the tacit knowledge of Native American Yup'ik people in Alaska. *Academic and Practical Intelligence, 14*(4), 183–207.

Gutiérrez, R. (2008). Realizing the potential of Chicanas/os and Native Americans: Engaging identity and power issues in teaching students mathematics and science. *Hispanic Outlook in Higher Education, 19*(18), 27.

Harris, P. (1991). *Mathematics in a cultural context: Aboriginal perspectives on space, time, and money.* Geelong: Deakin University.

Institute for Government Research. (1971). *The problem of Indian administration* (Report of a survey made at the request of Honorable Hubert Work, Secretary of the Interior, and submitted to him, February 21, 1928). New York, NY: Johnson Reprint. (Original work published in 1928.) Also known as the Merriam Report.

Jacobson, S. A. (1984). *Yup'ik Eskimo dictionary*. Fairbanks, AL: University of Alaska Fairbanks, Alaska Native Language Center.

Kieren, T. (1995). Creating spaces for learning fractions. In J. Sowder & B. Schappelle (Eds.), *Providing a foundation for teaching mathematics in the middle grades* (pp. 31–65). Albany, NY: SUNY Press.

Kisker, E. E., Lipka, J., Adams, B. L., Rickard, A., Andrew-Ihrke, D., Yanez, E. E., & Millard, A. (2012). The potential of a culturally based supplemental mathematics curriculum to improve the mathematics performance of Alaska Native and other students. *Journal for Research in Mathematics Education, 43*(1), 75–113.

Kristine Moore et al. v. State of Alaska, 3AN-04-9756 (Superior Court for the State of Alaska Third Judicial Division 2010).

Lakoff, G., & Núñez, R. E. (2000). *Where mathematics comes from: How the embodied mind brings mathematics into being.* New York, NY: Basic Books.

Lancy, D. F. (1981). The indigenous mathematics project: An overview. *Educational Studies in Mathematics, 12*(4), 445–453.

Lave, J. (1977). Cognitive consequences of traditional apprenticeship training in West Africa. *Anthropology & Education Quarterly, 8*(3), 177–180.

Lave, J. (1982). A comparative approach to educational forms and learning processes. *Anthropology & Education Quarterly, 13*(2), 181–187.

Lee, C. D. (1995). A culturally based cognitive apprenticeship: Teaching African American high school students skills in literary interpretation. *Reading Research Quarterly, 30*(4), 608–630.

Lipka, J., Andrew-Ihrke, D., Koester, D., Zinger, V., Olson, M., Yanez, E., & Rubinstein, D. (2015). Indigenous knowledge provides an elegant way to teach the foundations of mathematics. In T. G. Bartell, R. T. Putnam, K. Bradfield, & H. Dominguez (Eds.), *Proceedings of the 37th Annual Meeting of the North American Chapter of the International Group for the Psychology of Mathematics Education* (pp. 2–18). East Lansing, MI: Michigan State University Press.

Lipka, J., Andrew-Ihrke, D., & Yanez, E. E. (2011). Yup'ik cosmology to school mathematics: The power of symmetry and proportional measuring. *Interchange, 42*(2), 157–183.

Lipka, J., Hogan, M. P., Webster, J. P., Yanez, E., Adams, B., Clark, S., & Lacy, D. (2005). Math in a Cultural Context: Two case studies of a successful culturally based math project. *Anthropology & Education Quarterly, 36*(4), 367–385.

Lipka, J., Parmelee, J., & Adams, R. (2004). *Picking berries: Connections between data collection, graphing, and measuring.* Calgary: Detselig.

Lipka, J., Sharp, N., Adams, B., & Sharp, F. (2007). Creating a third space for authentic biculturalism: Examples from math in a cultural context. *Journal of American Indian Education, 46*(3), 94–105.

Lipka, J., Sharp, N., Brenner, M., Yanez, E., & Sharp, F. (2005). The relevance of culturally based curriculum and instructions: The case of Nancy Sharp. *Journal of American Indian Education, 44*(3), 30–54.

Lipka, J., & Yanez, E. (1998). Identifying and understanding cultural differences: Toward culturally based pedagogy. In J. Lipka, G. Mohatt, & The Ciulistet Group (Eds.), *Transforming the culture of schools: Yup'ik Eskimo examples* (pp. 111–137). Mahwah, NJ: Lawrence Erlbaum and Associates.

Lipka, J., Yanez, E., Andrew-Ihrke, D., & Adam, S. (2009). A two way process for developing effective culturally based math. In B. Greer, S. Mukhopadhyay, A. B. Powell, & S. Nelson-Barber (Eds.), *Responsive mathematics education.* New York, NY: Routledge, Taylor and Francis Group.

Lipka, J. W., Mohatt, G. V., & The Ciulistet Group. (1998). *Transforming the culture of schools: Yup'ik Eskimo examples.* Mahwah, NJ: Lawrence Erlbaum Associates.

Mohatt, G., & Erickson, F. (1981). Cultural differences in teaching styles in an Odawa school: A sociolinguistics approach. In H. Trueba, G. Guthrie, & K. Au. (Eds.), *Culture and the bilingual classroom: Studies in classroom ethnography* (pp. 105–119). Rowley, MA: Newbury House.

Moll, L., Amanti, C., Neff, D., & González, N. (1992). Funds of knowledge for teaching: A qualitative approach to developing strategic connections between homes and classrooms. *Theory into Practice, 31*(2), 132–141.

Moses, R., & Cobb, C. J. (2001). *Radical equations: Math literacy and civil rights.* Boston, MA: Beacon Press.

Nasir, N. S., Hand, V., & Taylor, A. V. (2008). Culture and mathematics in school: Boundaries between "cultural" and "domain" knowledge in the mathematics classroom and beyond. *Review of Research in Education, 32*(1), 187–240.

National Council of Teachers of Mathematics. (2000). *Principles and standards for school mathematics.* Reston, VA: Author.

National Mathematics Advisory Panel. (2008). *Foundations for success: The final report of the National Mathematics Advisory Panel.* Washington, DC: US Department of Education.

National Research Council. (2001). *Adding it up: Helping children learn mathematics* (J. Kilpatrick, J. Swafford, & B. Findell, Eds.). Washington, DC: National Academy Press.

Pothier, Y., & Sawada, D. (1983). Partitioning: The emergence of rational number ideas in young children. *Journal for Research in Mathematics Education, 14*(5), 307–317.

Pothier, Y., & Sawada, D. (1990). Partitioning: An approach to fractions. *Arithmetic Teacher, 38*(4), 12–16.

Reyhner, J. (2009). Indigenous education in global contexts. *International Studies in Education, 10*(1), 6–9.

Saxe, G. B. (1988a). Candy selling and math learning. *Educational Researcher, 17*(6), 14–21.

Saxe, G. B. (1988b). The mathematics of child street vendors. *Child Development, 59*(5), 1415–1425.

Sternberg, R., Lipka, J., Newman, T., Wildfeuer, S., & Grigorenko, E. L. (2006). Triarchically-based instruction and assessment of sixth-grade mathematics in a Yup'ik cultural setting in Alaska. *Gifted and Talented International, 21*(2), 6–19.

Sternberg, R., Nokes, K., Geissler, P. W., Prince, R., Okatcha, F., Bundy, D. A., & Grigorenko, E. L. (1998). The relationship between academic and practical intelligence: A case study in Kenya. *Intelligence, 29*(5), 401–418.

Thompson, P. W., & Saldanha, L. A. (2003). Fractions and multiplicative reasoning. In J. Kilpatrick, W. G. Martin & D. Schifter (Eds.), *A research companion to principles and standards for school mathematics* (pp. 95–114). Reston, VA: National Council of Teachers of Mathematics.

Tzur, R., & Simon, M. (2005). Distinguishing two stages of mathematics conceptual learning. *International Journal of Science and Mathematics Education, 2*(2), 287–304.

United States Department of Education. (2002). *No child left behind.* Retrieved April 5, 2007, from https://www2.ed.gov/nclb/landing.jhtml

Washburn, D. K., & Crowe, D. W. (1998). *Symmetries of culture: Theory and practice of plane pattern analysis.* Seattle, WA: University of Washington Press.

Watson-Gegeo, K. (1994). Keeping culture out of the classroom in rural Solomon Islands schools: A critical analysis. *Educational Foundations, 8*(2), 27–55.

White, L. A. (1959). *The evolution of culture: The development of civilization for the fall of Rome*. New York, NY: McGraw-Hill.

White House Conference on Indian Education. (1992). *The final report of the White House Conference on Indian Education* (Executive summary). Washington, DC: Author.

Wynn, T. (2002). Archeology and cognitive evolution. *Behavioral and Brain Sciences, 25*(3), 389–402.

Zehr, M. A. (2008). Evidence of culture-based teaching called thin. *Education Week, 27*(17), 8.

JENNIFER S. THOM

12. BENDING AN EAR IN THE QUEST OF "WHAT IS IT?"

INITIAL THOUGHTS: (DIS)ORIENTATION

A few months before I received the invitation to contribute to this book, I was rereading *The Wayfinders: Why Ancient Wisdom Matters in the Modern World* by Wade Davis (2009). The published edition of Davis's cross-Canada Massey Lectures reveals the rich diversity of Indigenous cultures around the world while also exposing the rapid loss of Indigenous languages, the eroding of local knowledge, and the devastating effects of these events on the earth's biosphere. This time, as I read Davis's book, I was intrigued by his composition on the ancient art of navigation called wayfinding. Davis describes his experience with the crew and wayfinder Nainoa Thompson aboard the *Hōkūle'a*. The *Hōkūle'a*, a 62-foot-long doubled-hulled open-decked catamaran, is a replica of a traditional Polynesian voyaging canoe. In the composition, Davis also details stories about the years Nainoa spent with master navigator Mau Piailug, learning the ancient tradition of wayfinding.

As I prepared to write my chapter for this book, the title—*Living Culturally Responsive Education: Mathematics Education with/in Indigenous Communities*, immediately gave rise to the following question: What is it to live a culturally responsive mathematics curriculum and pedagogy? Now stuck between the two notions of curriculum and pedagogy, I found myself in a conceptual space that demanded an immediate answer—my answer.

Just then, I remembered that such a question does not ask for a simple answer (van Manen, 1986). Rather, "what" questions, as van Manen suggests, ask us to "dwell on the different meaning dimensions of that object … getting to know what that thing really *is,* what it is in its *whatness* and *thatness*" (1986, p. 38). I returned to the question, and instead of focusing on producing an answer, I inquired into the possible and different meanings that the question incites. For example, if cultures are as Davis portrays them—diverse and irreducible—then the members, communities, and geographical places of which they are a part must surely impact the meanings we make of mathematics, curricula, and teaching. As I continued thinking about the issues, I realized that there would likely be multiple answers to my question, each of them, no doubt, complex. I then wondered what implications arise when one desires to live such a curriculum and pedagogy where the knowledge and ways of knowing come from a culture of which one is not a member and lacks firsthand experience. Certainly, this must be a situation in which many if not all mathematics educators

 | DOI: 10.1163/9789004415768_012

find themselves, at some point, in their practice. Here, re-viewing the question in this manner created an open space for me to develop deeper understandings about these issues, while the traditional art of wayfinding offered a completely unfamiliar context for me to situate my learning.

In this chapter, I share what emerged for me as I moved beyond Davis's composition about wayfinding and explored navigation as mathematics and navigating as mathematical activity. Conscious of my point of view—that is, as someone who is not a member of Polynesian culture—I considered the implications that wayfinding poses as a form of navigation and navigating, and then, possible meanings that might be made about such knowledge and knowing.

The chapter includes a series of six conversations featuring: wayfinders Nainoa Thompson, Ka'iulani Murphy, and Chad Baybayan; ethnobotanist and anthropologist Wade Davis; anthropologists Ben Finney, Charles Frake, and Sam Low; English professor Dennis Kawaharada; fictional characters Ida Deal, Cy Phact, and Thea Rey; and mathematician Keith Devlin. I created the dialogues using direct quotations from interviews, films, compositions, journal articles, and books. The three fictional characters first appeared in a play that I wrote and performed with two curriculum educators (see Oberg, Blades, & Thom, 2007). The play, which focused on cultural issues in education, included a cast of characters who possessed social-cultural stereotypes and served as caricatures of Antoinette Oberg, David Blades, and me. Including Ida, Cy, and Thea in this chapter not only gave the characters new life, doing so also allowed me to participate in the conversations as both speaker and listener, which, in turn, enabled me to explore the topics as they emerged in recursive and complementary manners. As speaker, I resituated the ideas of Ida, Cy, and Thea from the play to the context of navigation and, more specifically, wayfinding. Here, I was able to listen and re*cognize*[1] the ideas as they took shape, this time, within mathematics education and in relation to the responses from the other speakers.

To be clear, the conversations are the convergence of similar and differing perspectives as well as my inquiry into the associated conceptions regarding cultural forms of navigation and navigating. My reason for including the conversations in the chapter was not simply to leave a conceptual path for the reader to follow but also to invite the reader to engage in his or her own exploration of the issues at hand. In this manner, the conversations are not dialogues or representations of complete or linear trains of thought; they are co-emergent metalogues (Bateson, 1972) that provide generative and nonlinear spaces in which to ponder, pose questions, and develop new ways of thinking—about mathematics, the nature of it, and mathematical ways of knowing. The discussion that follows each of the six conversations details the understandings that transpired.

CONVERSATION ONE: WHAT IS IT?

Wade Davis (Ethnobotanist and Anthropologist):

> If you could take all the brilliance that has allowed us to put a man on the moon, and apply it to an understanding of the ocean, what you would get would be Polynesia. Here, the ancient skill of navigation called "wayfinding" linked people on thousands of islands spread over 25 million square kilometres, nearly a quarter of the surface of the planet. It's an amazing story. (Davis, Gregg, & Gregg, 2009)

Ka'iulani Murphy (Apprentice Wayfinder):

> A modern ship might have a magnetic compass on board and steer their course by looking at that compass ....When we are sailing our voyaging canoes, we're just going solely off of clues in nature. (Patterson, 2010, para. 1)

Ben Finney (Anthropologist):

> The best-known traditional stellar navigation system in the Pacific is the one employed by a few dozen traditional navigators who still practice this ancient art on remote atolls in the central Caroline Islands of Micronesia—including Satawal, Mau Piailug's island. Carolinian navigators are particularly noted for what has been called their "star compass," which they visualize as a series of bearings along the horizon known by the rising and setting points of the key stars and constellations. That writers commonly call this conception a "star compass" is perhaps unfortunate for it is not a physical instrument like a magnetic compass. It is a purely mental construct, a conceptual direction system by which the navigator mentally divides the horizon surrounding him. (Finney, 1994, p. 55)

Charles Frake (Anthropologist):

> Although our understanding of the cognitive basis of Pacific Islanders' navigation is certainly still imperfect and incomplete, it becomes increasingly apparent that their skills, great as they are, are based on cognitive processes shared by all humans. There is nothing especially weird, wonderful, or "non-western" in the mental form of the "star compass" as a directional framework. It is basically similar to other directional systems developed by seafarers elsewhere in the world. (Frake, 1995, p. 157)

Nainoa Thompson (Wayfinder):

> [But] you cannot look up at the stars and tell where you are. You only know where you are in this kind of navigation by memorizing where you sailed from. That means constant observation. You have to constantly remember your speed, your direction and time. You don't have a speedometer. You don't have a compass. You don't have a watch. It all has to be done in your head. It is easy in principle but it's hard to do. (Thompson, n.d., para. 3)

Wade Davis:

> It is one thing, for example, to measure the speed of the *Hōkūle'a* with a simple calculation: the time a bit of foam or flotsam, or perhaps a mere bubble, takes to pass the known length separating the crossbeams of the canoe. Three seconds and the speed will be 8.5 knots; fifteen seconds and the vessel slogs at a mere 1.5 knots …. But is quite another to make such calculations continually, day and night, while also taking the measure of stars breaking the horizon, winds shifting both in speed and direction, swells moving through the canoe, clouds, and waves. (Davis, 2009, p. 60)

Nainoa Thompson:

> [Now] consider that you have to remember those … for a month—every time you change course, every time you slow down. (Thompson, n.d., para. 4)

DISCUSSION ONE: MAKING A DIFFERENCE TO THE NOTION OF DIFFERENCE

Immediately, the speakers' descriptions of the ancient traditions of wayfinding draw me into the conversation. Once inside, it is the idea of difference that catches my attention; that is, how the perspectives of the speakers discursively shape the conversation, orienting me as I bend an ear to listen and make sense of it. Below, I examine the distinctions that the speakers make about wayfinding, the manners in which they communicate these, and how the conceptual differences give rise to distinct meanings of navigation and navigating.

Difference as Different

For Ka'iulani Murphy, other types of navigation such as those that use a magnetic compass, are fundamentally unlike wayfinding, because the ancient skill does not use external devices. Rather, the positioning and the setting of a course depend on the wayfinder's constant observations and making sense of that which is sensed intimately and directly from the natural environment, such as the location and movement of the stars, winds, clouds, and waves. Ka'iulani argues that this difference alone separates wayfinding from other modern types of navigation.

French philosopher Gilles Deleuze (e.g., see Deleuze & Parnet, 1987) described such a difference as a difference in *kind*. Differences in kind are characteristics that reveal an entity or being as qualitatively distinct from another regardless of identity. Therefore, it does not matter that instrument-based sea travel and wayfinding are both types of navigation. The fact that the former uses external devices while the latter does not render each as different.

Ben Finney's distinction between the physical presence of the magnetic compass and the lack thereof regarding the stellar compass gives rise to other features that make the two types of navigation different in kind. For example, the magnetic compass is also fixed and external, whereas the star compass is a mental construct. The star compass is completely dependent upon the conception and continuous (re)creation of the wayfinder. Moreover, because the stellar compass is based on the rising and setting positions of particular stars, it is too in this way dynamic—brought into being and inseparable from the wayfinder and the places in which it is articulated.

These differences in kind between the magnetic and stellar compass impact how I conceptualize navigating and relationally, the role of the navigator. For example, to use a magnetic compass largely involves the navigator's procedural understanding, or "know-what," to interpret the compass, identify one's position, and chart a course. It is reasonable to assume that if Nainoa Thompson employed the required methods for using a magnetic compass, he would be able to navigate the *Hōkūle'a* without wayfinding. The converse to this situation, however, seems unlikely; that is, that an instrument-based navigator could copy Nainoa's process of wayfinding. This is not to say that there are not other considerations that the instrument-based navigator must take into account which extend beyond the use of the compass. Local currents, direction and velocity of the wind, and the fact that a navigator would not direct a vessel to travel through a landmass in order to reach a destination are obvious examples of how instrument-based navigation also depends on the immediate context for important information. However, the wayfinder does not use set procedures. She or he must draw on an intuitive and conceptual understanding of the stars as well as their varying locations to imagine and make use of the stellar compass. This requires a great deal of know-how (Varela, 1999)—*know*ledge to be able to continually identify which stars to observe, interpret *how* the position of certain stars in the cycle of rising and setting relate to a particular place, as well as the wayfinder's conception of the star compass as a whole.

Difference as Same

It is not without irony that Charles Frake also compares the instrument-based navigation to wayfinding and arrives at an opposite conclusion: wayfinding generally and therefore essentially, is the same as other navigational constructs both in process and product. In the conversation, Charles makes it clear that cultural knowledge, practices, and identities do not define wayfinding as distinct—neither in terms of the concept of navigation nor the artifacts that it produces. Instead, the logic upon which Charles considers wayfinding to be based and which he refers to as the star compass is consistent with what he understands of other cultures' directional frameworks.

As I listen to Charles, I imagine each of the ideas he raises in the conversation as descriptors for particular scales that gauge how wayfinding and its associated activities compare to other types of navigation. Here, Charles's account contrasts in another way to the other conceptualizations; rather than difference as difference in kind, difference is a measure or what Deleuze considered as differences in *degree* (e.g., see Deleuze & Parnet, 1987). He described difference in degree as qualities that can be reduced, quantified, categorized, and expressed in mechanistic extremes of more or less. Consequently, when I think of wayfinding and instrument-based navigation in terms of more or less, I too disregard the qualitative differences identified by the other speakers. I see, as Charles does, the star compass as essentially the same as the Medieval Western Wild Rose (see, for example, Frake, 1995, p. 151) and magnetic compass. It does not matter whether these are mental or physical, external or internal, Western[2] or non-Western, subjective or objective, dynamic or static, or even abstract or contextual, because navigation and the act of navigating is ultimately about identifying one's position and direction. In the end, the outcome or result is the same.

Temporal Differences: Timeless, in Time

For Charles, it is the final product—magnetic compass, star compass, or Medieval Western Wild Rose—against which all else should be compared and measured, including the logic on which they operate. As listener, my attention *fix*at*es* on *object*ively de*lineat*ing meanings for navigation and navigating. Navigation for certain, is a matter of objectively determining one's position and direction based on the magnetic compass as an external, abstract, and static entity. Consequently, navigating is reduced to a procedure or application of knowledge, whereby the magnetic compass with its absolute and universal directions exists independent of the navigator. Timeless, such methods and instrument(s) can be used anytime, anywhere, and by anyone.

The other speakers, however, reveal wayfinding as an emergent event; one that evolves as the wayfinder continuously observes, measures, and calculates the natural world all at once, moment to moment, and context by context. Navigating in this way is an ongoing and recursive relationship *with* the environment. Moreover, as the mathematical knowledge that arises from wayfinding is not pre-existent, it also cannot be predetermined. Everything relies on and relates to everything else. These differences in kind reveal a temporal dependence that weaves past, present, and future into an ever-changing now (Maturana, 2001). Deleuze asserted that differences in kind possess temporal duration and are never pre-existent, fixed, static, or complete. Differences in kind are indeterminate and irreducible. Not completely unlike dead reckoning—"the process of calculating one's position, especially at sea, by estimating the direction and distance travelled" ("Dead reckoning," 2017), wayfinding happens precisely *in* time *with* the environment as continuously changing contexts that occur "now."

The distinctions which Ka'iulani, Ben, Charles, Wade, and Nainoa observe about wayfinding and other types of navigation expose similar and varied meanings about the idea of difference itself, that is, as differences in kind, of degree, and with respect to temporality. These ways of making sense of difference make a difference to the meanings that define navigation and navigating. Here another challenge emerges: resisting the urge to definitively discern one way or the other, whether wayfinding is the same *or* different from other forms of navigation. Instead, engaging in making sense of the ways in which they are the same *and* different occasions opportunity to understand how particular conceptions of difference make a difference to whether certain meanings are even possible when considering, comparing, and relating wayfinding to other forms of navigation.

CONVERSATION TWO: FOR REAL

Wade Davis:

> Polynesians never thought of a voyage as a journey. They believed that the canoe stood still and that the islands came out of the sea to greet it. (Davis, Gregg, & Gregg, 2009)

Nainoa Thompson:

> [And] when you have a black night where the ocean gets black and that's when you can't navigate with your eyes anymore, then you have to go someplace else, and that discipline is something very difficult for me to learn. It's almost as if I have to wash away that dependence on what I can see and what I can believe and trust other senses. And that's one level, of what I say is, knowing without knowing how you know … I'm not very good at it. I'm still a virtual kindergartner. I mean, and I say that, because I know the difference between myself and my teacher [Mau Pialiug] where that deeper understanding of trusting other instincts is [sic] nurtured when he was a young child and he was dependent on it for survival. (Davis, Gregg, & Gregg, 2009)
>
> This is where traditional navigators like Mau are so skilled. Lying inside the hull of the canoe, he can feel the different wave patterns as they come to the canoe, and from them tell the canoe's direction. I can't do that. I think that's what he started learning when he was a child with his grandfather, when he was placed in tide pools to feel the ocean. (Thompson, n.d., para. 7)

Dennis Kawaharada (English Professor):

> The stars are available about 20 percent of the time as clues to direction. Daylight and cloud-cover at night hide them from the navigator during the other 80 percent of the time. (Kawaharada, n.d.a, para. 30)

Ka'iulani Murphy:

> A cloudy night is probably [sic] most difficult time for navigation. During the day, if it's cloudy, you can at least see the ocean. (Patterson, 2010, para. 7)

Nainoa Thompson:

> [Once] the sky was so black, I couldn't see the moon, but I could feel where it was …. I knew where to go. I directed the canoe on a new course and then, just for a moment, there was a hole in the clouds and the light of the moon shone

> through—just where I expected it to be. I can't explain it, but that was one of the most precious moments in all my sailing experience. I realized there was some deep connection I was making, something very deep inside my abilities and my senses that goes beyond the analytical, beyond seeing with my eyes. I cannot explain what this is from a scientific point of view .... Before that happened, I tended to rely on math and science because it was so much easier to explain things that way. I didn't know how to trust my instincts. They were not trained enough to be trusted. Hawaiians call it *na'au*—your instincts, your feelings, rather than your mind, your intellect. (Thompson, n.d., paras. 42–43)

Ida Deal (Fictional Character):

> Think of how powerful introducing a variety of different cultural perspectives in our courses would be! It would enlarge our appreciation of the topics we study. That's why we need to make space for multiple ways of knowing. (Oberg, Blades, & Thom, 2007, p. 119)

Cy Phact (Fictional Character):

> I think it'd be inappropriate to start bringing in ... different cultural views .... It's like I tell my students: "Welcome to science education—leave your cultural bias at the door." ... Let me give you an example. If you drop something off a cliff, we can measure the rate of fall, and how it falls has nothing to do with what culture you belong to. It's going to fall at a determined rate—that's a scientific fact. It's going to fall at a different rate on Venus than it is on the Earth—another scientific fact .... You see the good thing about science is that it's a level playing ground for students. No one has to worry about what culture they're from. They can just come to my course and study the content. I just tell them that in my classroom, everything is equal and fair. (Oberg, Blades, & Thom, 2007, p. 116)

DISCUSSION TWO: FOR REAL?

In this conversation, certain ontologies emerge and become obvious. In some instances, more than one and even complementary realities surface. The speakers communicate what is "real," and as they do so, they demonstrate how certain beliefs are embedded in our worldviews and how these serve as our taken for granted ways of being in the world. For the speakers, these manners of being not only ground their participation in the conversation, but also in*form* what navigation and navigating mean or could mean.

"The Canoe Stood Still and ... the Islands Came out of the Sea to Greet It"

This comment of Wade's (in Davis, Gregg, & Gregg, 2009) bends my ear in a completely opposite direction than Charles's (1995) earlier remark that there is nothing "special or unique" about the wayfinder's cognitive basis compared to the directional systems of other cultures. Identifying this feature of the wayfinder's mind, Wade makes important the role that cultural languages play in determining how we know what we know (Bowers, 1997, 2017). Each culture's language identifies what manners of knowing, acting, and being its members will and will not distinguish as important by grounding these characteristics within the culture's spoken, written, symbolic, temporal, and spatial language. Recognizing the complexity of language as more than what is spoken or written includes the metaphorical nature of language itself[3] and allows for inquiry into how culturally embodied structures shape an individual's or entire group's perceptions and experiences and, importantly, how we bring forth a world of *sign*ificance with others (Maturana & Varela, 1987). "Language … is a flash of the human spirit, the vehicle by which the soul of each particular culture comes into the material world. Every language is an old-growth forest of the mind, a watershed of thought, an ecosystem of spiritual possibilities" (Davis, 2009, p. 3). Here, as ecological {[(biological) and social] and cultural} beings we do not so much choose a reality by which to live our lives, but rather, through the very social and cultural nature of language and our ways of languaging, we immerse ourselves and are immersed in the living forth of a particular reality or realities with others. Situated within this conception of culture and language, it is possible to understand how the magnetic compass, the Medieval Western Wild Rose and the stellar compass, as artifacts and activities, are produced and carried forth within the socially-culturally specific languages that bring them into being (Barton, 2008).

The Canoe Moves, the Canoe Stands Still

The contrast between Charles's earlier comment—that the star compass "is basically similar to other directional systems developed by seafarers elsewhere in the world"—and Wade's comment in this conversation identifies their remarks as arising from two very different ontologies. In the same manner, while it may seem strange or even impossible to imagine that the canoe does not move and the rest of the world does, so too is the idea that the canoe does move and everything else remains still, depending on which reality one's way of being is embedded in. The two disparate conceptions create a compelling case concerning the effect that a culture's metaphorical language has on specifying the kind of mathematics and mathematical ways of knowing that are possible and impossible. While in contrast with one another, each reality arises from certain beliefs, that is, one from an ontology that creates a static, unchanging, timeless, independent, and universal world and the other from an ontology that creates a dynamic, ever changing, temporally situated, interdependent, and locally contextualized world. Each worldview gives rise to certain ways of knowing while at the same time obscures or even makes other kinds of knowing completely imperceptible. Further still, just as we can gain an appreciation for how the two ontologies complement each other by moving back and forth between the mathematics each enables, it also seems important to consider the ways in which each disables particular kinds of mathematics and mathematical ways of knowing.

For example, magnetic compass-based navigation allows one to appreciate the universality of navigation and navigating. Here, mathematics rests on the idea of absolute directions of north, south, east, and west, and these directions are independent of geographical location and external to the boat. The canoe moves while all other entities remain at static locations upon which a fixed circle of directions is transposed. The movement of the canoe and the compass directions become most important in establishing where one is and the charting of a course. The thinking carried forth is that of an out-there, inanimate, predictable, and eternally "extra-human" world that exists regardless of us (Rotman, 1987). Navigation in this sense emphasizes reference points and distances (Barton, 2008). As such, navigation can be defined as "the process or activity of accurately ascertaining one's position and planning and following a route" ("Navigation," 2017).

In contrast, the very notion that the canoe remains still and it is the islands that come out of the sea to greet it occasions a completely different comprehension of the world and a radically different meaning for wayfinding than "planning and following a route." Rather, the world and the act of finding a way in it are imagined as living and ever-changing phenomena. The world as animate implies right here and now, integral, unpredictably emergent, and locally specific. This necessitates recognizing the world and ways of being in a world that are more-than-human (Abram, 1996; Apffel-Marglin, 2011; Cajete, 1999; Ishizawa & Rengifo, 2011). Different from instrument-based navigation, which exists as "something" one does to determine location and direct travel, wayfinding presents as "doing." As such, wayfinding engenders dynamic forms of geometry that emphasize pathways and speed to find the direction of the next sea mark, whereas static forms of navigation focus on universal reference points and distance in order to ascertain one's position and plot the route to a final destination (Barton, 2008).

"You Have to Go Someplace Else ... I Cannot Explain What this Is from a Scientific Point of View"

In this statement by Nainoa Thompson (Davis, Gregg, & Gregg, 2009), his difficulty in describing characteristics of wayfinding that distinguish it from other kinds of navigation moves the conversation toward a familiar and ongoing dilemma in mathematics education, namely, how we conceive and articulate mathematics and mathematical ways of knowing in manners that elucidate sentient, embodied, and contextual qualities which are indescribable from a Western scientific perspective. Traditionally, conceptions of mathematics and mathematical knowing are portrayed as disembodied, abstract "heirlooms of logic and rational knowledge … [that are] neutral and [live] without regard for us … our beliefs … our values … our actions … our cultural ways" (Thom, 2012, p. 74). Consequently, there is no genuine space in this conceptual realm to describe or expand our meanings of mathematics and these other ways of knowing. One is forced to choose either to discard aspects of embodiment as subjective interpretations or errors in logic or, as Nainoa explains, go someplace else and trust other senses that give rise to knowing without knowing how you know. Trusting other senses and instincts requires a different way of being than thinking logically, rationally, and analytically. Here, the body plays a central role, necessitating more than a material body but one that is alive (flesh) and occasions consciousness, knowledge, subjectivity, and so on, through processes of self-affection (Henry, 2000).

Mathematics in and of the Flesh

Nainoa's description of traditional wayfinding as a kind of knowing that cannot be objectified or disconnected from the act itself enables wayfinding to be a type of navigating in which "all doing is knowing and all knowing is doing" (Maturana & Varela, 1987). While this aphorism makes it possible to reveal the very personal and contextual characteristics of wayfinding, it also elucidates the dialectic nature by which knowing and doing emerge. Knowing and doing as necessary parts of each other means that conceptions and embodied knowings presuppose and constitute each other *as* bodily experiences of realization (e.g., Roth & Thom, 2009; Thom, 2011)

Therefore, it is not necessary for abstract conceptions to be understood as separate or simply metaphorical extensions of bodily schema, but rather, as potential provocations of experiences that effectively and metonymically expand conceptions as wholes. Each and every articulation made by the wayfinder is immersed in and inherently part of the natural world where knowing emerges. The sense made and communicated by the wayfinder can only originate and evolve with the body. Any and all conceptions then exist as enactions; *em*bedded and *bodied* activity that is the wayfinder's dynamic, collective and thus, (em)bodied comprehension of sense made. Wayfinding is not some abstract idea or process; it is knowing and doing "in and of the flesh" (Thom & Roth, 2011; Thom, Roth, & Bautista, 2010)—mindful knowing and doing that require the wayfinder's entire being as living, breathing, sentient, emotional, and spiritual self.

Knowing without Knowing How You Know

Taking seriously the perspective of mathematics in and of the flesh means that knowing the direction of the waves or the canoe, or even the location of the moon without relying on visual perception entails the continuous experiencing of lived and living bodies, that is bodies of knowers, knowing, and knowledge which in*form* perceptions and actions. This ongoing making sense of sensation, rightly cognition, is grounded in and arises from immediate contexts. In contrast to abstract constructs, feeling the waves or the movement of the canoe are kinesthetic-sensible events integrated by living bodies that not only give rise to such events in the first place but also serve as the means by which events are recursively (re)enacted across time and space (Merleau-Ponty, 1945) and that which enable the development of mathematical processes and objects.[4] This is the very work of mathematics—imbued with affect, emotions, and all else that the wayfinder feels and is in the act of wayfinding—that constitutes the development of mathematical knowing and doing as these are expressed, metaphorized, and metonymized.

"We Need to Make Space for Multiple Ways of Knowing ... that's a Scientific Fact"

Ida Deal claims that there are multiple ways of knowing and that the curriculum should make room for all of these. For Cy Phact, quoted above (in Oberg, Blades, & Thom, 2007, p. 116), however, it makes sense that only the Western scientific form of knowledge be presented. Here, there is opportunity to examine not only the contrasts between the two seemingly opposite realities and how other realities in turn might be regarded and treated but also possible reasons for why other forms of mathematics and ways of knowing might be included in or excluded from a given curriculum. Although both speakers acknowledge the presence of multiple ways of knowing, how this is taken up can be seen in Ida's and Cy's particular ways of being.

Ida believes that other realities do exist and that they are constitutive in nature. This is because she acknowledges and accepts other ways of knowing and acting as a natural consequence of one's participation in culture(s) (Maturana, 2005). She regards these as legitimate and valid. For Ida, this is why all ways of knowing should be part of a curriculum. In contrast, Cy acknowledges that while other cultural views exist, he privileges a Western scientific reality. His reality is grounded in facts, ones that are generalizable "for any" or "for all" situations, and is therefore formulaic. Cy's reality is essentialist in nature, whereby universal laws exist as undeniable truths and as such, it is this reality that surpasses all other ways of knowing. Further, with an essentialist reality like Cy's, any events that cannot be explained through scientific facts, such as Nainoa's account of Mau's skills as navigator or the time when Nainoa could feel but could not see where the moon was, are regarded as unfounded, subjective accounts or perhaps even mystical experiences and as such, could not be what really happened. For Cy, only scientific facts are accepted as coherent, and therefore all other views must be denied.

Moving beyond the distinctive manners in which cultural forms of knowing are regarded within constitutive and essentialist realities, both Ida's and Cy's responses to how to approach the issue of different cultural ways of knowing in the classroom have to be considered to be impositions, albeit complementary ones. The two approaches, regardless of their realities, both assume solutions of an either-or mindset. For Cy, it is a matter of one reality over all others, while for Ida it is a matter of including all realities over just one. Neither solution can be considered inclusive or fair because both deny the existence of the other's reality or, at best, delegitimize the reality of the other. Cy and Ida strive to be inclusive and fair; however, in the end, their contradictory one-size-fits-all solutions serve as prescriptions. What results is missed opportunity to understand how our ontologies impact the existence of certain worlds when we choose all or particular mathematics and mathematical ways of being over others as a response to the situation.

CONVERSATION THREE: DWELLING ON ITS WHATNESS

Thea Rey (Fictional Character):

> Theoretically speaking, is it even possible to "represent" another culture and its perspective? (Oberg, Blades, & Thom, 2007, p. 116)

Wade Davis:

> Nainoa explained: as the Earth rotates, every star comes up over the eastern horizon, describes an arc through the sky, and then sets on a westerly bearing. These two points on the horizon, where a specific star rises in the east and sets in the west, remain the same throughout the year, though the time at which a star emerges changes by four minutes every night. Thus, as long as one is able to commit to memory all the stars and their unique positions, the time at which each is to appear on a particular night, and their bearings as they break the horizon or slip beneath it, one can envision a 360-degree compass, which the Hawaiians divide conceptually into the thirty-two star houses, each segment on the horizon named for a celestial body. Any one star is only dependable for a time, for as it arcs through the star breaking the horizon, again on a bearing known to the navigator. Over the course of a night at sea—roughly twelve hours in the tropics—ten such guiding stars are enough to maintain a course. To steer, the crew at the helm, instructed by the navigator, takes advantage of the canoe itself, positioning the vessel so that a particular star or celestial body remains framed, for example, within the angle subtended between the top of the mast and stays that support it. (Davis, 2009, p. 57)

Ben Finney:

> Like their Western counterparts, Carolinian navigators divide the horizon around them into thirty-two points or bearings, as did, according to evidence from the Cook Islands … at least some Polynesian navigators …. Unlike, however, the Western practice … these do not rise and set exactly at the points derived from the successive bisections, charting the Carolinian conception on paper (or by coral rocks placed on a mat) yields irregularly spaced bearings. (Finney, 1994, p. 55)[5]

Charles Frake:

> This apparent difference suggests that we take another look at the Pacific …. Micronesian Islanders are reputed to think about directions in another way ….

> Not only do reciprocal courses not have the same name at either end, but, because the segmentation of the horizon is not equal, there will not even be a named compass point in the opposite direction of one's current course .... These conclusions assume we have interpreted the Micronesian system correctly. One must keep in mind the possibility that investigators have here committed the same error of misplaced concreteness that so often distorts our views of the mental schemes of other peoples. (Frake, 1995, p. 152)
>
> ... Consider the displays of their directional system used by Micronesian navigators themselves when teaching novices. An eight-rayed octagon is laid out on a mat with sticks. Eight large coral stones are placed on the end of each ray, and then equally spaced between each of these are three smaller stones. The mid-stone is placed at each corner of the octagon, the large stones at the centre of each side of the octagon. This construction ... not only yields 32 *equally* spaced points, but the eight main points are visually distinguished from the eight subsidiary points and the 16 final points (Goodenough & Thomas, 1987. These pairs of reciprocals are explicitly taught as part of navigation training. (Frake, 1995, p. 152, emphasis added)[6]

The diametrically opposite points of this compass are seen as connecting in straight lines through a central point. A navigator thinks of himself or of any place from which he is determining directions as at this central point. Thus, whatever compass point he faces, there is a reciprocal point at his back. If this is the case, then the actual rising and setting positions of the stars *cannot* mark the directions the stars name. My argument is that the stars provide the names, *not* the positions, for abstract conceptual segmentations of the horizon circle into 32 equally spaced points.

Micronesian navigators did use stars for finding directions. They used the whole changing visual pattern of the night sky for doing so. They also used the sensory pattern of sea swells passing under the canoe and its outrigger. They used birds, sea marks, clouds, and even mythical creatures. They used anything they could get. But when they found their bearings they referenced them to a conceptual system of abstract named directions that equally segmented the horizon circle into 32 points (Frake, 1995, p. 155).

DISCUSSION THREE: IS IT EVEN POSSIBLE TO "RE-PRESENT" ANOTHER CULTURE AND ITS PERSPECTIVE?

This conversation focuses solely on the stellar compass and moves for the most part in the same direction until Charles turns it completely around. As I reflect on what the speakers say, while at a loss as to how to reconcile the contradictory ideas about the stellar compass, I return several times to Thea's initial question and wonder if it is at all possible to represent another culture's mathematics and mathematical ways of knowing.

"Everything Said Is Said by Someone"

The quotation (Maturana & Varela, 1987, p. 27) refers to the constant everyday activity of making distinctions that we do, for the most part, with little or no conscious thought. Maturana and Varela (1987) explain distinction making as the means by which we indicate some being or thing from the context of which it is part. At the same time as we make a distinction either explicitly or implicitly, we reveal not only the criteria that gave rise to the distinction but also the particular qualities that the being or object demonstrates in relation to the criteria.

Therefore, in the conversation, what the speakers say of the stellar compass cannot be taken as re-presentations of it but rather, their particular personal and collective meanings of the compass. If what I assumed to be representations are now conceived as distinctions made by another, and if distinguishing people, places, objects, events, and so on can only arise in language—that is, through our verbal, written, symbolic, spatial, and temporal forms of communication—then all that is distinguished and the act of making distinctions itself cannot be assumed to be a neutral or fixed activity but must be considered as non-neutral and ever changing. Both the manner in which an event or entity is distinguished and the distinctions that are made of it are, then, necessarily shaped by the agents, their culture, and their cultural manners of languaging.

After once more considering what Wade, Ben, and Charles said, it is clear that it is not possible to capture a true or exact representation of "what" the stellar compass is or is not, what wayfinding is or is not, or even what the Polynesian perspective is or is not. Despite the fact that these three people are knowledgeable, trustworthy, and indeed all talking about the stellar compass, the distinctions they make do not collapse neatly onto one another. In fundamental ways, their articulations about the star compass are different and even contradictory. Yet, if it were possible to somehow take all of their interpretations and comprehend them in a coherent manner, our conception of the stellar compass would still be incomplete. This is because Wade, Ben, and Charles are not re-presenting the stellar compass. They are making distinctions about it as observers, and each of their observations is embedded in their own social-cultural understandings.

Given this, it does not make sense to think that teaching another culture's mathematics and mathematical ways of knowing means that that is what is actually being taught. For example, in the conversation, each anthropologist distinguishes different aspects he sees as relevant to certain mental or material manifestations of the star compass. Wade points out the arcs created by the rising and setting of the stars on the horizon, the circular image of a compass divided into 32 sections, and how the canoe is steered by being lined up with the stars. Ben speaks of the 32 points or bearings as irregularly spaced around the circumference of the circle. And Charles asserts that the star names on the stellar compass are placed, not representatively in a concrete way but arbitrarily in an abstract manner, equidistantly around the circle of the compass or octagon of the teaching diagrams. What is being taught and what is being represented here is not Polynesian mathematics and mathematical ways of knowing but observed distinctions of mathematics and mathematics knowing from Polynesian culture based on three observers' perspectives from their specific realities. When we neglect to consider how we explain our experiences, we trivialize the significance that lies in knowing how we know what we know, while keeping hidden the fact that everything said is said by someone (Maturana, 2005).

CONVERSATION FOUR: DWELLING ON ITS THATNESS

Wade Davis:

> So the canoe becomes the actual needle of the compass. (Davis, Gregg, & Gregg, 2009)

Ka'iulani Murphy:

> Yeah, that's right. (Davis, Gregg, & Gregg, 2009)
>
> The basic aspect of the navigation is this idea of a star compass. You have your circle of horizon. Different points along this circle mark star paths. So you're watching for stars that rise from these points on the horizon, and that's how you know your direction. (Patterson, 2010, para. 3)
>
> The star compass is a mathematical way of dividing up the horizon. (Patterson, 2010, para. 4)
>
> We learn what stars rise and set. So when we see one star, we have one given direction, and we can orient ourselves based on that. (Patterson, 2010, para. 5)

Dennis Kawaharada:

> Nainoa developd [sic] a star compass with thirty-two equidistant directional points around the horizon, each point 11.25° from the next point (11.25° x 32 points = 360°). Each point is the midpoint of a house of the same name, and each house is 11.25° wide (11.25° x 32 houses = 360°). [This] Star Compass … [is] loosely based on the Micronesian Star Compass of Mau Piailug. (Kawaharada, n.d.a [Hawaiian Star Compass], para. 1)[7]

Nainoa Thompson:

> The star compass also reads the flight path of birds and the direction of waves …. This mental construct of the star compass with its Hawaiian names is from Mau. The genius of this construct is that it compacts a lot of information and enables you to make decisions based on that information. (Thompson, n.d.b, para. 2, 4)

Sam Low (Anthropologist):

> Nainoa's is a hybrid system, our Polynesian ancestors did not divide their world into lines of longitude and latitude, for example, nor did they have charts. Nevertheless, many of the observations he used to find his way must have been available to ancient Polynesians as well. The star compass, which still is used in Micronesia, is one such technique as is Mau's reliance on wind and waves for direction and his use of birds, flotsam and sky coloration. What ties Nainoa's system to the ancient one is a similar painstaking observation of the natural world—seeking clues to location in the night sky and the world around him. Polynesian navigators undoubtedly noticed the shape of the sky changing as they moved across Earth—the rising and setting points of stars, their arc across the sky, the height they rose to, perhaps even relationships such as equal distances and simultaneous setting. (Low, 2003 [A Hybrid System])

Thea Rey:

> Seriously .... Does each culture have only one perspective? (Oberg, Blades, & Thom, 2007, p. 116)

DISCUSSION FOUR: THE GENIUS OF IT

In this conversation like the one before, the speakers reiterate that there is no singular cultural perspective of the stellar compass or wayfinding but several personal, social, and cultural distinctions that can be made of them. Perhaps even more interesting is what the conversation further reveals about the nature of the star compass and wayfinding.

The stellar compass as *a* conceptual structure is specific to each wayfinder and, thus, relies on the conceptualizations of the wayfinder. For example, for Mau's compass included bird paths and wave directions, 32 points with star names that are not equidistant points, and depicted in an east, south, west, and north orientation. In contrast, Nainoa's compass involves star houses and degree equivalents, 32 equidistant points, and north, east, south, and west orientations.[8]

Yet even with these variations, there is coherence that cuts across the individual conceptions and enables the star compass and the tradition of wayfinding to transcend from one navigator to another and from one generation to the next. The articulation of particular stars, their paths, the direction of waves, and so on are all forms of knowledge not created or located within any one individual or event but arise from generations of cultural experience. The star compass and wayfinding are the collective intelligence that lives forth as socially and culturally shared patterns, recursively and intergenerationally created and re-enacted. This is evident even during this short conversation, as it was Mau's father and grandfather who were wayfinders and who trained him. It was Mau who taught and mentored Nainoa for several years, and it is Ka'iulani who is a student of Nainoa.

Also, evident in the conversation and discussions thus far is that the stellar compass arises in compatible but not identical manners for the three wayfinders. In one sense, the diversity reiterates the coexistence of multiple ways of knowing discussed earlier in the first three conversations. In another sense, the diversity can be considered as that which effectively enables the expansion of intelligence in the form of new concepts and conceptions prompted by the star compass and the tradition of wayfinding. Here, intelligence and agency are not attributes of an individual but distinct qualities of a(ny) (eco)system as a whole (Bowers, 2009).

For example, through his own research and experiences in navigating, Nainoa developed the art of wayfinding by integrating Mau's traditional methods with ideas founded in Western mathematics and science. Three such concepts include determining one's latitude and the two which are mentioned in the conversation as "equal distances" and "simultaneous setting" (Low, 2003). These are based on patterns that Nainoa observed in the positions and movements of certain stars. First, the horizon at the equator appears parallel to the earth's axis. Polaris is observed as if it were resting on the surface of the ocean, and it rises a single degree above the horizon at one degree north latitude. This pattern is also true for the rest of the stars. For each degree of change in latitude, there is an equivalent rising or sinking of the stars. Therefore, when moving one degree north, each northern star rises one degree and each southern star sinks one degree. When moving one degree south, one will observe the southern stars rising and the northern stars sinking by one degree each. However, because these stars move, their altitude must be identified in order to determine one's latitude. This is done by measuring the highest point in a star's arc, that is, when it crosses the meridian. For instance, the highest point for Acrux, a southern star, is when it crosses the meridian at 27 degrees. Thus, by moving one degree north, Acrux's highest point would be at 26 degrees. Nainoa finds the altitude of the stars by measuring it with his hand. Holding his hand out to the sky, he stretches his thumb so that it rests along the horizon. He then determines the altitude of the stars by using benchmarks on his hand, which is also extended (Finney et al., 1986). Here, Nainoa knows that if a star that rises to the height of his second knuckle, its altitude is 15 degrees. Second, Nainoa noticed that particular pairs of stars set in a specific sequence when he is in certain places at sea, and when he is in other locations, the stars set at the exactly the same latitude. Nainoa explored the observation further and found that there are many other pairs of stars that set in the same manner. He calls this concept "simultaneous setting." Third, Nainoa realized that the distance between certain "top" stars and "bottom" stars within constellations is the same as the distance between the bottom star(s) and the horizon (see Figure 12.1). By knowing and using this relationship, Nainoa is able to determine his latitude from many different locations around the world. He calls this concept "equal distance."

The star compass and wayfinding can be understood as resulting from the languaging process of cultural "drift" (Maturana & Varela, 1987) or, in other words, collective ways of knowing and intelligence that evolve in a nested manner with individuals, communities, and the natural world (Bowers, 2003). They are not static

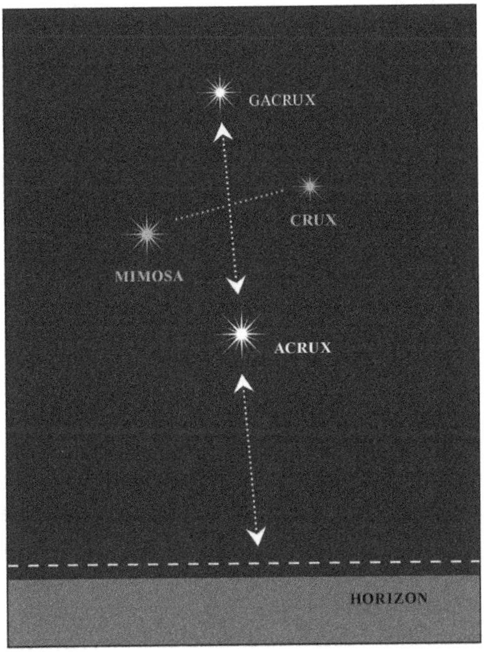

Figure 12.1. Diagram of the Southern Cross shows the distance between the stars Gacrux and Acrux as equal to the distance between Acrux and the horizon

forms of knowledge passed down from generation to generation. The concepts and conceptions are conceived through intergenerational re-enactments, always existing "now" and therefore, dynamic and ever changing.

The recursive interaction of different individuals, different generations, and different cultures nested within cultures—including ones that might be considered disparate (e.g., Polynesian ways of knowing and Western mathematics and science)—and the natural world demonstrates how the star compass and wayfinding have endured and, in radical ways, are living mathematical embodiments of intersecting, co-evolving cultures. Even in this brief fourth conversation, the complexity of the star compass and methods of wayfinding can be appreciated. Today's stellar compass and practice of wayfinding are hybrids that simultaneously embody an in-here and out-there sensibility: a "fish's eye" view of the arcs, heights, positions, and relationships of stars' paths, the flight of birds, the direction of waves, and the shape and colour of the sky, that connect with a geometric and trigonometric "beyond the stars" view of mathematical objects such as circles, longitude, latitude, altitudes, and so on.

Moreover, as a concept, the stellar compass lives as the wayfinders' conceptions in relation to the environments in which it arises and is enacted and re-enacted. This makes it impossible for the stellar compass to be a thing or object but, instead, to be an emergent and dynamic event in which none of its articulated meanings can be understood as inputs into a collection of parts. Rather, making sense of the star compass entails understanding how the meanings of it *part*icipate on different levels and in different ways to give rise to new meanings for it and for wayfinding. After reflecting on the third conversation again in relation to the dilemma of representing the stellar compass: another reason that this might be so is precisely because the stellar compass is a live and growing concept that changes as it lives forth and is articulated by wayfinder to wayfinder and within the circumstances in which they find themselves.

CONVERSATION FIVE: WHAT AND HOW

Nainoa Thompson:

> On my first voyage in 1980 (Thompson, n.d., para. 15), Mau … said, "turn the canoe around and follow the bird." I was really puzzled. I didn't know why …. Mau said, "you wait one hour and you will find the island you are looking for." And after about an hour, Mau, who is about twenty years older than me—my eyes are physically much more powerful than his—… gets up on the rail of the canoe and says: "The island is right there." And we all stood up and we climbed the mast and everything and we just couldn't see it. Vision is not so much about *what* you do—but *how* you do it. It's experience. Mau had seen in the beak of the bird a little fish and he knew that the birds were nesting. They had flown out to sea before sunrise and were taking food back at mid-morning to feed their young, before they flew out to sea again to feed themselves. (Kawaharada, n.d.b. [Land-Based Seabirds], para. 3, emphasis added)

DISCUSSION FIVE: DOING AND BEING

Nainoa realizes that the brilliance of what Mau did, as a navigator, was completely dependent on how he did what he did when he navigated. Here, Nainoa speaks directly to the complexity and mindfulness of Mau as wayfinder. The shift in Nainoa's awareness from a focus on "doing" (what Mau did to navigate) to also include "being" (how it is that Mau navigated the way he did) addresses not only the intuitive sensibility required by the wayfinder but also particular knowledge and ways of knowing the world. Here, two important questions arise: How might this sensibility be described in terms of the kind of navigating and the ways of navigating that are occasioned by the wayfinder? And, how might such an understanding inform the way(s) in which I, as other, regard this culture's knowledge and ways of knowing?

Light, Love, Kindness, and Compassion

At first glance, the four virtues might appear to be out of place and have little or no relevance to the topic of this chapter's discussion. However, they are in fact the responsibilities given to the wayfinders by Mau when he inducted them into Pwo on March 18, 2007. Pwo is the ninth of fifteen degrees in the Weriyeng School of Navigation of Micronesia (Flannery, 2010). Mau also explained to his graduates, "If there are conflicts, the navigator must resolve them; if there is sickness, the navigator's responsibility is to heal; if there is damage, the navigator must repair it" (cited in Flannery, 2010, para. 16). Truly, navigating for the wayfinder is much more than identifying one's position and direction on the surface of the earth! The following discussion focuses on Maturana's ideas about emotions, particularly the emotion of love, and how these conceptualizations informed my further thinking about Nainoa's explanation of Mau as a navigator.

Emotions as Orienting What We Know

In addition to our social and cultural existence in language and languaging, Maturana and Verden-Zöller (2008) explain that it is our emotions that define what is possible in the relations, interactions, and world(s) we live forth with others. Emotions are not conceived as feelings but arise in relationships as behaviours or dynamics that effectively orient us with all others, moment to moment, context to context. In this way, our emotions determine what can and cannot arise in our interactions. Whenever our emotion changes, the manners in which we relate to others change.

While each emotion orients us to the other in different ways, love is considered to be the most expansive emotion because it is the only one that accepts the other as a legitimate other (Bunnell, 2004). Love as a relational dynamic then, enables the other to coexist with others as itself and vice versa. Even respect, which might be viewed as a similar emotion to love, engenders restraint or distance, whereas love creates closeness. Given this, it is possible to appreciate how love and respect enable different potentialities for future events. The emotion of anger, in greater contrast, does not enable the other to be acknowledged as legitimate and sets up dynamics that might potentially involve power and dominance of one over the other.

Follow the Bird

Mau's advice to Nainoa in light of the above discussion rings true in the sense that love defined the manners in which he navigated. For Mau, there seemed to be no separation between the wayfinder and the natural environment. The wayfinder and the environment coexist in what Indigenous scholars and philosopher and cultural ecologist David Abram (1996) refer to as a "more-than-human" world. And in this manner, love orients the more-than-human world and wayfinder in relations and interactions where the natural world is regarded as intelligent and genuine in its own right, that is, other animals, plants, ocean, stars, wind, and everything else visible and invisible are taken to be unique and legitimate entities. Moreover, not only would it be undesirable to disregard being human, but being human becomes vital to understanding how being human, as living, breathing, sentient, and emotional self in the flesh, creates meaning and potential for the wayfinder.

Love as demonstrated by Mau and observed by Nainoa reveals how Mau's emotional orientation engendered the kinds of relations and interactions he had with the environment, the knowledge that arose, and the navigating he enabled. In contrast to instrument-based navigation that reduces and abstracts methods, wayfinding oriented by love can be considered as that which expands intelligence and creativity through its intimate connection with the immediate environment (Maturana & Bunnell, 1998). What is kept whole is a complex conversation with the more-than-human world that the wayfinder must perceive and interpret. This conversation involves patterns of position, movement, formation, appearance, and rhythms of the sun, moon, stars, wind, clouds, currents, ocean, birds, and marine life—all of which provide continuous, meaningful information moment to moment and completely unscripted. Mau once told Nainoa, "If you can read the ocean, you will never be lost" (in Thompson, n.d., para. 7). This ability to read the (con)texts of the natural world intimately and openly implies that "language … is not a specifically human possession, but is a property of the animate earth, in which we humans participate" (Abram, 2010, p. 11). As such, this kind of participation is not a conversation in which the wayfinder listens to the (con)texts of the earth, compares what is heard against what she or he already knows to be true, and then speaks abstractly about the natural world. Rather, it is a conversation that requires the wayfinder to listen and respond directly back to and with the earth, always seeking to comprehend more of what the other is saying even if the other does not speak with words.

What Does Love Have to Do with It?

Now, in reference to my second question: How might an understanding of emotions, particularly love, inform the way(s) in which I, as other, regard the knowledge of wayfinding and ways of knowing? I realize that to consider this from my cultural orientation—that is, how it fits within my world(s)—makes it impossible for me to conceive wayfinding as genuine in its own right. When I consider how wayfinding measures up against my cultural manners of knowing, doing, and being, there is no need for me to bend my ear to attend in a way that expands what I know or could know. Whatever the result, it simply reinforces what I already know because I am not listening in a genuine way to the other as a legitimate other. In fact, when I listen, I do not hear the other, I hear only myself (Poerksen & Maturana, 2006).

However, when I take wayfinding as unique and legitimate, there is opportunity to re*cognize* how mathematical distinctions that emerge from particular contexts arise meaningfully for the wayfinder, such as shape of the clouds, their formations, movements, or their positions in the sky as foretelling approaching winds and weather, or the reverberation of the waves predicting the distance to land or even how a change in the course of the canoe might be identified through the pitch and roll of the waves (Davis, 2009). All of these patterns exist as necessary relationships that connect wayfinder with earth and earth with wayfinder. They do not resemble what I think of as familiar, abstract, and formulaic mathematics, because the mathematics of wayfinding never seems to be abstracted or disconnected from the contexts in which it is sensed, made sense of, and lived forth. Of course, this is not to say that it cannot be, should not be, or has not been done. However, important questions to ask include: What might be reasons for wishing to do so? What new meanings might be achieved? And, what meanings might be lost in the process of doing so? Wayfinding is not a failed attempt at being truly mathematical. It is not a less sophisticated form of mathematics or mathematical activity. Rather, wayfinding is a genuine and unique expression of complex mathematics and mathematical work—a live kind of mathematics in and of the flesh that necessitates and occasions earthly ways of knowing.

CONVERSATION SIX: COME AGAIN, WHAT IS IT?

Ka'iulani Murphy:

> I believe we were lucky to have Mau Piailug to show us how ancestors may have sailed. The way he navigates, it's very intuitive; he's very close with nature. The way he looks at the world, the eyes he looks through are very different than the eyes we look through the world with. In our developing of our system of wayfinding that works for us today, it's incorporated a lot of the things we learned. (Patterson, 2010, para. 11)

Nainoa Thomspon:

> When everything is going right … I get into a zone, a special place in which all of my relations with the canoe, the natural world and the crew are integrated. You have to be in that special place to navigate well …. You have the star patterns in mind and you seem to know where you are even when the sky is cloudy and you can't see the stars …. It's an awesome feeling but it's hard to describe. It is like being inside the navigation, participating from the inside. (Low, 2003, para. 22)

Sam Low:

> The modern instrument navigator inhabits a world of abstractions. To find his place in the world, he reduces a star's altitude to a number, then consults his astronomical tables, does some math, and plots his position on a chart. He then forgets his navigational tasks until it's time for another peek through his sextant.
>
> But the noninstrument navigator never ceases to regard nature. The world all around offers a continual "heads up display"—of stars, swells, and wind …. Nainoa … constantly senses the direction of his vessel—in the soles of his

> feet, in the feel of wind on his cheek, in the sound of the rushing wake. He is continually, as he describes it, "in the navigation." (Low, 2003, paras. 1–2)

Keith Devlin (Mathematician):

> When we approach mathematics as a purely human endeavor, we focus almost exclusively on the conscious performance of computational processes—numerical, algebraic, geometric, etc.—often carried out with the aid of a pencil and paper, or these days some form of electronic computational device such as a calculator or computer. Those kinds of activity [sic] are certainly part of mathematics, but if you start from the fact that mathematics—the science of patterns as I like to call it—is about recognizing and manipulating patterns, then viewing the paper-and-pencil stuff we humans do as being all there is to mathematics is like saying that flying is about having wings and flapping them up and down. True, that is how birds fly, but if you take that as being what flying is about you exclude all those jet aircraft that fly around the globe every day. *Flying* is more fundamental than either birds or airplanes; it is about leaving the ground and moving through the air for extended periods of time. Feathered wings that flap and metal wings engineered by Boeing (that hopefully do not flap) are just two particular *ways of performing* that activity …. Once you get away from the pencil-and-paper view of mathematics we all get from our school days, and you think about the more fundamental activity that those school methods provide *just one way of doing*, you find that math is all around us. (Devlin, 2005, para. 21, 22, emphasis in original)

Ka'iulani Murphy:

> Beyond the navigation, what we've learned with our voyages, is that we learn how to better live on our island by the lessons we learn by living on our canoe. (Patterson, 2010, para. 13)

Chad Baybayan (Wayfinder):

> The land and the sea and all life on our planet are interconnected. So if we take care of even the smallest portion of land, or ocean or the smallest creature, we take care of ourselves. (Low, 2002 [A Quarter-Century of Achievement], para. 5)

Wade Davis:

> Margaret Mead said it so well—what this is really all about—If we are not careful, we may wake up one day as from a dream and forget that there were ever any other possibilities for life other than our own (Davis, Gregg, & Gregg, 2009). The point is not to ask or suggest which perspective is right or wrong …. What matters is the potency of a belief, the manner in which a conviction plays

> out in the day-to-day lives of a people, for in a very real sense this determines the ecological footprint of a culture, the impact that any society has on its environment .... The full measure of a culture embraces both the actions of a people and the quality of their aspirations, the nature of the metaphors that propel them onward. (Davis, 2009, pp. 122–123)

CONCLUDING THOUGHTS: FINDING A WAY

What began as a simple question of "What is it?" that then transformed into a series of conversations about wayfinding has proven to be a generative space to explore issues deeply relevant to mathematics education. In a profound way, it is the traditional art of wayfinding which emerges as a powerful metaphor for guiding exploration into what it means to live a culturally responsive mathematics curriculum and pedagogy by occasioning curricula and pedagogies that are living and life-giving.

This chapter on its own and in relation to the book, illustrates the challenge that comes when meanings of *whatness* and *thatness* are taken as final and definitive answers. Rather differently and more akin to wayfinding, meaning making depends on insights that emerge from the ongoing act of voyaging or, quest(ion)ing.[9] My participation as listener and speaker in the six conversations, my further thinking, new understandings, as well as the questions that arose from these is evidence of this.

In the same way that a voyage is unpredictable and happens moment to moment, so too should our quest(ion)s be open, the answers as yet unknown. It makes no sense to chart a course before beginning an inquiry or to locate a final destination prior to actually reaching it. Approaching and staying with an open question means meeting it head-on, but in such a way "to suspend judgment … go gentle … [and] have the patience and forbearance … to contemplate the question" (Francisco Varela, in Hoen & Reichle, 2008). Staying with an open question also necessitates letting the question shift and change course while allowing such change to shift and change us in the process. Doing so enables the possibility that genuinely different kinds of knowledge and ways of knowing can emerge that can then, in turn, inform the field of mathematics education—even our everyday manners of being. This kind of intuitive and open sensibility that grows only from love enables *re*cognizing mathematics, teaching, and learning as more-than-human abstractions. It is a sensibility that acknowledges the need for a more-than-human world to occasion intelligence and creativity. It is a way of being that incites understanding of how mathematics, curricula, and pedagogy happen precisely *because* of us—because of who we are, our emotions, and the realities we live forth with others.

Living a culturally responsive curriculum and pedagogy as wayfinding implies "participating from the inside." Any mathematics curricula, teaching, and learning based on pedagogical dead reckoning then, suggests that we know where we are only by knowing where we have been and how we got to where we are (Davis, 2009). To delve deeper into the historicity and legitimacy of our own and others' mathematics and ways of knowing is another way that the emergence of new understanding and action might be provoked. And just as the actual voyage of the *Hōkūle'a* depends on the crew who depend on one another who also depend on the weather, waves, stars, and so on, mathematics curricula and pedagogies depend on the relationships, environments, and realities created with others. Like the stellar compass, curriculum and pedagogy are not fixed objects, static and external to us but inherently dynamic and inseparable.

Finally, as I consider the questions at the beginning of this chapter still once more, I understand better that living a culturally responsive curriculum and pedagogy involves enriching what we know, the ways we know, and what is possible for us to know about mathematics and the world. It is about taking opportunities to explore who we are and how our mathematical ways of knowing the world are intimately connected. Responsive forms of mathematics education cannot be those that assume difference as clarity that comes from separating what is from what is not. Doing so entirely misses the point of the conversations and discussions in this chapter. In contrast, being responsive involves conceiving difference as arising from and distinguished in relationship with one another (Deleuze, 1988; Heidegger, 1962). This meaning of difference is vital to re*cognizing* mathematics curricula and pedagogy as co-emergent and living relations—relations that are constantly changing, changing us, and at the same time, compelling inquiry into how it is that we bend an ear in the quest of "What is it?"

NOTES

1 This word comes from the Latin word, *recognoscere*. Its meaning here is not the term's common definition which is to "identify as already known" but its second meaning, to "know again" as in "to know anew."

2 In this chapter, *Western* is taken to mean colonial ways of thinking following the Industrial Revolution. *Non-Western*, then, includes all other ways of thinking.

3 For example, "objects" of knowledge or "web" of life.

4 Here, the emergence and development of mathematical activity and objects also enables them to be made available to oneself and to others through (re)enactions.

5 For an example of a Carolinian Star Compass, see figure 1 in Finney et al. (1986, p. 45).

6 For an example of this Micronesian Star Compass as displayed in teaching diagrams, see figure 3 in Frake (1995, p. 155).

7 See also http://pvs.kcc.hawaii.edu/ike/hookele/star_compasses.html for diagrams of Nainoa's and Mau's compasses.

8 See http://archive.hokulea.com/ike/hookele/star_compasses.html for different conceptualizations of a stellar compass.

9 The word *quest* comes from the Latin word *quaerere*, which means "to seek, gain, and ask."

REFERENCES

Abram, D. (1996). *The spell of the sensuous: Perception and language in a more-than-human world.* New York, NY: Vintage Books.

Abram, D. (2010). *Becoming animal: An earthly cosmology*. New York, NY: Pantheon.

Apffel-Marglin, F. (2011). *Subversive spiritualities: How rituals enact the world.* Oxford: Oxford University Press.

Bateson, G. (1972). *Steps to an ecology of mind.* New York, NY: Ballantine.

Barton, B. (2008). *The language of mathematics: Telling mathematical tales.* New York, NY: Springer.

Bowers, C. A. (1997). *The culture of denial: Why the environmental movement needs a strategy for reforming universities and public schools.* Albany, NY: SUNY Press.

Bowers, C. A. (2003). *Mindful conservatism: Rethinking the ideological and educational basis of an ecologically sustainable future.* Lanham, MD: Roman and Littlefield.

Bowers, C. A. (2009). *Educating for ecological intelligence: Practices and challenges.* Retrieved from https://scholarsbank.uoregon.edu/xmlui/handle/1794/9268

Bowers, C. A. (2017). An ecojustice approach to educational reform in adult education. *New Directions for Adult and Continuing Education, 2017*(153), 53–64.

Bunnell, P. (2004). Reflections on the ontology of observing. *Cybernetics & Human Knowing, 11*(4), 72–84.

Cajete, G. (2000). *Native science: Natural laws of interdependence.* Santa Fe, NM: Clear Light.

Davis, W. (2009). *The wayfinders: Why ancient wisdom matters in the modern world.* Toronto: House of Anansi.

Davis, W., & Gregg, A. (Producers), & Gregg, A. (Director). (2009). *Light at the edge of the world: The wayfinders* [Motion picture]. Washington, DC: Smithsonian Networks.

Dead reckoning. (2017). *English Oxford living dictionaries.* Retrieved from https://en.oxforddictionaries.com/definition/dead_reckoning

Deleuze, G. (1988). *Bergsonism.* New York, NY: Zone Books.

Deleuze, G., & Parnet, C. (1987). *Dialogues.* New York, NY: Columbia University Press.

Devlin, K. (2005). *Mathematics: A natural pursuit.* Retrieved from https://www.maa.org/external_archive/devlin/devlin_03_05.html

Finney, B. R. (1994). *Voyage of rediscovery: A cultural odyssey through Polynesia.* Berkeley, CA: University of California Press.

Finney, B. R., Kilonsky, B. J., Somsen, S., & Stroup, E. D. (1986). Re-learning a vanishing art. *The Journal of Polynesian Society, 95*(1), 41–90.

Flannery, J. (2010). Master navigator shared his secrets. *Soundings Magazine.* Retrieved from http://www.soundingsonline.com/features/in-depth/264552-master-navigator-shared-his-secrets

Frake, C. O. (1995). A reinterpretation of the Micronesian "star compass." *The Journal of the Polynesian Society, 104*(2), 147–158.

Goodenough, W. H., & Thomas, S. (1987). Traditional navigation in the Western Pacific: A search for pattern. *Expedition: The Magazine of the University of Pennsylvania, 29*(3), 3–14.

Heidegger, M. (1962). *Being and time* (J. Macquarrie & E. Robinson, Trans.). New York, NY: Harper.

Henry, M. (2000). *Incarnation : une philosophie de la chair.* Paris, France: Seuil.

Hoen, M. (Producer), & Reichle, F. (Director). (2008). *Monte grande: What is life?* [Motion picture]. Brooklyn, NY: Icarus Films.

Ishizawa, J., & Rengifo, G. (2011). Revitalizing the ecological intelligence of Andean Amazonian communities: The way back to respect. In C. A. Bowers (Ed.), *Perspectives on the ideas of Gregory Bateson, ecological intelligence, and educational reforms* (pp. 163–184). Eugene, OR: Eco-Justice.

Kawaharada, D. (n.d.a). *Hawaiian voyaging traditions: Holding a course.* Retrieved from http://pvs.kcc.hawaii.edu/ike/hookele/holding_a_course.html

Kawaharada, D. (n.d.b). *Summary wayfinding, or non-instrument navigation.* Retrieved from http://www.pvs-hawaii.com/navigation/summary.htm

Low, S. (2002). *Voyages of awakening: 25 years of* Hōkūle'a. Retrieved from http://www.samlow.com/sail-nav/awakening.html

Low, S. (2003). *Star navigation.* Retrieved from http://www.samlow.com/sail-nav/starnavigation.htm

Maturana, H. (2001, May). Our genome does not determine us. In P. Bunnell (Chair.), *Remaining human.* Public Forum conducted at the meeting of the American Society for Cybernetics, Vancouver, BC.

Maturana, H. (2005). The origin and conservation of self-consciousness: Reflections on four questions by Heinz von Foerster. *Kybernetes, 34*(1–2), 54–88.

Maturana, H., & Bunnell, P. (1998). *Biosphere, homosphere, and robosphere: What has that to do with business?* Retrieved September 15, 2011, from http://www.solonline.org/res/wp/maturana/index.html
Maturana, H., & Varela, F. (1987). *The tree of knowledge: The biological roots of human understanding.* Boston, MA: Shambhala.
Maturana, H., & Verden-Zöller, G. (2008). *The origin of humanness in the biology of love.* Charlottesville, VA: Imprint Academic.
Maurice, M. P. (1945). *Phénoménologie de la perception.* Paris, France: Gallimard.
Navigation. (2017). *English Oxford living dictionaries.* Retrieved from https://en.oxforddictionaries.com/definition/navigation
Oberg, A., Blades, D., & Thom, J. S. (2007). Untying a dreamcatcher: Coming to understand possibilities for teaching students of Aboriginal inheritance. *Educational Studies, 42*(2), 111–139.
Patterson, L. (2010, October 25). Ka'iulani Murphy's star compass helps navigate oceans. *Earth Sky.* Retrieved from http://earthsky.org/human-world/kaiulani-murphys-star-compass-helps-navigate-oceans
Roth, W.-M., & Thom, J. S. (2009). Bodily experience and mathematical conceptions: From classical views to a phenomenological reconceptualization. *Educational Studies in Mathematics, 70*(2), 175–189.
Rotman, B. (1987). *Signifying nothing: The semiotics of zero.* Stanford, CA: Stanford University Press.
Thom, J. S. (2011). Nurturing mathematical reasoning. *Teaching Children Mathematics, 18*(4), 234–243.
Thom, J. S. (2012). *Re-rooting the learning space: Minding where children's mathematics grow.* Rotterdam, The Netherlands: Sense Publishers.
Thom, J. S., & Roth, W.-M. (2011). Radical embodiment and semiotics: Toward a theory of mathematics in the flesh. *Educational Studies in Mathematics, 77*(2–3), 267–284.
Thom, J. S., Roth, W.-M., & Bautista, A. (2010). In the flesh: Living growing conceptual domains in a geometry lesson. *Complicity, 7*, 77–87.
Thompson, N. (n.d.a). Non-instrument navigation. *Wayfinders: A Pacific odyssey.* Retrieved from https://www.pbs.org/wayfinders/wayfinding2.html
Thompson, N. (n.d.b). On wayfinding. *Hawaian Voyaging traditions.* Retrieved from http://archive.hokulea.com/ike/hookele/on_wayfinding.html
van Manen, M. (1986). *The tone of teaching: The language of pedagogy* (2nd ed.). New York, NY: Routledge.
Varela, F. J. (1999). *Ethical know-how: Action, wisdom, and cognition.* Palo Alto, CA: Stanford University Press.

CYNTHIA NICOL, JO-ANN ARCHIBALD Q'UM Q'UM XIIEM AND FLORENCE GLANFIELD

13. CONCLUSION

Culturally Responsive Pedagogies to Re-Storying Mathematics Education

Culturally responsive pedagogies provide opportunities to re-story mathematics education for students, teachers, and community members. In discussing Indigenous storywork principles, Archibald (2008) refers to the "synergy of story threads that became a basket" (p. 152). We use this idea of story-thread synergy to consider how living culturally responsive mathematics education can create openings to extend and deepen—in other words, to re-story—what students, teachers, and community members in Indigenous communities typically experience as mathematics education. What story threads appear while living culturally responsive mathematics education that invite students to experience learning mathematics as connected, familiar, and empowering?

Examining the critical and social perspectives in mathematics education research, Atweh and Brady (2009) acknowledge the "strong rejection of the dominant view that mathematics is a singular, objective and value free discipline that is isolated from human interest" (p. 267). However, for many students, their experience with mathematics is just that: disconnected from their lives and communities, and seen as acultural. Culturally responsive pedagogies support re-storying students', teachers', and community's relationships to and experiences with mathematics. As a result, rigid structures of learning mathematics are challenged, what counts as doing mathematics and where mathematics problems live is reconsidered, and views of who does and who can do mathematics are transformed. Often mathematics or a particular view of mathematics has been and continues to be revered. Asking questions about whose perspective of mathematics is privileged and why, whose mathematical perspective is excluded and why, and how mathematics can be inclusive and meaningful to all learners may open learning environments.

Culturally responsive pedagogies in re-storying mathematics education point to the importance of place/land within Indigenous knowledges. King (2013), an Indigenous storyteller of the Cherokee Nation, writes of the interconnectedness of land and culture:

> Land has always been a defining element of Aboriginal culture. Land contains the language, the stories, and the histories of a people. It provides water, air,

 | DOI: 10.1163/9789004415768_013

> shelter, and food. Land participates in the ceremonies and the songs. Land is home. (p. 218)

It is through place or connections to the land that experiences are storied, remembered and re-told, and cultural histories are lived again (Cajete, 2000; Deloria & Wildcat, 2001; Little Bear, 2000; Peat, 2002). Culturally responsive pedagogies have the potential to reconnect histories and knowledges to place, and to open discussions that challenge the view there is only one way to know the world mathematically. Barton (2008) speaks of this one way of knowing mathematics as Nearly Universal and Conventional (NUC). Its universality tends to position NUC mathematics as above other mathematical knowledge systems; the argument posits that the knowledge system is more powerful when it is more abstract and less dependent on local place. Yet such an argument ignores the localness of knowledges created and the pedagogies created by that place. Culturally responsive pedagogies can nudge us to acknowledge that different kinds of economies, politics, relations, and languages are informed by different knowledge systems and that all knowledge systems are local and placed. Thom's (Chapter 12) exploration of culturally responsive education in the context of Polynesian wayfinding provides one example of uncovering and revitalizing placed mathematical knowledge. The Xavante Indigenous peoples of Central Brazil provide another excellent example.

In her research with Indigenous communities within Central Brazil focusing on mathematical thinking and mapping, Ferreira (2015) draws upon her experiences teaching and living in a Xavante community to highlight the nature of Xavante mathematics. Xavante worldviews of interdependent relationships and reciprocity or gift-giving structure solutions to mathematical problems. To illustrate this, Ferreira tells the story of how when she first arrived on Kuluene land, the Xavante people were curious as to why she arrived alone and without family. She was referred to as "*Mitsire*," where "*mitsi*" is the name for the number one (1) and "*re*" means "little one." With the name Mitsire, Ferreira was known as the "lonely self," one with "nobody to share with" (p. 175). Extending this dualistic thinking, Xavante people use *Maparané* to stand for the number two (2), meaning "the two of us together" (p. 175) or the "union of the lonely halves" (p. 177).

A Xavante worldview, of pairings or groupings that express relationships between people, lineages, clans, animals, and more than humans, thus brings forth a dualistic mathematical system. Pairs form a unit. Groups without pairs, such as a woman without a man, "[are] nothing but half of a unit, because man and woman together form the fundamental binary unitary value" (Ferreira, 2015, p. 176). As numbers are named after social relationships, and as part of a Xavante worldview, Ferreira (2015) refers to this as a Xavante socionumerical system. In this system, the Xavante, "count, measure, order and classify time, space, and other dimensions and aspects of their universe using categories and norms that are governed to a large extent by their kinship system" (p. 170). As social relationships, and in particular dualism, are fundamental structuring principles of the Xavante cosmology, it is understandable

Table 13.1. Xavante representation of numbers from 1 to 6 (from Ferreira, 2015, pp. 177, 181)

Numeral	*Xavante language*
1	*Mitsi* means that the element is "alone" or "on its own"
2	*Maparané* is the unitary base for counting because it is the union of the "lonely halves"
3	*Tsi'umdatõ* starts with the prefix *tsi* meaning alone, on its own—implying that it is an odd number
4	*Maparané tsiuiwanã* represents 2 groups of 2
5	*Imrotõ* means "without a mate" (*imro* means "husband" or "mate"; *tõ* means "without")
6	*Imropõ* stands for "the one who has found his mate" (*imro* means "husband"; *põ* means "with")

then that a dialectical dualism is also the main feature of the Xavante socionumerical system.

Counting to six (6) for Xavante people involved indicating quantities as having mates or not.

Ferreira (2015) articulates this sequence of counting as quite different from the "natural order" where we might begin with one (1), followed by two (2), and then three (3) and so on where the next "natural number" is one more than the previous. The Xavante philosophy instead challenges this idea and uses two (2) as the basic unit, and counts follow as expressions related to the unit two (2) or to pairings with mates. Perhaps a more cyclical, rather than linear, representation better depicts this system paralleling Xavante's cyclical expressions of time, body and relationships to the environments (p. 183). Thus counting and place value for the Xavante people are determined by organizing principles of social relationships. Xavante philosophy shapes mathematical actions indicating there are multiple ways of engaging mathematically and that "there is more than one universal grammar of number" (p. 50), more than one universal story of mathematical activity.

Understanding other systems of mathematical knowing, like the one enacted by the Xavante people of Brazil, acknowledges other ways of knowing and relating mathematically. This means that mathematics curriculum experienced in relation is open to multiple ways of knowing. Culturally responsive pedagogies pay attention to how different kinds of economies, politics, relations, and languages are informed by different knowledge systems and to how all knowledge systems are local and placed. In this way, culturally responsive pedagogies re-story mathematics education, which challenges the view there is a one way to know the world mathematically. Considering how mathematics can be enacted in community, in social relation, and in the context of human history and culture could work toward rooting mathematics in place and toward what Johnson (2012) calls "healing placelessness" in mathematics education.

Culturally responsive pedagogies in re-storying mathematics education highlight the need for commitments to building and sustaining relationships that connect community, culture(s), land/place and mathematics. This work at relationship-building requires on-going effort and commitment, is actualized in different ways, and is responsive and relational to the local context (place/land). To engage with community requires engaging with the local place.

Culturally responsive pedagogies also re-story mathematics education by providing openings for mathematics educators, both Indigenous and non-Indigenous, to examine personal taken-for-granted assumptions, privileges, and implicit biases of teaching and learning mathematics with and within Indigenous communities. This could require educators to "confront fear and ignorance" as Potawatomi-Lenapé scholar Dion (2016) says, to re-story their own understandings of colonization, Indigenous knowledges, mathematics and historical and contemporary cultural practices. Culturally responsive pedagogies could frame the "ethical space" imagined by Indigenous scholar Ermine (2007) or the "space between" articulated by Dion (2016) to allow for this kind of critical dialogue to occur.

Culturally responsive mathematics education experienced in relation and responsive to community and cultural interests includes: (1) dynamic learning environments focusing on relationships with mathematics with/in communities; (2) acknowledging place/land for opening possibilities for meaningful activities that consider other ways of being mathematical; (3) recognizing ongoing required commitment for building and sustaining reciprocal relationships at multiple levels and across contexts; and (4) creating ethical spaces to embrace critical stances that can open up alternative ways of being in relation.

Living culturally responsive mathematics education attends to the role of place, land, and community in acknowledging and valuing other knowledge systems. Different understandings of place root different ways of interacting with others and with land, that then inspire different mathematical actions. Our communities guide us in our cultural responsiveness. This connection to place and land, as evidenced throughout the chapters in this book, emphasizes the importance of local context in living culturally responsive mathematics education. It is this connection to place and land that sets culturally responsive mathematics education with/in Indigenous communities apart from other pedagogies, and reminds us that the promotion of one cultural perspective can't be universally applied or seen as more sophisticated than other perspectives.

Yet this localness, this situated-ness, also raises questions from educators, researchers, and policy-makers. How can research and teaching so dependent on a particular place or context be scaled up, be transferable, or, as Māori scholar McKinley (2005) writes, be "translatable"? Is there some kind of universality of culturally responsive mathematics education? What can we learn from mathematical engagement of youth and Indigenous Elders and community members in particular places such as Alaska, New Zealand, Central Brazil, and from the West Coast to the

East Coast of Canada? Indigenous communities are diverse with different linguistic, cultural and historical roots. Although different, it is the story of living connections between community, culture, and mathematics are what threads inter-connectedness. Across contexts responsiveness to relations and place highlight an ethical disposition of listening, relating and engaging. Across geographical contexts, we learn of the possibilities and struggles that students and teachers face when re-storying and resisting challenges of deficit thinking. As the chapters in this book portray, living culturally responsive mathematics education is promising for communities across multiple contexts. However, we need more stories of living culturally responsive mathematics education—the possibilities and the challenges.

What does the future hold for culturally responsive mathematics education with/in Indigenous communities? Living mathematics education in relational, responsive, and reciprocal ways in our communities is an ongoing endeavour. We need to remember that there isn't one definition of culturally responsive education and that there are different strategies to address living mathematics in our communities that can work toward revitalization of culture and language, renewed interest and passion toward teaching and learning mathematics, and reconciliation for powerful placed community engagement and well-being. Living culturally responsive education implies change, growth, renewal, and emergence along a less-worn path. Indigenous writer and activist Noel Pearson (2011) in Australia and philosopher Jonathan Lear (2006) in the United States write of the need for "radical hope" for a future not yet imagined, a path not yet taken. Drawing upon the story of Plenty Coups, the last chief of the Crow Nation, Pearson and Lear call for a type of hope that is more than optimism. Perhaps culturally responsive education offers a place for this active radical hope threaded with courage. Living culturally responsive education is a call to re-storying mathematics education with radical hope for a future in which all students participate in responsive, relational mathematical actions for interpreting and transforming our local and global worlds.

You've now reached the end of particular journey and are at the beginning of a new journey. In arriving at this place, we are reminded of the teachings of Elders and knowledge keepers in that human beings live in cycles. Each day we wake as an infant and pass through stages from infant, to youth, to young adult, to older adult similar to the stages we live as human beings. We begin the day as an infant in the sense that we have never had any experience of that day previously. We travel through the day and are always having new experiences. We mature and gain wisdom through each experience in the day. By the end of the day we've gained wisdom that we did not have at the beginning of the day. With our new wisdom we can wake up the next day and begin to live the new day in a way that we could not have lived the previous day.

We invite you to live your new journey with the maturity and wisdom you've gained by traveling through this book by exploring questions such as:

- In what ways is culturally responsive mathematics education practiced and studied with/in Indigenous communities in the geographical locations of South and Central America, Europe, Asia, and Africa?
- What spaces for courage and vulnerabilities exist in living culturally responsive mathematics education?
- What ontological, epistemological, and axiological stances are associated with Indigenous worldviews? In what ways might these stances re-story the ways in which mathematics is studied, taught, and experienced in schools?
- Imagine a mathematics teacher education program in which culturally responsive pedagogies were privileged. What might be the experiences of teacher candidates in such a program?
- In what ways might culturally responsive pedagogies in mathematics education be practiced within urban contexts?
- How might culturally responsive mathematics education invite all learners/ teachers/communities (Indigenous and non-Indigenous) to imagine a future that is not yet imagined?

REFERENCES

Archibald, J. Q'um Q'um Xiiem. (2008). *Indigenous storywork: Educating the heart, mind, body, and spirit.* Vancouver: UBC Press.

Atweh, B., & Brady, K. (2009). Socially response-able mathematics education: Implications of an ethical approach. *Eurasia Journal of Mathematics, Science & Technology Education, 5*(3), 267–276.

Barton, B. (2008). *The language of mathematics: Telling mathematical tales.* New York, NY: Springer.

Cajete, G. (2000). *Native science: Natural laws of interdependence.* Santa Fe, NM: Clear Light Publishers.

Deloria, V., & Wildcat, D. (2001). *Power and place: Indian education in America.* Golden, CO: Fulcrum.

Dion, S. (2016). Mediating the space between: Voices of Indigenous youth an voices of educators in the service of reconciliation. *Canadian Review of Sociology, 53*(4), 468–473.

Ermine, W. (2007). The ethical space of engagement. *Indigenous Law Journal, 6*(1), 193–204.

Ferreira, M. L. (2015). *Mapping time, space, and the body: Indigenous knowledge and mathematical thinking in Brazil.* Rotterdam, The Netherlands: Sense Publishers.

Johnson, J. (2012). Place-based learning and knowing: Critical pedagogies grounded in Indigeneity. *GeoJournal, 77*, 829–836.

King, T. (2013). *The inconvenient Indian: A curious account of native people in North America.* Minneapolis, MN: University of Minnesota Press.

Lear, J. (2006). *Radical hope: Ethics in the face of devastation.* Cambridge, MA: Harvard University Press.

Little Bear, L. (2000). Jagged worldviews colliding. In M. Battiste (Ed.), *Reclaiming Indigenous voice and vision* (pp. 77–85). Vancouver: UBC Press.

McKinley, E. (2005). Locating the global: Culture, language and science education for indigenous students. *International Journal of Science Education, 27*(2), 227–241.

Pearson, N. (2011, October 16). *Radical hope book: A talk by Noel Pearson* [Video file]. Retrieved from https://www.youtube.com/watch?v=Uux8CoXULn0

Peat, D. (2002). *Blackfoot physics: A journey into the Native American universe.* Grand Rapids, MI: Phanes Press.

INDEX

www.ingramcontent.com/pod-product-compliance
Lightning Source LLC
Chambersburg PA
CBHW071429230926
51846CB00001B/160
9789004415744